T0133115

E-SYSTEMS FOR THE 21ST CENTURY

Concept, Developments, and Applications

Volume 1

E-Commerce, E-Decision, E-Government,
E-Health, and Social Networks

E-SYSTEMS FOR THE 21ST CENTURY

Concept, Developments, and Applications

Volume 1

E-Commerce, E-Decision, E-Government, E-Health, and Social Networks

Edited by
Seifedine Kadry, PhD
American University of the Middle East, Kuwait

Abdelkhalak El Hami, PhD
INSA of Rouen, France

APPLE ACADEMIC PRESS

Apple Academic Press Inc. | Apple Academic Press Inc.
3333 Mistwell Crescent | 9 Spinnaker Way
Oakville, ON L6L 0A2 | Waretown, NJ 08758
Canada | USA

©2016 by Apple Academic Press, Inc.
Exclusive worldwide distribution by CRC Press, a member of Taylor & Francis Group
No claim to original U.S. Government works
Printed in the United States of America on acid-free paper
International Standard Book Number-13: 978-1-77188-265-1 (Hardcover)

International Standard Book Number-13: 978-1-77188-263-7 (eBook)

Library and Archives Canada Cataloguing in Publication

E-systems for the 21st century : concept, developments, and applications.

Includes bibliographical references and indexes.
Contents: Volume 1. E-commerce, e-decision, e-government, e-health, and social networks / edited by Seifedine Kadry, PhD, American University of the Middle East, Kuwait ; Abdelkhalak El Hami, PhD, INSA of Rouen, France.

Issued in print and electronic formats.
ISBN 978-1-77188-265-1 (volume 1 : hardcover).--ISBN 978-1-77188-255-2 (set : hardcover).--ISBN 978-1-77188-263-7 (volume 1 : pdf)
1. Computer networks. 2. Computer systems. 3. Electronic systems.
I. Kadry, Seifedine, 1977-, editor II. El Hami, Abdelkhalak, editor
TK5105.5.E89 2016   004.6   C2016-901696-X   C2016-901697-8

CIP data on file with US Library of Congress

CONTENTS

LIST OF CONTRIBUTORS

Mourine Achieng
Cape Peninsula University of Technology, Cape Town, South Africa

Ebtisam Al-Harithi
Southampton University, London – UK

Mohammed Zuhair Al-Taie
UTM Big Data Centre and Faculty of Computing, Universiti Teknologi Malaysia (UTM), Skudai, 81310 Johor, Malaysia

Rasim Alguliyev
Director of the Institute of Information Technology of ANAS, Baku, Azerbaijan

Linus Atorf
Institute for Man-Machine Interaction, RWTH Aachen University, Germany

Ramatsetse Boitumelo
Department of Industrial Engineering, Tshwane University of Technology, South Africa

María Del Carmen Caba-Pérez
Department of Economía y Empresa Universidad de Almería, Carretera de Sacromonte s/n 04120 Almería (Spain)

Te Fu Chen
Department of Business Administration, Lunghwa University of Science and Technology, Taoyuan Taiwan

Robert Costello
Graduate School University of Hull, UK

P. Devika
Department of Communication, PSG College of Arts and Science, Coimbatore- 641014, Tamil Nadu, India

Abdelkhalak El Hami
National Institute of Applied Sciences, Rouen, FRANCE

Seyyed Muhamad Mutallebi Esfidvajani
Entrepreneurship Department, University of Tehran, Tehran, Iran

Leila Esmaeili
Computer Engineering and Information Technology Department, Amirkabir University of Technology, Tehran, Iran

Harry Fulgencio
Leiden Institute of Advanced Computer Science, Niels Bohrweg 1 Leiden, 2333CA, The Netherlands

María del Mar Gálvez-Rodríguez
University of Almeria, Spain

Seyyed Alireza Hashemi Golpayegani
Computer Engineering and Information Technology Department, Amirkabir University of Technology, Tehran, Iran

Norman Gwangwava
Department of Industrial and Manufacturing Engineering, National University of Science and Technology, Bulawayo, Zimbabwe

Maryam Haghshenas
Media Management, University of Tehran, Iran

Nurussobah Hussin
Faculty of Information Management, Universiti Teknologi MARA (UiTM) Malaysia

Seifedine Kadry
American University of the Middle East, Kuwait

Adeyeri Michael Kanisuru
Department of Industrial Engineering, Tshwane University of Technology, South Africa

Mpofu Khumbulani
Department of Industrial Engineering, Tshwane University of Technology, South Africa

Shahla Mardani
Computer Engineering and Information Technology Department, Amirkabir University of Technology, Tehran, Iran

Shahla Mardani
Computer Engineering and Information Technology Department, Amirkabir University of Technology, Tehran, Iran

Nasim Matar
Applied Science University, Amman – Jordan

N. Mathiyalagan
Department of Communication, PSG College of Arts and Science, Coimbatore- 641014, Tamil Nadu, India

Chi Man Mui
Department of Information Technology, Chinese YMCA College, Hong Kong SAR, China

Alaa Amir Najim
Mathematical Department, Science College, Basrah University, Basrah, Iraq

Safa Amir Najim
Computer Science Department, Science College, Basrah University, Basrah, Iraq

Mojtaba Nassiriyar
IT Management, University of Tehran, Iran

Makinde Olasumbo
Department of Industrial Engineering, Tshwane University of Technology, South Africa

Rachid Oumlil
Department of Management, ENCG-Agadir, Ibnou Zohr

Ashis Pani

Information Systems Area, XLRI Jamshedpur, Jharkhand, India

Malte Rast
Institute for Man-Machine Interaction, RWTH Aachen University, Germany

Jürgen Rossmann
Institute for Man-Machine Interaction, RWTH Aachen University, Germany

Ephias Ruhode
Cape Peninsula University of Technology, Cape Town, South Africa

Abouzar Sadeghzadeh
Electronics & Telecoms Engineering, University of Bradford, UK

Alejandro Sáez-Martín
University of Almeria, Spain

Wan Satirah Wan Mohd Saman
School of Information Management, Universiti Teknologi MARA, Malaysia

Michael Schluse
Institute for Man-Machine Interaction, RWTH Aachen University, Germany

Roghayeh Shahbazi
IT Management, Alzahra University, Tehran, Iran

Hamid Reza Shahriari
Computer Engineering and Information Technology Department, Amirkabir University of Technology, Tehran, Iran

Siti Mariyam Shamsuddin
UTM Big Data Centre and Faculty of computing, Universiti Teknologi Malaysia (UTM), Skudai, 81310 Johor, Malaysia

Dalibor Stanimirovic
University of Ljubljana, Faculty of Administration, Gosarjeva ulica 5, 1000 Ljubljana, Slovenia

Rakesh Tiwari
SSP India Private Limited, Gurgaon, Haryana, India

Watcharapol Wiboolyasarin
Suan Dusit Rajabhat University, Bangkok, Thailand

Farhad Yusifov
Department of Information Society, Institute of Information Technology of ANAS, Baku, Azerbaijan

LIST OF ABBREVIATIONS

ACMS	Asthma Care Mobile Service
AGFI	Adjusted Goodness-of-Fit Index
AHP	Analytic Hierarchy Process
ANOVA	Analysis of Variance
ATT	Attitude
AUTO DCR	Automating Development Control Rules
AVE	Average Variance Extracted
BI	Behavioral Intention
BSNL	Bharat Sanchar Nigam Ltd.
CAP	Common Alerting Protocol
CBSEM	Covariance-Based Structural Equation Modeling
CCMC	Coimbatore City Municipal Corporation
CERT	Computer Emergency Response Team
CFI	Comparative Fit Index
CLIPS	C Language Integrated Production System
CMS	Course Management Systems
COTS	Commercial Off the Shelf
CR	Composite Reliability
CRB	Cockpit Voice Recorder
CRM	Customer Relationship Management
CRT	Court Recording and Transcribing
CSCL	Computer-Supported Collaborative Learning
CSCW	Computer Supported Cooperative Work
DMT	Dedicated Machine Tools
DOFs	Degrees of Freedom
DOI	Diffusion of Innovations
DRG	Diagnosis Related Groups
DSM	Design Structure Matrix
EHCR	Electronic Health Care Records
EPR	Electronic Patient Records
EST	Environmentally Sound Technology

EU	European Union
EWS	Early Warning Systems
FAA	Federal Aviation Administration
FAM	Female-Focused Acceptance Model
FARR	Full-Automatic Reconfiguration for Reconfigurable Systems
FMS	Flexible Manufacturing Systems
FTA	Fault Tree Analysis
GDP	Gross Domestic Product
GFI	Goodness-of-Fit Index
GIS	Geographical Information Systems
GP	General Practitioner
hAOP	Health Portal
HIS	Health Information System
HIS	Hospital Information System
HIT	Healthcare Information Technology
hNET	Health Network
HOQ1	House of Quality
IBT	Interpersonal Behavior Theory
ICA	International Council on Archives
ICR	Internal Consistency Reliability
ICRIER	Indian Council for Research on International Economic Relations
ICT	Information and Communication Technologies
IDI	ICT Development Index
IDT	Innovation Diffusion Theory
IFR	Instrument Flight Rules
INT	Intention
IS	Information Systems
ISMS	Information Security Management Standard
ISO	International Standards
IT	Information Technologies
JKSM	Malaysia Shariah Judiciary Department
KBES	Knowledge-Based Expert System
KGISL	Kg Information Systems Limited
KM	Knowledge Management

LMS	Learning Management Systems
MC	Mass Customization
MD	Mean Difference
MES	Multi-Linear Events Sequencing
MHS	Mobile Healthcare Systems
MMRS	Metropolitan Medical Response System
NASA	National Aeronautics And Space Administration
NBiS	Network-Based Information Systems
NeGP	National E-Governance Plan
NEIC	National Earthquake Information Center
NGM	Next-Generation Manufacturing
NIC	National Informatics Centre
NTSB	National Transportation Safety Board
PEOU	Perceived Ease of Use
PH	Provincial Hospital
PHR	Personal Health Records
PLS	Partial Least Squares
PMFHS	Project Management and Financing of Health Sector
POH	Pilot Operating Handbook
PPP	Public Private Partnership
PU	Perceived Usefulness
QFD	Quality Function Deployment
RFID	Radio Frequency Identification
RMMA	Reconfigurable Manufacturing Machine Advisor
RMR	Root Mean Square Residual
RMS	Reconfigurable Manufacturing Systems
RMSEA	Root Mean Square Error of Approximation
RMT	Reconfigurable Machine Tools
SCT	Social Cognitive Theory
SEM	Structural Equation Modeling
SIG	Special Interest Group
SME	Small and Medium Enterprises
SMS	Short Message Service
SNS	Social Networking Sites
SSTs	Self-Service Technologies
TAM	Technology Acceptance Model

TAM2	Extension of Technology Acceptance Model
TPB	Theory of Planned Behavior
TRA	Theory of Reasoned Action
TT	Technology Transfer
TTDSS	Technology Transfer Decision Support Systems
UTAUT	Unified Theory of Acceptance and Use of Technology
VFR	Visual Flight Rules

PREFACE

E-based systems and computer networks are becoming standard practice across all sectors, including health, engineering, business, education, security, and citizen interaction with local and national government. They facilitate rapid and easy dissemination of information and data to assist service providers and end-users, offering existing and newly engineered services, products, and communication channels. Recent years have witnessed rising interest in these computerized systems and procedures, which exploit different forms of electronic media to offer effective and sophisticated solutions to a wide range of real-world applications.

With contributions from researchers and practitioners from around the world, this two-volume book discusses and reports on new and important developments in the field of e-systems, covering a wide range of current issues in the design, engineering, and adoption of e-systems. *E-Systems for the 21st Century: Concept, Developments and Applications* focuses on the use of e-systems in many areas of sectors of contemporary life, including commerce and business, learning and education, health care, government and law, voting, and service businesses.

The two-volume book offers comprehensive research and case studies addressing e-system use in health, business, education, security, and citizen interaction with local and national government. Several studies address the use of social networks in providing services as well as issues in maintenance and security of e-systems as well.

This collection will be valuable to researchers at universities and other institutions working in these fields, practitioners in the research and development departments in industry, and students conducting research in the areas of e-systems. The book can be used as an advanced reference for a course taught at the undergraduate and graduate-level in business and engineering schools as well.

Seifedine Kadry, PhD
Abdelkhalak El Hami, PhD

ABOUT THE EDITORS

Seifedine Kadry, PhD, has been an associate professor with the American University of the Middle East in Kuwait since 2010. He serves as editor-in-chief of the *Research Journal of Mathematics and Statistics* and the *ARPN Journal of Systems and Software*. He worked as head of the software support and analysis unit of First National Bank, where he designed and implemented the data warehouse and business intelligence. He has published several books and is the author of more than 50 papers on applied math, computer science, and stochastic systems in peer-reviewed journals. At present his research focuses on system prognostics, stochastic systems, and probability and reliability analysis. He received a PhD in computational and applied mathematics in 2007 from the Blaise Pascal University (Clermont-II) – Clermont-Ferrand in France.

Abdelkhalak El Hami, PhD, is a full professor at the National Institute of Applied Sciences in the Rouen area of France as well as Deputy Director and Head of LOFIMS Laboratory and mechanical chair of the National Conservatory of Arts and Crafts. He is the author of many articles in international journals and books on optimization and uncertainty software and has presented at many conferences. He is also an IEEE Senior Member as well as on the editorial boards of several journals. He has published several books and is the author of more than 500 papers published in international journals and conferences. He received a doctorate in engineering sciences from the University of Franche-Comté in France (1992). He received his Habilitation diploma to supervise research (HDR) in 2000.

INTRODUCTION

E-systems are an integral part of the increasingly prevalent complex, pervasive, embedded, and ubiquitous computing solutions that have been, or are being, developed.

Recent years have witnessed the rising interest in these computerized systems and procedures, which exploit different forms of electronic media in order to offer effective and sophisticated solutions to a wide range of real-world applications. Initially, the impetus toward e-systems uptake was prompted by convenience (for the customer) and cost savings (for the merchant) of e-commerce transactions. Subsequently, a rapidly growing number of government services and information sources are now available online, providing fast, reliable, and convenient global access to e-government. Innovation and research development continue to sustain this rapidly evolving area and are increasing its scope into many more areas, for which we adopt the term 'e-systems engineering.' Such developments have typically been reported on as part of cognate fields such as information and communications technology, computer science, systems engineering, and social science and engineering. The fundamental elements of e-systems engineering, based on shared standards, web services, service-oriented architecture (SOA), and distributed data, are changing the way computer and software systems are designed, architected, delivered, and consumed, arguably meriting e-systems engineering as a separate and self-contained field of study.

The developments and innovations of e-systems are becoming prevalent in many diverse domains including business, education, security, and governance. This usually involves the use and implementations of Internet technology to reproduce inter/intra organizational procedures and frameworks that are conveniently available for the end-user, offering existing and newly engineered services, products, and communication channels.

This book is a collection of chapters that could completely cover the huge coverage that e-systems encompass. The chapters are draw from different science fields like e-commerce, e-decision, e-government, e-health, social networks, e-learning, e-maintenance, e-portfolio, e-system and e-voting.

The book is organized into two volumes of 12 chapters each. Each volume is divided into five sections. Volume I sections are: e-commerce, e-decision, e-government, e-health, and social networks. Volume II sections are: e-learning, e-maintenance, e-portfolio, e-system, and e-voting.

PART 1:

E-COMMERCE

CHAPTER 1

A SURVEY OF TRUST IN SOCIAL COMMERCE

SEYYED ALIREZA HASHEMI GOLPAYEGANI,[1,a] LEILA ESMAEILI,[2] SHAHLA MARDANI,[3,b] and SEYYED MUHAMAD MUTALLEBI ESFIDVAJANI[4]

[1]*Assistant Professor, Computer Engineering and Information Technology Department, Amirkabir University of Technology, Tehran, Iran, Tel.: +982164542726, E-mail: Sa.Hashemi@aut.ac.ir*

[2]*PhD Candidate, Computer Engineering and Information Technology Department, Amirkabir University of Technology, Tehran, Iran, Tel.: +982164542726, E-mail: Leila.Esmaeili@aut.ac.ir*

[3]*MSc, Computer Engineering and Information Technology Department, Amirkabir University of Technology, Tehran, Iran, Tel.: +982164542726, E-mail: Shahla.Mardani@aut.ac.ir*

[4]*MSc Student, Entrepreneurship Department, University of Tehran, Tehran, Iran, E-mail: M.Mutallebi@ut.ac.ir*

CONTENTS

[a]Seyyed Alireza Hashemi Golpayegani is corresponding author.
[b]These authors contributed equally to this work.

ABSTRACT

The increasing popularity of social networks in recent years has brought about many novel opportunities for e-commerce. Social commerce (S-commerce) is a new stream of e-commerce that, through social networks as well as other social media, interconnects users and facilitated the commerce process. Irrespective of their mechanisms, there are essential elements, such as trust, to any type of commerce. Trust is considered more vital especially in commerce, where it may affect users behavior and influence their purchase intention. By the same logic, it is the center of attention in s-commerce because a higher risk and uncertainty lays in the nature of its transactions. The issue of trust in cyberspace is not new and many researches have been carried out to cover and fully understand the phenomenon; however, the concept of e-commerce has just recently been introduced to the literature and requires special attention. To fulfill this end, this chapter explores previous studies to review the range of studies as well as give insights for future studies.

1.1   INTRODUCTION

The failure of many of the online companies in the dot-com boom of 2000 and 2001 [1] forced companies to rethink theirs strategies to improve and implement their virtual distribution channels and develop a better understanding of consumer behavior as well as customer loyalty and customer retention [2]. Therefore, to build loyalty and retain customers, web based businesses began to implement new tools and operations in their websites.

In addition, online shopping by means of some very special and interactive methods was provided to consumers, this was made possible by support of the emerging Web 2.0 technologies and their widespread use in social network service markets. Consequently, e-commerce made its stride towards the emergence of s-commerce [3]. In general, s-commerce refers to the use of Web 2.0 technologies in traditional e-commerce [4–6] have explained the impacts of Web 2.0 on e-commerce.

Empirical evidence shows the use of a set of features by web-based businesses to increase the social interaction and sense of cooperation among users in late 90s [3, 7]; however, an official use of the term "social commerce" has been introduced to the literature by the first use of the term by Yahoo in 2005 [7–9] and it was first appeared in a scientific paper in 2007 [10, 11]. S-commerce has ever since drawn a considerable attention [7, 9] and has been known rather by its practical applications and not academically [12].

Despite lack of a standard definition for "s-commerce," this term generally refers to providing and carrying out e-commerce activities and transactions in the context of social media, and social networks in particular; therefore, by definition, s-commerce could be considered as a subset of e-commerce [14] or as its evolved form Refs. [7, 13] and it mainly constitutes a combination of social activities and commerce [14]. Marsden [15] has collected 22 different definitions of social commerce. They include a range of s-commerce features such as word of mouth advertising, trusted consultation, and assisted shopping. S-commerce is defined in respect to different fields of study like marketing, computer science, sociology, and psychology. In marketing, for instance, s-commerce regards the prominent trends in online markets; businesses use social media or Web 2.0 as a direct marketing method in order to support customers' decision-making process and behavior [16]. Regarding computer science, Lee et al., [17] Introduce s-commerce as an online application that combines Web 2.0 technologies like Ajax [18] and RSS [19] with interactive platforms such as online social media websites and content communities in a commercial environment. Regarding sociology, s-commerce is about employment of web based social communities by e-commerce companies. It mainly focuses on the effects of social influence, which shapes the interaction among consumers [4]. Finally, Marsden [20] from a perspective of psychology science, defines s-commerce as social shopping. In social shopping people intend to shop online under the influence of the prominent information in a social network. Based on the mentioned and other definitions, in this chapter we define s-commerce as an Internet based commercial application that makes use of Web 2.0 technologies and social media; and it supports user created content and social interactions. S-commerce considers the network of buyers and sellers as a single platform that includes the buying/selling activities and all related interactions. All of these commercial activities and transactions are carried out inside online communities and markets.

Regardless of the type of commerce, electronic or not and social or not, there are essential elements to them. Trust (see Section 1.2.1) is one of these elements without which conducting financial transactions proves problematic. Trust is the willingness of one party to put himself in a position of harm by another party, expecting that the other party will do actions that are important to him, even in the absence of control and monitoring [21]. In general, trust is considered an essential factor in all financial and social circumstances, especially in commerce, due to its impact on consumers' behavior and its role in strengthening their purchase intention. The importance of trust is regarded even more vital [22]. Furthermore, the vital importance of trust is further stressed in e-commerce when the risk of transactions is higher than normal situations and customers face an uncertain environment [23] (see Section 1.3).

To improve trust in e-commerce and reduce uncertainties in that environment, researchers have recommended the relationship among people and knowledge transfer as an effective solution [24–26]. Knowledge transfer is now make possible in social media platforms, and as previously discussed, combined with e-commerce, social media shape the concept of s-commerce.

As discussed, s-commerce is a novel phenomenon and research topics of a bibliography study of 2004 to 2013 mainly cover: study of customer/ consumers behavior [27–29], s-commerce acceptance [30, 31], and design of s-commerce websites [3, 5]. In the first two topics, the issue of trust is a main concern of research. Over all, researches on this topic cover two subjects: (i) identification of effective factors on consumers' trust, and (ii) studying the effects of various features of s-commerce on consumers' trust. In addition, another stream of researchers have studies the methods for improving consumers' trust on a s-commerce platform like word of mouth marketing and user experience. This chapter, therefore, is the result of a survey study on previous researches on trust in e-commerce to identify factors affecting trust and those factors affected by it. In the end, along with a summary of each chapter, a classification and comparison will be listed and as a contribution, open discussions and avenues for future research are also provided.

Basic concepts and definitions for trust (see Section 1.2.1) and trust in cyberspace (see Section 1.2.2) as used in this chapter are discussed in Section 1.2. Section 1.3 reviews trust concepts in e-commerce. Effects of virtual communities and social networks on virtual trust are studies in Section 1.4. Subsequently, Section 1.5 covers trust in s-commerce along with the research methodology classifications and paper reviews. This is followed by ideas for future research. Finally, Section 1.6 concludes the chapter.

1.2   BASIC CONCEPTS AND DEFINITION

Trust is constantly apparent in people lives and it shapes their behavior. The broad nature of this concept makes it challenging to give a simple explicit description of trust. Literature reviews reveal a variety of definitions for trust, each of which sheds light on one or more aspects of trust from different perspective and knowledge fields. Here, we intent to explore this concept and study its dimensions and features. This section is concluded by a description of trust in the virtual world.

1.2.1   TRUST

As explained, trust is a complex concept and there are various definitions to it [32]; nonetheless, there are general definitions [33]: reliability trust and decision trust.

Definition 1: reliability trust: trust is a subjective probability and based on it, one party expects another peer to do an action in a way that satisfies the first party, the truster.

Here, the trusted party is central to the definition and the trust of trusting party depends on reliability of the trusted party. However, in real life, trust does not exclusively depend upon the reliability of the trusted party.

Imagine a situation that a person is imprisoned in a deep well with a single rope hanging down the well, in such circumstances, people may not sufficiently trust that rope, but they may try to use it to escape the current situation and trust the uncertain rope. Falcone and Castelfranchi discuss the same topic in Ref. [34] and do not consider high reliability as a sufficient factor for decision making. Therefore, another definition for trust is suggested.

Definition 2: decision trust is the willingness of one party to trust another person or thing in a particular situation with a relative sense of security and awareness of the probability of negative consequences.

In this definition, trust and reliability as well as results and benefits are relative. In fact, here an attitude of risk is introduced. This definition gives a more comprehensive definition compared with definition 1. Its implied uncertainty considers all possible consequences.

1.2.1.1 Trust and Reputation

Trust and reputation are interlinked. To eliminate misunderstandings, the different aspects of these two concepts are discussed.

Definition 1.3: reputation: the belief, view, and judgment of the public about the quality of a general characteristic of a person or thing [35].

The readers should make the distinction by now that belief in a person in the definition of reputation is based on the belief and ideas of other people, whereas in trust, which is a subjective phenomenon formed by various factors, personal experience has more effects that others in shaping trust [33]. In the absence of personal experiences, trust could be built based on the recommendations. This of trust is based on reputation.

1.2.1.2 Trust Characteristics

Complex as it is, trust possesses some generic characteristics [36, 37]. Trust is:

- Directed: trust is a directed relationship between truster and trustee [37].
- Subjective: trust is a personal belief and a subjective concept shaped by different factors with various levels of importance [37].

- Context specific: trust is context specific, the truster may trust the trustee in only a specific area and context [36]. For example, one may trust a person as a dentist and not as a baby sitter.
- Measurable: the amount of trust could be measured and compared against the amount of trust of another peer. By measuring trust, trust evaluation and modeling is possible [37].
- Dependent on history: previous experiences could affect the current level of trust [37].
- Multi-faceted: even in the same context and with the same content, trust varies based on the trustee's capabilities. For instance, when receiving a service, its price, quality, and support are evaluated. The overall trust depends upon each of the mentioned aspect. These aspects could assist in deciding whether or not to trust [36].
- Dynamic: trust may be improved or decreased over time and by gain of experience [36]. It may cease to exist or be recreated [37].
- Conditionally transferable: information based trust could be sent or received through a chain or network of recommendations. Conditions mostly refer to context and objectives of the truster [37].
- Constitutionality: trust is a combination of different characteristic: reliability, dependability, honesty, truthfulness, security, competence, and timeliness [37].
- Asymmetric: when two people trust one another, they do not necessarily hold the same level of trust against each other. This arises from their difference in experiences, psychological background, and histories [38].

1.2.1.3   Trust Types (Objective)

Based on the various parameters, there could be varying classifications to trust. A trust classification by Grandison and Sloman is based on trust objectives [39].

- Provision trust: it is a type of trust that concerns supplier of resources o services. Truster, in this type of trust seeks protection against unreliable and ill-purposed suppliers.
- Access trust: the trust about the managerial performance of the management of the resources, this type of trust, is related to the principle of access control.
- Delegation trust: the trust on an agent who makes decisions on behalf of truster.

- Identity trust: this type of trust is to put faith and believe the identity of a person. The trust systems performing on this usually consider using identification patterns like X.509 and PGP.
- Context trust: truster trusts a system or institute based on the assumption that the trusted system could securely manage to support the transactions.

1.2.2 TRUST IN VIRTUAL WORLD

Trust is a persistent characteristic of social interactions. Researches reveal that people wish to reduce their social uncertainties. This means that they wish to understand, predict, and even sometimes control other people behavior [40]. Therefore, trust, as a means to reduce uncertainties, holds an important position in social interactions. In some interactions, like commerce where there is a matter of profit, the importance of trust is more stressed. In real world, people have face to face contact and that makes it easier to trust. However, in the virtual world, where "nobody knows if you're dog," it is challenging to trust. In e-commerce, customers and sellers do not know each other, there are no means of face to face interaction and the product or service is not tangible, and trust is difficult to shape. Consequently, trust building and acceptance is a totally different story in the virtual world. Trust must suit the nature of the net. An electronic trust!

In the following, trust and e-trust are used interchangeably.

1.2.2.1 Trust Management

Trust management includes collecting information to make decisions about a trust relationship, evaluation of criteria about trust relationship, monitoring and reevaluation of existing trust relationships, and automation of the process [41]. Two main approaches towards trust management are policy based approach and reputation based approach.

1.2.2.1.1 Policy Based Trust Management

This approach is used in open and distributed service architectures to solve the issue of authorization access control in open systems. In this approach policy languages and engines are used to identify and reason rules for trust. In other words, this seeks to understand if an unknown user can be trusted [42].

1.2.2.1.2   *Reputation Based Trust Management*

This approach emerged in e-commerce systems; it has been used to distributed conditions for trust in public key certificates, p2p systems, mobile ad-hoc networks, and semantic web. In this trust management approach, by using histories, trust computation model estimates the trust level and a peer [42].

1.2.2.2   Trust Modeling

Trust management needs to develop and use trust models. A trust model is the method of specifying, evaluating and setting up links among peers to compute trust. Trust modeling is a technical approach for the purpose of digital processing [37]. Scholars try to recreate real life trust in cyberspace and to reflect the characteristics of the real life trust on the Internet. The subjective nature of trust is a challenging issue in trust modeling. Therefore, the interpretation of concept of the subjective trust to machine-readable language is the main objective of trust modeling. To do this job, many models have been introduced, each of which employs a distinctive methodology and a different approach. In a classification by Yan and Holtmanns [37], the available trust models are presented (Figure 1.1).

1.3   TRUST IN E-COMMERCE

The nature of e-commerce refrains sellers and customers to have a direct face-to-face transaction, and customers are not able to touch and feel the products. Face to face relations give both sides of the deal an opportunity to evaluate and learn about each other. Online transactions lack such opportunity and financial transactions usually carry a sense of uncertainty and risk. This uncertainty and risk is more of a problem when there has been no prior history for transaction between the peers. Social exchange theory states that people participate in transactions based on trust [43].

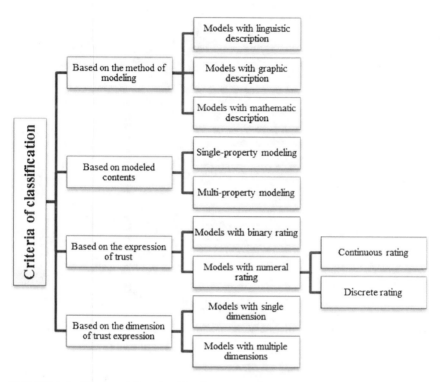

FIGURE 1.1 Trust modeling classification.

Therefore, as offline commerce requires trust. In online businesses, likewise, trust is an essential ingredient to reduce uncertainty. As previously discussed (see Section 1.2.2.1), to do so, the transaction history of a peer is scrutinized and the trust value is then computed. Predicted trust is a criteria to reduce risk and eliminate uncertainties. In the following, reputation based trust systems are reviewed.

1.3.1   REPUTATION BASED TRUST SYSTEMS

Most reputation based trust systems use the following four phases to achieve their goals (Figure 1.2) [44]:

1. Collecting information about a particular peer in the society by learning the beliefs and recommendations of other peers.

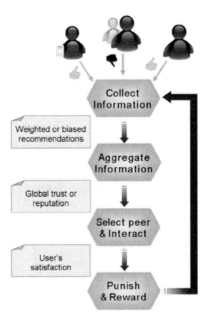

FIGURE 1.2   Main phases in trust and reputation systems [44].

2. Aggregating all the information and giving each peer in the system a score.
3. Selecting the most trustworthy or most reputed peer in the society, providing a particular service, interacting with the selected peer, and comparably assess users' satisfaction of using that service.
4. Based on level of satisfaction, a punish or reward phase should be implemented and the level of public reputation or trust of the service provider should be computed.

To better understand the mechanism of such systems, we will further look into the architecture of these systems and their trust computation methodology.

1.3.2 REPUTATION-BASED SYSTEMS ARCHITECTURE

Architecture here means the method of relating reputation scores and ratings between sub-systems. The two principal architectures are distributed architecture and centralized architecture.

1.3.2.1   Centralized Reputation Systems

In centralized systems, when community members establish direct contact with a peer and gain some experience, they give a score to that peer, all of these scores are then aggregated by a central authority and a reputation score is assigned to each peer. These scores are published to the public. Members could decide, based on these reputation scores, if a peer is trustworthy for a transaction or not. After each transaction, each participant in the transaction scores the peer. The reputation center aggregates the ratings and updates the overall scores.

Centralized reputation systems have two main aspects [33].

- Centralized communication protocols which aggregates ratings for the central authority and fetches the scores.
- Reputation computation engine which is used by the central authority to compute the reputation score for each peer.

1.3.2.2   Distribute Reputation System

In these systems, there is no central authority to receive, compute, and send the reputation scores; instead, there are distributed spots that aggregate the scores; or alternately, the peers themselves express their belief about their experience with other peers and transfer these information once requested by other peers in the network. Therefore, if any peer should want to learn another peer's score, he should either look into the distinction spots or find the members with prior experience with that peer and inquire their belief about that particular peer [33].

Distributed reputation systems have two main aspects [33]:

- Distributed communication protocol, which aggregates score from other members; and
- Reputation computation method, which is used by the central authority to compute the reputation score for each peer.

1.3.3   REPUTATION COMPUTATION ENGINE

Trust and reputation are computed based on private information, public information, or both. Private information is the primary information that a user gains through experiencing a contact with another peer; and public information is the secondary kind of information received from third parties [33]. Various reputation computation methods are explained in following.

1.3.3.1   Sum or Average Scores

The most straightforward reputation computation method is the sum of positive and negative scores. It is also possible to consider the average of score as the reputation. A more advanced method is to give each rater a value and compute reputation based on the weighted average.

1.3.3.2   Bayesian Systems

Bayesian systems take binary scores as input and compute reputation by updating beta probability density function. The inductive reputation score is computed by mixing the primary reputation score as well as all new scores [33].

1.3.3.3   Discrete Trust Models

Some researchers have presented discrete trust models. They reason that statements are more comprehensible [33]. For example, instead of a score, they use "very certain," "certain," "uncertain," and "very uncertain" to describe the level of trust for a peer. Consequently, this way both the person who gives the scores and the member who uses the ratings fully understands the provided reputation score.

1.3.3.4   Belief Models

The belief theory framework is related to the probability theory; but in belief theory, unlike the probability theory, the sum of all possible probabilities does not necessarily equal 1. The remaining probabilities are interpreted as uncertainty [33].

1.3.3.5   Fuzzy Models

Trust and reputation could be converted into fuzzy concepts. The fuzzy logic provides rules to deduct fuzzy criteria [33].

1.3.3.6   Flow Models

Flow model systems compute trust and reputation by transitive interaction through long or circular chains [33].

1.4   EFFECTS OF PRESENCE OF VIRTUAL COMMUNITIES ON VIRTUAL TRUST

Trust is always shaped as a result of interactions. The interactions may or may not be direct. Unlike traditional shoppings, in online purchases user or customer has no means to interact with other people. In traditional offline commerce, customers walk into the store and talk to other customers and establish social contact. This contact with other customers makes a basis for deciding whether or not to trust that store. And once trust is formed, chances for financial transactions to take place increase. In online environments, how can customers trust when they cannot see sellers and other customers? Trust models in s-commerce mean to, in a way, transfer the experience of other peers to the buyer. The challenge here would be to trust the beliefs of people with whom you have had no previous interactions. Obviously, we put a better faith on the words of people we know and with whom we have a history. Likewise, these people have a better chance of persuading us to purchase a particular product. The emergence of Web 2.0 and social networks that gather a big community of family members and friends has created a great opportunity to overcome this challenge. It is now possible to form a community of friends that was formerly unavailable on the net. In the following, the relationship between social networks and trust is discussed.

1.4.1   SOCIAL NETWORKS AND TRUST

Information about a peer is usually collected from a group of people and is shared by the same people. People make decisions based on such shared information. Use of social networks is one of the means through which people easily discuss and share information. Now, social networks have turned into an important service on the net and provide a wide range of applications such as collaborative work, collaborative service rating, resource sharing, and searching for new friends [45]. Social networks are made of a set of nodes, users, interconnected by edges, the relationships. Therefore, social networks have created a suitable platform for people to aggregate the information they need to evaluate trust and reputation. Of course, the information shared in social networks is raw and needs further processing to be used in trust models. As previously discussed, trust is a concept related to context and content; this should be considered when trying to exploit shared information. Another important issue is the existence of social structure; in a society people possess different characteristics (such as sexuality, race, education, and feelings) that distinguish each individual. These individuals have relationships, they interact and share their information; however, people do not intent to have relations with just any other people. They select other individuals with common interests and beliefs as themselves. Such issues help identify the structure of social networks as well as assist perfectly evaluate trust.

1.4.2   RELATIONSHIPS IN SOCIAL NETWORKS

Awareness of relationships in social networks is very important when deciding to use them in trust and social networks. Here, we give a brief explanation on the structure of relationships in social networks.

Edges in social networks resemble relationships which could be either rooted in real life relationships, like work or school relationships or kinship or be based on the relations originally shaped in the virtual world. Just like real life, edges or relationships on the net are also created by the existence of similarities between peers. As a result, if we consider a social network as a graph, we would see a high density of edges in some parts and few of them in other areas. High-density areas in the graph represent a set of people who are closely linked due to their similarities. In the study of social networks, these edges that have close relationships with each other are called a community. The formation of communities is closely related with Homophily principle. Based on this principle, people have a tendency to have relationships [46] with others whom have the same interests as themselves.

Recognizing these facts makes it easier to use social networks to create trust in e-commerce and produces better results.

1.5   TRUST IN S-COMMERCE

In this section, the previous literature on s-commerce about trust is reviewed, based on data collection and analysis methods [47], we selected researches in 2004–2013 time period to find the researches in e-commerce and particularly to review papers on the topic of trust. Exploring these papers reveals emerging topics, challenges, and opportunities as well as gives insights on the literature.

1.5.1   METHODOLOGY

To extract papers on trust in s-commerce, Google Scholar search engine was employed. Google Scholar, besides its simple search tools, provides researchers with a more advanced search options of looking into papers, theses, books, and publications. Some main resources include IEEE, ACM, Springer, and Science Direct [48]. We used "s-commerce" and "trust" keywords in 2004 to 2013 in the publications field. The results were further refined to exclusively list full-text publications in English. Since there was a limited number of results (5 journal papers), papers on trust in commercial social networks or those focusing on the effects of social networks on trust in e-commerce were also added to the list of review papers. A total of 14 papers were selected and reviewed in our survey study (Table 1.1).

1.5.2   CLASSIFICATION OF RESEARCHES

There are different classifications in e-commerce research [49–52]. Exploring 275 published papers between 1993 and 1999 [51] has identified

TABLE 1.1   List of Journal and Conference Papers on Trust in S-Commerce

Title	Number of Publications	Type
Behavior and Information Technology	1	Journal
Decision Support Systems	1	
Electronic Commerce Research and Applications	3	
Information and Management	1	
Information Technology Journal	1	
International Journal of Information Management	1	
International Journal of Information Science and Management	1	
Journal of Applied Sciences	1	
Journal of Economics, Business and Management	1	
International Conference on Computational Intelligence and Security	1	Conference
ACM conference on Electronic commerce	1	
IEEE International Conference on Communications Workshops	1	

four main topics of discussion in e-commerce: (i) technical issues such as security, networking technology, and support systems; (ii) applications including sales, marketing, and advertisement, (iii) implementation and support such as public policy and corporate strategies; and (iv) other issues like fundamental, introductory, and public purposes. In Ref. [52], Wareham et al., identified important issues in e-commerce based on journal papers between 1997 and 2003, and also extracted methodologies employed in those papers. In Ref. [14] the author has proposed a research framework in s-commerce that could be used to classify and determine this distinguishing finding of researches in this new research field. Based on this framework, each research has six elements: commercial activities, social media, research themes, research methods, theories, and outcome measures. Refs. [14, 51, 52] are classified and shown in Table 1.2. Table 1.2 gives an overview of all the papers used in our survey along with their main topics, research methodology, modeling method and analysis, data set, and data collection techniques. Section 1.5.3 explores each paper.

TABLE 1.2 Classification of Research Paper on Trust in S-Commerce

ID	Author; Year	Research Theme	Research Method	Modeling/Analyzing Method	Dataset; Size	Data collection method
1	Gan et al.; 2009[53]	Graphical modeling of trust in	N/A	Graphical representation	N/A	N/A
2	Lu et al.; 2010[55]	Trust in virtual communities	Survey	Research model, SEM	Taoboa virtual community 1 – Offline: Wuhan university students 2 – Online: university students in Taoboa; 373 users	Manual; distributing online and offline questionnaires
3	Chuang; 2010[54]	Quantitatively measure trust	N/A	Algorithm (Probability theoretic framework), Distributed decision	N/A	Systematic; no detail
4	Guo et al.; 2011[56]	Price of trust	Triads Direct closure Process	N/A	Taoboa members with at least one commercial interaction, for two months; 1 million users	Systematic;the first one million users with at least one commercial interaction are restored and analyzed

TABLE 1.2 Continued

ID	Author; Year	Research Theme	Research Method	Modeling/Analyzing Method	Dataset; Size	Data collection method
5	Utz et al.; 2012[57]	Effects of online reviews on trust	Experimental research	Regression analysis	University of Amsterdam students; 100 students for the first experiment and 131 users for the second experiment	Manual; no detail
6	Kim and Noh; 2012[60]	Influencing factors on consumers' trust and trust performances	Survey	Research model, SEM	Korean s-commerce users; 466 users	Manual; online, offline, telephone and email methods
7	Gorner et al.; 2013[61]	Improving trust modeling of agents	Experimental research	Algorithm	Simulation; 1 buyer 80 adviser 100 seller	Systematic; based on imaginary data; N/A
8	Leeraphong and Mardjo; 2013[63]	Trust and risk in purchase intention	Focus group	Research model	Working adults with 25 to 34 years of age; 15 users	Manual; no detail
9	Li et al.; 2013[66]	Recommender system	Experimental research using survey	Framework, AHP	Yahoo! Shopping customers in Taiwan; 424 participants	Manual; sending e-questionnaires

TABLE 1.2 Continued

ID	Author; Year	Research Theme	Research Method	Modeling/Analyzing Method	Dataset; Size	Data collection method
10	Kim and Park; 2013[64]	Effects of s-commerce factors on trust and trust performance	Survey	Research model, PLS, SEM	Users living in Korea; 371 users	Manual; online and offline questionnaires, email, telephones, and random distribution of 2000 questionnaires among Korean users
11	Hajli et al.; 2013[65]	Word of mouth and trust	Survey	Research model, SEM	Forum users, virtual communities, and social networks; 295 users	Manual; sending invitations through email to users in various networks in 2 months
12	Shin; 2013[27]	Customer behavior	Survey (in-depth interview, focus group, final survey questionnaire)	Research model, SEM	S-commerce users in Korea; 329 users	Manual; placing survey invitations in blogs, forums and communities and collecting data in 6 months
13	Ng; 2013[28]	Culture and intention to purchase	Survey	Research model, Covariance-based Structural Equation Modeling (CBSEM)	Facebook social network; 248 users	Manual; sampling using snowball technique

1.5.3   LITERATURE REVIEW: A CLOSE LOOK

In Section 5.2 are further explored in this section. It is worth mentioning that papers, which only present a model and do not include research findings, such as Ref. [53], are excluded. For each paper, the main topic, objectives, methodology, and important findings are provided.

• **Researches in 2010**

Chuang in Ref. [54] introduces a theoretical framework based on probabilities to quantitatively measure trust and also model customers and sellers behavior in an e-commerce system by measuring trust. The objective of this paper is to provide a model to identify trustworthy and non-trustworthy users through mathematical reasoning and probabilities. In his paper, Chuang first identifies trust antecedents and finally implements an algorithm that is reliable against malicious sellers. He considers user confidence as the foundation for e-commerce trust and states four characteristics for trust: (i) trust is a set of states (trust = {trustworthy, non-trustworthy}); (ii) trust is a system for mathematical reasoning; (iii) trust is a statistical decision resulting from sellers certainty level and consumers' benefit; and (iv) trust results from the credit (reputation) of seller and cooperation with networks. Therefore, trust is a statistical measure in {0, 1}. Chuang then provides a network model in which any transaction between users is illustrated. By carefully studying the transactions we would be able to learn about trustworthiness of a user. Nonetheless, in any circumstances, there is a chance of system error; and misbehavior of users may lead to mistakes in identifying trustworthy users. In the end, to illustrate the improvements and advantages of the propose algorithms, some Yahoo user accounts were used by different algorithms (weighted-sum algorithm and belief propagation) to detect misbehaving users; the results proved the improvement of the proposed algorithm over the older methods.

Exploring and analyzing the effective factors on trust building in virtual communities and effects of this trust on purchase intention in Ref. [55], Lu et al., conclude that trust in virtual communities affects trust in the website or vendor. The main contribution of his work is relating virtual communities to C2C e-commerce. Lu et al., have studied the conversion of virtual communities to C2C buyers; in his paper, Lu names ability, benevolence, and integrity as the three constructs of "trust in website or vendor" and ability and benevolence/integrity as the two constructs of "trust in members." The interconnection of these elements and trust has also been studied. To test the hypotheses, they used the database of Taoboa, a Chinese website. Data collection included three different forms: distributing questionnaires among students and graduates of Wuhan University; placing e-questionnaires on the e-market website of Huazhong University for two weeks; and distributing questionnaires in Taoboa community. Some important findings of this paper are: (i) familiarity has a direct relationship with trust in members, but this trust only concerns the aspect of ability and not benevolence or integrity; (ii) perceived familiarity has a direct relationship with trust in members. Evidence shows that the effect of familiarity is greater on benevolence and integrity than on ability; (iii) trust disposition has a direct relationship with trust in member and the ability and integrity aspects of trust in website or vendor; and 4- trust in members has no proved effects on intention to obtain information. The intentions to obtain information and purchase intention are both affected by trust in members, or more specifically, trust in members' ability.

• **Researches in 2011**

Guo et al., in an article in 2011 [56], "the role of social networks in online shopping information passing, price of trust, and consumer choice," quantitatively measured information passing by triads and directed closure process. They also went further to put a price on trust by rating stores. The logic behind estimating trust price is how much more are customers willing to pay to purchase their products from a trusted seller; the extra payment is considered as the price paid for trust. Guo, in his paper, by analyzing performance of a set of perceived features, shows how social graph (the seller-buyer graph) helps better understand customers' behavior. The advantage of Guo's research is the inclusion of all activities of the first million users who have had at least one commercial interaction in Taoboa, a Chinese e-commerce company. One of the hypotheses of Guo's study is that all commercial activities are founded on social networks and the direct relationship between the number of exchanged messages and their commercial activities supports this hypothesis. Furthermore, the relationship between social proximity and the probability of doing commercial activity between users is supported by results. This shows that the increase in the number of mutual friends leads to an increase in the probability that a commercial activity is going to take place. The findings of Guo's work support Granovetter studies. Granovetter believes that economic transactions are rooted in social network streams and that each person's social graph determines the people with whom he establishes commercial links.

• Researches in 2012

In Ref. [57] to study the effects of online store review on consumers' trust in that store, Utz et al., conducted two separate experiments: experiment 1 concerns use of information about social relationships like consumer reviews to judge the trustworthiness of an online store, compared with deciding about its trustworthiness on its overall reputation; and experiment 2 assesses assurance seals. Furthermore, role of disposition to trust has been studied. Trust has a positive relationship with trust disposition and perceived trusting beliefs. Mcknight et al., [58, 59] developed some measurements for evaluating perceived trustworthiness. Utz et al., to evaluate the same factor used the signaling theory. Signaling theory states that in an environment of information asymmetry, signals must be notified. Another principle used as the theoretical foundation is warranting principle. It states that a signal is more effective in shaping trust if it is harder to manipulate (user reviews in their research). In the first experiment, one hundred students from University of Amsterdam were invited; each of these students were taken to separate rooms equipped with computer sets. The instructions were given through the computer. Respondent were asked to imagine that they have entered an e-shop to buy an iPad. Pictures of the store were then shown to the respondents followed by questions about trustworthiness of that store as well as demographic questions. Trust disposition of each respondent was also measured by 8 items. Perceived trust was the dependent variable of this experiment measure by 8 item questions developed by McKnight. In the second experiment, 131 respondents took part and trust disposition was measured. Respondents were selected through snowball selection technique. Questions began with demographic questions and then their trust disposition was measured. Like the previous experiment, respondents were asked to imagine that they want to buy an iPad set and pictures of the online store were displayed. The independent variables in this experiment were assurance seals, user reviews, and disposition to trust. In the latter experiment three states of positive, negative, and neutral were assumed for user reviews. Unlike the previous experiments, there were no significant effects of store reputation its perceived trustworthiness. Utz et al., suggest that special attentions must be paid to transactions based on social relationships such as user reviews in online stores. Assurance seals also did not have a significant effect on improving trustworthiness of online stores.

Kim and Noh in [60] studied the effects of various trust antecedents (reputation, information quality, communication, and company size) on consumers' trust in s-commerce. They also studied the moderating effects of consumers' experience on the relationship between these antecedents and between trust and trust performance in s-commerce. Trust antecedents and hypotheses were mostly extracted from the literature and others were identified from interviews. This paper defines trust as the willingness of users to trust the capacity, integrity, benevolence, and predictability of sellers based on their beliefs. The sample population consisted of s-commerce users in Korea. Different data collection techniques such as online, offline, telephone, and email methods were employed. Respondent had 12–24 months of experience in using s-commerce sites. All hypotheses were supported by the findings. Therefore: (i) proposed trust antecedents have a positive effect on consumers' trust; (ii) the moderating effect of experience on consumers' trust is supported; and finally, (iii) the positive effect of trust on trust performance is proved.

• Researches in 2013

Shin in "user experience in social commerce: in friends we trust" [27], analyzes consumers' behavior, emphasizing on the role of social impacts in s-commerce. He develops a model to evaluate the relationship between subjective norms and trust, social support, attitude, and intention. To formulate the research model and research hypotheses, Shin uses the theory of planned behavior (TPB) and Technology Acceptance Model (TAM). It along with the original dimensions, includes perceived trust, social support, social norms, usefulness, and enjoyment. The research methodology is survey with four phases. First in an interview with 10 users, questions about theirs s-commerce experience were asked to gain a clear big picture of the subject. Then assisted by s-commerce service providers, 5 focus groups were selected and questions about their experience and influential factors were asked to validate survey questions and gain an insight about influential factors on users' behavior. Based on focus group interviews, the final survey questionnaire was then created. Finally, the questionnaire were distributed between 30 users for pilot test. The finalized questionnaire were then distributed electronically between users. Shin uses AMOS modeling method to analyze results. The results proved all hypotheses of the paper and the key effect of social norms on perceived trust and social support was significantly proved. Findings also emphasize the influential role of social norm on user Intention to Use s-commerce.

Gorner et al., in [61] advance the current modeling techniques like TRAVOS [62], and Personalized Trust by two improvements. The first improvement is limiting the size of network advisors. This method defines a threshold for trustworthiness and buyers get advice exclusively from advisors with a predefined level of trustworthiness. The other improvement is putting a limit on the number of advisors. By the second improvement, buyers are given access to advisors' ratings to a particular seller and may use this information to determine a public reputation score for that seller. To experiment the proposed algorithm, a set of data including 80 advisors, 100 sellers and one buyer was used in a simulation. The stimulation was conducted and results proved the superiority of the proposed model over previous algorithms.

Leeraphong and Mardjo in [63] developed a model that takes risk, trust, social norm and online shopping experience as factors affecting online purchase intention of consumers between ages of 25–34. The four mentioned factors are extracted from the literature. These factors shape the primary hypotheses and the interrelationship of these factors is proposed by this research. In fact, they first presented the primary research model based on the literature and then revisited the model after analyzing the results of focus group interviews. The focus group consisted of 15 working adults between 25 and 34. After analyzing the results, word of mouth advertisement was added to the factors and a few alterations were made to the relationship between the proposed factors. Findings suggest the following relationships that are considered a contribution of this paper to the literature: the relationship between social norm and trust, social norm and risk, word of mouth advertisement and trust, and also word of mouth and risk.

In their article "effects of various characteristics of social commerce on consumers' trust and trust performance" [64], Kim and Park have identified the principal s-commerce factors affecting trust (reputation, company size, information quality, security of transactions, communications, economic feasibility, and word of mouth referrals). They have then studied the effects of these factors on users' trust. The research model has been extracted from literature on trust and s-commerce and informal interviews. The s-commerce users in Korea have been the main target population of this study; however, to make it more generalizable, users of different countries (China, Japan, USA, and some European countries) living in Korea were also added to the target population. Various online and offline survey methods, phone calls and email were used to collect data. PLS approach was employed to analyze the model. All hypotheses of the research (except for the positive effect of financial feasibility on users' trust) were supported by the data. Findings show that despite the importance of discounts in commerce, it is not the most important factor for all users when deciding the trustworthiness of s-commerce.

In order to develop a research model to study the effect of consumers' Internet contact (including word of mouth referrals) on trust building mechanism, Hajli et al., in [65] studied the theory of s-commerce constructs and explored the new product development and trust concepts. In this work, two dimensions (benevolence and credibility) are assumed for trust. Recommendations, ratings, and user reviews as well as user communities and forums have defined as s-commerce constructs. To test the hypotheses, Hajli et al., used survey method. Survey questions were extracted from previous research. E-questionnaires were sent to users in different networks and a total of 295 valid responses were received. SEM data analysis technique was used to learn that the s-commerce constructs have a positive effect on trust. This finding shows that consumers prefer to learn about the products from the information and content provided by other consumers. Therefore, social links between people through s-commerce introduces word of mouth referrals as a valuable source of trust.

S-P Ng studied purchase intention in s-commerce considering the moderating effects of culture [28]. Also the mediating effect of trust on the relationship between social interaction and the intention to purchase was proposed. To test the hypotheses, the theories of trust transference, social interaction and Hofstede's cultural dimension were introduced and employed. To study the effects of culture on s-commerce this research uses a dataset from Facebook. In his experiment See and Ng put up an experimental s-commerce site on Facebook social networking website and monitored the user activities. The first level users were chosen from different geographical zones (Latin America and East Asia) to control the effects of culture. The experimental site was online for two weeks and using snowball-sampling method, e-questionnaires were sent to customers of the experimental s-commerce site. Data were analyzed using the Covariance-based Structural Equation Modeling (CBSEM) and findings suggest that closeness, familiarity (the two antecedents of social interaction) and trust predict purchase intention in social network community.

S-commerce acceptance and the collective social aspect of s-commerce have been the interest of study for

Li et al., in [66] introduced a recommender system that is an improvement over previous e-commerce recommender systems. They designed this system to be used in s-commerce websites where a peer's previous purchases and others' similar purchases as well as the relationships of that peer are of importance when deciding which products to buy or which brands to purchase from. Unlike traditional e-commerce sites, by definition, in s-commerce an important tenant of the system is the social relations each member has; Li et al., have tried to fill this gap by developing a new recommender system that incorporates preference similarity, recommendation trust, and social relations. The provided recommender system is made of four different modules, considering different aspects of the system; namely preference similarity module, recommendation trust analysis module, social relations analysis module, and personalized product recommendation module. The authors conducted an experiment by inviting Yahoo Shopping users and collected the data. Of the 1075 invitations sent, 424 participants joined the experiment. Using the MCDM and APH methods the improvement of the proposed algorithm over the previous ones was supported.

1.5.4 DISCUSSION

Regarding the trend in Table 1.2, apparently the researches in this area are carried out after 2009 and mainly in 2013; this reveals the importance of this topic and the gap in literature. In most researches, the research model is based on literature review and data set to evaluate the model is collected via questionnaires (online or offline) from a limited population. Therefore, some of these studies [28, 57, 65] may be not very generalizable.

To conclude the reviewed papers, Table 1.3 briefly shows, based on the provided models and evaluations, what factors affect trust in s-commerce. In addition, it gives a picture of what elements are affected by trust in s-commerce. In Table 1.3 papers (including Refs. [53, 54, 56, 61, 66]), which did not study the effects of trust or factors influencing trust, are not listed.

1.5.5 FUTURE STUDIES

Based on the research framework in [14], and research classification in Section 1.5.2, avenues for future research are classified into six categories.

• **Commercial activities:** commercial activities cover a large range of activities. Generally, from the perspective of customers, they are divided into three streams: before purchase, amid purchase, and after purchase activities. Some commercial activities may belong to more than one category. Marketing, advertisement, WoM, and recommenders are before purchase activities. Transactions are considered amid purchase activities. Customer service, CRM, and rating are considered after purchase activities. The issue of trust could be studied in any of these activities with the objective of exploring its effect on their performance. Of course, the importance of trust varies in each of these activities. Viral marketing, for instance, which influences customers' choice in s-commerce could be studies regarding how trust influences it [56]. As another example, studying the effects of trust-based Internet communications of consumers on socially developing word of mouth referring could be a good research topic [65].

• **Employed theories:** many different theories have been used in s-commerce researches. They include theories of social support [65], social influence, and social interactions. Influence of online virtual communities on s-commerce stakeholders is also worth discussing [55]. Future studies could broaden the topics to include more social and individual characteristics [60, 64, 27]. The Homophily principle also is another social theory that begs further research. For example, as an idea for future research, one could explore the effects of

TABLE 1.3   Summary of Findings on Factors Affecting Trust and Those Affected by it in S-Commerce

ID	Author; Year	Factors positively influencing trust	Factors with a negative effect on trust	Factors affected by trust
1	Lu et al.; 2010[55]	Familiarity with other members (affects trust in other members ability) Perceived similarity with other members (affects trust in other member' benevolence and integrity) Perception of structural assurance (affects trust in other members' ability) Trust propensity (affects trust in other members) Trust propensity (affects trust in website or vendor's ability and integrity) Trust in other members' ability (affects trust in the website or vendor)	N/A	(Trust in website or vendor's ability effects) purchase intention (Trust in other members' benevolence and integrity effects) purchase intention (trust in website or vendor's ability effects) intention to get information
2	Utz et al.; 2012[57]	Dispositional trust (influences perceived trustworthiness of the online stone) Positive user reviews (influences perceived trustworthiness of the online stone)	Negative user reviews	N/A
3	Kim and Noh; 2012[60]	Trust antecedents in a firm (reputation, size of the firm, information quality and communication). Experience has a moderating effect	N/A	Trust performance
4	Leeraphong and Mardjo; 2013[63]	Subjective norm Online word of mouth	N/A	Purchase intention
5	Kim and Park; 2013[64]	Size of the firm, information quality, transaction safety, communication, and word of mouth referrals	N/A	Purchase intention Word of mouth Intention (Trust performance)

TABLE 1.3 Continued

ID	Author; Year	Factors positively influencing trust	Factors with a negative effect on trust	Factors affected by trust
6	Hajli et al.; 2013[65]	Social commerce constructs (recommendations and referrals, ratings and reviews, forums and communities)	N/A	N/A
7	Shin; 2013[27]	N/A	N/A	Behavior (with moderating effect of social norm)
8	Ng; 2013[28]	Closeness to the social network Familiarity with a social network	N/A	Purchase intention The relationship between closeness and intention to purchase (mediating effect of trust) The relationship between familiarity and intention to purchase (mediating effect of trust) [culture moderates these mediating effects of trust]

familiarity with a store, the perceived similarity of websites and its effects on trust [55]. Social norms, due to their key role in trust should be more deeply studied in the context of s-commerce [27].

• **Research method:** based on Table 1.2, most studies have used survey method and questionnaires. As stated by [52] the other methodologies like data analysis, case study, and empirical and experimental methodologies could be employed along with real life or stimulated data.

• **Dataset:** in most of these studies, research data were collected using a limited number of questionnaires from a specific group of people living in a particular geographic location. Future studies could employ a broader dataset with more population and more diverse demographic and geographic characteristics to improve generalizability [27, 63, 64]. Furthermore, another way to increase generalizability would be to gather more data from different social platforms [65] and different world regions [28]. Monitoring customer behavior on s-commerce platform (e.g., controlling payment behavior [28]) and use of systematic methods for collecting individual data and costumer behavior instead of using questionnaires may lead to more accurate insights about the actual behavior of customers. Stimulating user behavior in s-commerce is another useful method for collecting research dataset.

• **Outcome measures:** research models could be developed to include more constructs such as consumers' purchase intention [65], intention to adoption of s-commerce, intention to have more participation in s-commerce or continuity of activities in s-commerce. Customer loyalty, customer satisfaction, website usage and similar issues are outcome measures that may be affected by trust.

• **Research theme:** the issue of trust could be studied in many research themes in s-commerce. So far, most researches have revolved around users' (customer/consumer) behavior; future research could search to find more factors affecting behavior and also focus on sellers' behavior. In addition, firm performance, networks analysis, business models, enterprise strategies, and system quality, information quality [28] social processes are other areas that deserve future research.

The current research models could be experimented in other environments to verify their comprehensiveness [65]. Nowadays, mobile social networks like WhatsApp, Line, and Viber have provided a chance for "mobile s-commerce"; and this makes trust also face the new challenges of mobile platform. Furthermore, trust, due to its dependability on time and its dynamic nature is always changing. Recognizing and understanding social behavior of people as well as their individual and social characteristics, which cause a change (increase/decrease) in trust is of importance in studying trust. Furthermore, since social networks are an important aspect of s-commerce and, as discussed in Section 1.4, trust in social networks depends on context, content and social structures, these elements are also generalizable to commercial platforms. Therefore, trust in a person for buying/selling digital equipment may vary with the trust in same person for buying/selling cars.

Future studies may, in addition, repeat the same studies in different cultural settings [28] or add other factors variable to their proposed model to enhance its effectiveness [66]. Some factors require a longer period of time to show their effects, it is suggested to carry out longitudinal studies to gain a better understanding of those factors [28].

1.6   CONCLUSION

S-commerce is a recent stream of e-commerce [7, 12, 13], which links users through social networks and other social media, and facilitates the commerce process [64]. The actual nature of s-commerce is sharing, experiencing, networking, and building long-term relationships. Many scientists consider s-commerce the center of the next wave of e-commerce [9, 14, 67]. Since the term "social commerce," after the emergence of Web 2.0 and the related technologies, was first used in 2005, the literature on this subject is still scarce and the academic research still requires more time and attention [9].

In any kind of commerce, financial transactions require trust and without it they undoubtedly face difficulties. Doing business on e-commerce platforms, due to the existence of uncertainty and risk, are in more need of trust building mechanisms. Therefore, researches have entered the subject and looked for methods to reduce this uncertainty by improving trust.

In the study of trust, decisions are made based on sharing of information about a certain peer by other people. Information sharing, user created content, and interacting with other peers, are the three very important features in social networks; and since s-commerce is a combination of Web 2.0 technologies (and especially social networks) and commercial activities in a social and collaborative platform, it has created an opportunity to study trust and to improve it. Doing commercial activities in a social platform provides the stakeholders (customer, seller, firms, and intermediaries) the information needed to evaluation the reputation and trust of users.

This chapter, after introducing the basic concepts of trust, explored the issue of trust in e-commerce. The effect of social networks and online virtual communities on trust in the literature were then studied. Using the survey method, conference and journal papers were studied. The main concern of these papers were trust in s-commerce or the effects of social networks on trust in e-commerce. The papers were classified and explored in detail. Finally, ideas for future research related to trust were presented in Section 1.5. Since there are a diverse range of factors influencing trust, and because trust also does influence other important factors that affect consumers' decision-making process in s-commerce, more systematic researches are required.

KEYWORDS

- **e-commerce**
- **s-commerce**
- **social media**
- **social networking websites**
- **survey**
- **trust**
- **Web 2.0**

REFERENCES

1. Cassidy, J. Dot.Con: The Greatest Story Ever Sold, Harper Collins Publishers, New York, 2002.
2. Machado, A. Drivers of shopping online: a literature review, *In Proceedings of the IADIS International Conference E-Commerce*, International Association for the Development of the Information Society Press, 2005.

3. Curty, R. G., Zhang, P. Website features that gave rise to social commerce: a historical analysis, *Electronic Commerce Research and Applications*, 12, pp. 260–279, 2013.

4. Kim, Y. A., Srivastava, J. Impact of social influence in e-commerce decision making, *In Proceedings of the Ninth International Conference on Electronic Commerce*, Minneapolis, MN, ACM Press, New York, NY, pp. 293–302, 2007.

5. Huang, Z., Benyoucef, M. From e-commerce to social commerce: A close look at design features, *Electronic Commerce Research and Applications*, 12, pp. 246–259, 2013.

6. Stephen, A. T., Toubia, O. Driving value from social commerce networks, *Journal of Marketing Research*, 47, 2, pp. 215–228, 2010.

7. Curty, R. G., Zhang, P. Social Commerce: Looking Back and Forward, *ASIST 2011*, October 9–13, New Orleans, LA, USA, 2011.

8. Yahoo! Social commerce via the Shoposphere and pick lists. 2005. Available at http://www.ysearchblog.com/2005/11/14/social-commerce-via-the-shoposphere-pick-lists/.

9. Zhou, L., Zhang, P., Zimmermann, H.-D. Social commerce research: An integrated view, *Electronic Commerce Research and Applications*, 12, pp. 61–68, 2013.

10. Wang, C. Linking shopping and social networking: Approaches to social shopping, *15th Americas Conference on Information Systems (AMCIS)*, San Diego, CA, USA, 2009.

11. Jascanu, N., Jascanu, V., Nicolau, F. A new approach to E-commerce multi-agent systems, *The Annals of "Dunarea De Jos" University of Galati: Fascicle III Electrotechnics, Electronics, Automatic Control and Informatics*, pp. 8–11, 2007.

12. Wang, C., Zhang, P., The evolution of social commerce: an examination from the people, business, technology, and information perspective, *Communication of the Association for Information Systems*, 31, 5, pp. 105–127, 2012.

13. Wang, C., Zhang, P., The evolution of social commerce: the people, management, technology and information dimensions, *Communications of the Association for Information Systems*, 32, pp. 1–23, 2012.

14. Liang, T.-P., Turban, E. Introduction to the special issue – Social Commerce: A Research Framework for Social Commerce, *International Journal of Electronic Commerce*, Vol. 16, No. 2, pp. 5–13, Winter 2011–12.

15. Marsden, P. Commerce gets social: How your networks are driving what you buy, *Social Commerce Today*, January 2011.

16. Constantinides, E., Fountain, S. J. Web 2.0: conceptual foundations and marketing issues, *Journal of Direct, Data and Digital Marketing Practice*, 9, 3, pp. 231–244, 2008.

17. Lee, S. H., DeWester, D., Park, S. R. Web 2.0 and opportunities for small business, *Service Business*, 2, 4, pp. 335–345, 2008.

18. Murugesan, S. Understanding Web 2.0. IT Professional, 9, 4, pp. 34–41, 2007.

19. Wigand, R. T., Benjamin, R. I., Birkland, J. Web 2.0 and beyond: implications for electronic commerce, *In Proceedings of the 10th International Conference on Electronic Commerce*, Innsbruck, Austria, ACM Press, New York, NY, 2008.

20. Marsden, P. How Social Commerce Works: The Social Psychology of Social Shopping, *Social Commerce Today*, Syzygy London, London, UK, 2009. Available at http://digitalintelligencetoday.com/how-social-commerce-works-the-social-psychology-of-social-shopping.

21. Mayer, R. C., Davis, J. H., Schoorman, F. D. An integrative model of organizational trust, *Academy of Management Review*, 20, pp. 709–734, 1995.

22. Jones, K. and Leonard, N. K. Trust in consumer-to-consumer electronic commerce, *Information & Management*, 45 (2), pp. 88–95, 2008.

23. Gefen, D., Karahanna, E., W. Straub, D. Trust and TAM in Online Shopping: An Integrated Model, *Management Information Systems Quarterly*, 27, 1, 2003.

24. Hajli, M. An Integrated Model for E-commerce Adoption at the Customer Level with the Impact of Social Commerce, *International Journal of Information Science and Management*, 16 (Special-Issue 2012 ECDC), pp. 77–97, 2012.

25. Hajli, M. Social Commerce: The Role of Trust, *Paper presented at the AMCIS 2012 Proceedings*, Paper 9, Seattle, 2012.

26. Hennig-Thurau, T., Malthouse, E. C., Friege, C., Gensler, S., Lobschat, L., Rangaswamy, A. and Skiera, Bernd. The Impact of New Media on Customer Relationships, *Journal of Service Research*, 13 (3), pp. 311–330, 2010.

27. Shin, D.-H. User experience in social commerce: in friends we trust, *Behaviour & Information Technology*, Vol. 32, No. 1, pp. 52–67, 2013.

28. Ng, C. S.-P. Intention to Purchase on Social Commerce Websites across Cultures: A Cross-regional Study, *Information & Management*, Vol. 50, Issue 8, pp. 609–620, 2013.

29. Kim, S.-B., Sun, K.-A., Kim, D.-Y. The Influence of Consumer Value-Based Factors on Attitude-Behavioral Intention in Social Commerce: The Differences between High- and Low-Technology Experience Groups, *Journal of Travel & Tourism Marketing*, Special Issue: Social Media, Vol. 30, Issue 1–2, 2013.

30. The, P.-L., P. K. A. Understanding social commerce adoption: An extension of the Technology Acceptance Model, *Management of Innovation and Technology (ICMIT), 2012 IEEE International Conference on*, pp. 359–364, 2012.

31. Suraworachet, W., Premsiri, S., Cooharojananone, N. The Study on the Effect of Facebook's Social Network Features toward Intention to Buy on F-commerce in Thailand, *Applications and the Internet (SAINT), 2012 IEEE/IPSJ 12th International Symposium on*, pp. 245–250, 2012.

32. McKnight, D. H., Norman, L. C. The Meanings of Trust, 1996.

33. Josang, A., Ismail, R., Boyd, C., A Survey of Trust and Reputation Systems for Online Service Provision, *Decision Support System*, Vol. 43, pp. 618–644, 2007.

34. Falcone, R., Castelfranchi, C. Social Trust: A Cognitive Approach, *Kluwer Academic Publisher*, pp. 55–99, 2001.

35. Oxford dictionary, Available at http://www.oxforddictionaries.com/definition/english/reputation.

36. Wang, Y., Vassileva, J. Trust and Reputation Model in Peer-to-Peer Networks, *IEEE Conference on P2P Computing*, Sweden, 2003.

37. Yan, Z., Holtmanns, S. Trust Modeling and Management: from Social Trust to Digital Trust, *book chapter of Computer Security, Privacy and Politics: Current Issues, Challenges and Solutions*, IGI Global, 2007.

38. Golbeck, J. Computing and Applying Trust in Web-based Social Networks, *PhD thesis, University of Maryland*, 2005.

39. Grandison, T., Sloman, M. A Survey of Trust in Internet Applications, *IEEE Communications Surveys and Tutorials*, 3, 2000.

40. Luhmann, N. Trust and power. London: Wiley; 1979.

41. Grandison, T., Sloman, M. A survey of trust in internet applications, *IEEE Communications and Survey*, 4th Quarter, 3(4), pp. 2–16, 2000.

42. Bonatti, P., Duma, C., Olmedilla, D., Shahmehri, N. An integration of reputation-based and policy-based trust management, *In Proceedings of the Semantic Web Policy Workshop*, 2005.

43. Thibaut, J. W. and Kelley, H. H. The Social Psychology of Groups, Wiley, New York, 1959.

44. Marmol, F., Perez, G. Security Threats Scenarios in Trust and Reputation Models for Distributed Systems, *Journal of Computer and Security*, 2009.

45. Domingo-Ferrer, J., Viejo, A., Sebe, F., Gonzalez-Nicolas, U. Privacy homomorphisms for social networks with private relationships, *Computer Networks*, 52, pp. 3007–3016, 2008.

46. McPherson, M., Smith-Lovin, L., M. Cook, J. Birds of a feather: Homophily in social networks, *Annual review of sociology*, pp. 415–444, 2001.

47. Chen, H., Chiang, R. H. L., Storey, V. C. Business intelligence and analytics: from big data to big impact, *MIS Quarterly*, 36, 4, pp. 1165–1188, 2012.

48. Available at http://scholar.google.com/intl/en-US/scholar/about.html

49. Ngai, E. W. T., Gunasekaran, A. A review for mobile commerce research and applications, *Decision Support Systems*, 43, pp. 3–15, 2007.

50. Turban, E., King, D., Lang, J. Introduction to Electronic Commerce, Edition 3, Pearson Education, Appendix A: Current Electronic Commerce Research, 2013.

51. Ngai, E. W. T., Wat, F. K. T. A literature review and classification of electronic commerce research, *Information & Management*, 39, pp. 415–429, 2002.

52. Wareham, J., Zheng, J. G., Straub, D. Critical themes in electronic commerce research: a meta-analysis, *Journal of Information Technology*, 20, pp. 1–19, 2005.

53. Gan, Z., He, J., Ding, Q., Varadharajan, V. Trust Relationship Modeling in E-Commerce-Based Social Network, International Conference on Computational Intelligence and Security, 2009.

54. Chuang, T.-Y. Trust with Social Network Learning in E-Commerce, *Communications Workshops (ICC), 2010 IEEE International Conference on,* pp. 1–6, 2010.

55. Lu, Y., Zhao, L., and, Wang, B. From virtual community members to C2C e-commerce buyers: Trust in virtual communities and its effect on consumers' purchase intention, *Electronic Commerce Research and Applications*, 9, pp. 346–360, 2010.

56. Guo, S., Wang, M., Leskovec, J. The Role of Social Networks in Online Shopping: Information Passing, Price of Trust, and Consumer Choice, *EC '11 Proceedings of the 12th ACM conference on Electronic commerce*, pp. 157–166, 2011.

57. Utz, S., Kerkhof, P., Bos, J. Consumers rule: How consumer reviews influence perceived trustworthiness of online stores, Electronic Commerce Research and Applications, 11, pp. 49–58, 2012.

58. McKnight, D. H., Choudhury, V., Kacmar, C. Developing and validating trust measures for e-commerce: an integrative typology, *Information Systems Research Special Issue: Measuring e-Commerce in Net-Enabled Organizations*, 13, 3, pp. 334–359, 2002.

59. McKnight, D. H., Choudhury, V., Kacmar, C. The impact of initial consumer trust on intentions to transact with a web site: a trust building model, *Journal of Strategic Information Systems*, 11, 3–4, pp. 297–323, 2002.

60. Kim, S., Noh, M.-J. Determinants influencing consumers' trust and trust performance of social commerce and moderating effect of experience, *Information technology journal*, 11 (10), pp. 1369–1380, 2012.

61. Gorner, J., Zhang, J., Cohen, R. Improving trust modeling through the limit of advisor network size and use of referrals, *Electronic Commerce Research and Applications*, 12, pp. 112–123, 2013.

62. Teacy, W. T. L., Patel, J., Jennings, N. R., Luck, M. TRAVOS: trust and reputation in the context of inaccurate information sources, *Autonomous Agents and Multi-Agent Systems*, 12, 2, pp. 183–198, 2006.

63. Leeraphong, A., Mardjo, A. Trust and Risk in Purchase Intention through Online Social Network: A Focus Group Study of Facebook in Thailand, *Journal of Economics, Business and Management*, Vol. 1, No. 4, pp. 314–318, 2013.

64. Kim, S., Park, H. Effects of various characteristics of social commerce (s-commerce) on consumers' trust and trust performance, *International Journal of Information Management*, 33, pp. 318–332, 2013.

65. Hajli, M., Hajli, M., Khani, F. Establishing Trust in Social Commerce through Social Word of Mouth, *International Journal of Information Science and Management*, Special Issue (ECDC 2013), pp. 39–54, 2013.

66. Li, Y.-M., Wu, C.-T., Lai, C.-Y. A social recommender mechanism for e-commerce: Combining similarity, trust, and relationship, *Decision Support Systems*, Vol. 55, Issue 3, pp. 740–752, 2013.

67. Nutley, M. Forget E-Commerce; Social Commerce Is Where It's At, Marketing Week, July 28, 2010, Available at http://www.marketingweek.co.uk/disciplines/digital/forget-ecommerce-social-commerce-iswhere-its-at/3016388.article.

PART 2:

E-DECISION

CHAPTER 2

E-DECISION SUPPORT IN RECONFIGURABLE MANUFACTURING SYSTEMS

NORMAN GWANGWAVA

Lecturer, Concurrent Engineering, Manufacturing Engineering Design, Manufacturing Information and Database Systems; Department of Industrial and Manufacturing Engineering, National University of Science and Technology, Bulawayo, Zimbabwe

CONTENTS

Manufacturing is one of the most important sectors of any economy. The high priority given to the sector by governments of different nations as well as from the management of the firms has also exposed the sector to stiff competition. Traditionally, manufacturers were producing for their local countries but the trend has drastically transformed to global focus, denoted as "Global Manufacturing." The transition led to the evolution of new design trends for manufacturing systems, which can be categorized into Dedicated Machine Tools (DMT), Flexible Manufacturing Systems (FMS) and Reconfigurable Manufacturing Systems (RMS). The revolution of the manufacturing industry can also be tracked from industry 1.0, industry 2.0, industry 3.0, to the current level, industry 4.0. Manufacturers are under pressure to offer cost effective products, improve the quality of manufactured products, achieve higher efficiency, improve communication, and completely integrate their processes. Large quantities of data are being collected in manufacturing factories, either on real-time or "post-mortem" basis. Numerous decisions need to be made using accumulated data so that factories can be more effective in their routine operation.

As the world population continues to grow, production output and production varieties in many companies continue to grow. Other factors of concern to manufacturers include shorter product life cycles, increasing product variation (Mass Customization – MC), volatile markets, cost reduction pressures, scarce resources, cleaner production, lack of skilled workforce and aging community. In order to cope with the pressure, modern day factories have big data repositories to analyze and make informed decisions.

Manufacturers are adopting RMS so as to handle a variety of manufactured goods through their own adjustable machine operations. RMS uses reconfigurable machine tools (RMTs) to achieve adjustability, customizability, scalability, and diagnosability. However, this comes with its own challenges. One major challenge faced by the manufacturing industry is the rapid reconfiguration of manufacturing systems to handle rapid change in business environment without human intervention. Reconfiguring existing manufacturing systems require detailed data about the capability of each machine and its current state is very necessary so that an optimal solution can be achieved. In order to minimize ramp-up time during the reconfiguration process, there should be less manual involvement. Automating the whole reconfiguration process minimizes the reconfiguration process overheads and also enables the manufacturer to meet demand within a short time frame. Full-Automatic Reconfiguration for Reconfigurable Systems (FARR) can be achieved through an agent based approach using ontology knowledge of the manufacturing environment so as to make decisions. This chapter presents different e-decision support tools that manufacturers adopting RMS systems can use to complement their processes.

2.1   RECONFIGURABLE MANUFACTURING SYSTEMS (RMS)

RMS is defined as a system designed at the outset for rapid change in structure, as well as in hardware and software components, in order to quickly adjust production capacity and functionality within a part family in response to sudden changes in market or regulatory requirements [1]. A system built with changeable structure provides scalability and customized flexibility and focuses on a part family, thus generating a responsive manufacturing system. The flexibility of RMS, although it is only "customized flexibility," provides all the flexibility needed to process that entire part family. RMS aims to achieve the cost effective and rapid system changes needed, by incorporating principles of modularity, integrability and scalability. In other words, RMS promises customized flexibility in a short time [2]. Drivers for the need and rationale for reconfigurable manufacturing systems arises from unpredictable market changes occurring with increased frequency in recent years [3]. These include:

- increasing frequency of introduction of existing products;
- changes in parts for existing products;

- large fluctuations in product demand and mix
- changes in government regulations (safety and environment) and,
- changes in process technology.

A typical application scenario of a reconfigurable manufacturing system can be illustrated by Figure 2.1. Customization, scalability and convertibility are the critical reconfiguration characteristics [4]. The other three, modularity, integrability and diagnosability allow rapid reconfiguration, but they do not guarantee modifications in production capacity and functionality. Customization is the most essential characteristic of RMS; it is based upon design for a part family or a product family [5].

2.1.1 COMPONENTS OF AN RMS

Two major components of an RMS are the reconfigurable machine tool (RMT) and the reconfigurable controller. By contrast to conventional CNCs, which are general-purpose machines, RMT design is for a specific customized range of operation requirements and may be cost effectively converted when the requirements change. RMTs are designed to produce a specific set of features for a specific range of cycle time. Some operation

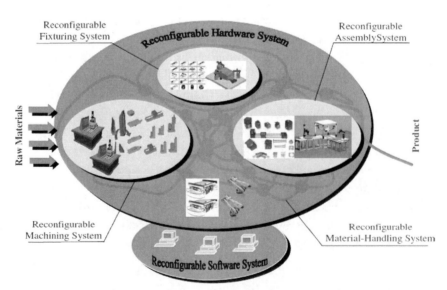

FIGURE 2.1 An application scenario for RMS [1, 6].

requirements will be constant over the lifetime of the machining system. The primary aim of the RMT is to cope with changes in the products or parts to be manufactured. The following possible changes must be taken into consideration:

- work piece size;
- part geometry and complexity;
- production volume and production rates;
- required processes;
- accuracy requirement in terms of geometrical accuracy and surface quality;
- material property, such as type of material, hardness, etc.

According to reconfigurable machines, tools comprise of four important modules to develop a reconfigurable control system as follows:

- Automatic Part Transfer System
- Automatic Part Clamping Rotating System
- Automatic Part Lifting System
- Automatic Tool Changing System.

The RMT must be designed and built into Reconfigurable Controller.

2.2   MASS CUSTOMIZATION (MC)

Mass Customization relates to the ability of a business entity to provide customized products or services in high volumes with short lead times through flexible processes at costs similar to standardized mass products [7]. Companies implementing MC often find themselves mired in a net of conflicts both strategically and operationally, for example the conflict between manufacturing costs and customization, and between responsiveness and customization. On the other hand it is asserted that MC is only viable for a very limited range of applications. Salvador and Forza [8] says innovative tactics and technologies like differentiation postponement, product family design (Group Technology GT), and product configuration systems have greatly mitigated the severity of these challenges. According to Koren [9], mass customized production is in practice realized with flexible production systems, which are able to deal with a variety of manufactured parts and adjustable assembly operations.

2.2.1 CONSOLIDATED KEY ACTIVITIES AND RESOURCES OF MASS CUSTOMIZATION

Key activities of MC include: elicit customer requirements; increase agility of supply chain; develop product variants and solution space based on MC dimensions; share and manage customer knowledge; and arrange efficient production. Implementation of MC is not feasible without application of some key resources. These resources might be physical, intellectual, human or financial resources. There are different studies focusing on required key resources to pursue MC. Flexible manufacturing processes and integrated information system as two main resources for MC, while some others focus also on other physical resources such as scanners and measurements systems [10]. Reconfigurable manufacturing system is considered one of the main resources for a company offering MC [11].

Table 2.1 illustrates the relationships that exist between the key resources and activities of MC. Wherever an alternative corresponds to a specific pillar of MC; this correspondence is illustrated by "X." For instance the alternative "develop product platform and modules considering commonality in modules and components" impacts on three pillars of MC. It supports the company to meet the needs of each individual customer due to the fact that diverse customized products can be configured by different customers through combinations of common modules. Moreover, applying modularity helps company to define a stable solution space and it is an enabler to control cost level (and consequently price level) since standard modules do not require high cost. Hence two MC pillars "stable solution space" and "adequate cost and price level" are covered by this alternatives in addition to "meeting the needs of each individual customer."

2.2.2 LEVELS OF MASS CUSTOMIZATION

Determining the level of individualization and characterizing truly mass-customized products is a major point of contention in the MC debate. Purists attributed the MC concept only to products that contemplate all requirements made by individual customers. Pragmatists suggest MC to be simply about delivering products following customer options, independent of the number of options actually offered. According to Hart [12], the solution for

TABLE 2.1 Key Activities and Resources of MC Pillars [10]

Key Activities	MC Pillars / Alternatives	Customer co-design	Meeting the needs of each individual customer	Stable solution space	Adequate price and cost level
Elicit customer requirements	Recognizing customer individuality	X	X		
	Collecting customer feedback from prototypes	X	X		
	Translating customer requirements through co-creation	X	X		
Develop product variants and solution space based on MC dimentions(axes)	Developing product platform and modules by considering commonality in modules and components		X	X	X
Increase agility of supply chain	Undertaking variety management	X	X	X	X
	Integrating with supply chain partners in processes and in sharing information	X	X	X	X
	Reaching customers through efficient logistics services		X		X
Share and manage customer knowledge	Managing customer knowledge		X		X
Arrange efficient production	Applying efficient and flexible manufacturing system			X	X
	Implementing postponement strategy		X	X	X

TABLE 2.1 Continued

	MC Pillars / Alternatives	Customer co-design	Meeting the needs of each individual customer	Stable solution space	Adequate price and cost level
Key Activities	Increase flexability of manufacturing system				
	Automated manufacturing and assembly		X	X	X
	Automated materials handling system			X	X
	Automated inspection system				X
Increase reconfigurability of manufacturing system	Reconfigurable machine tools			X	X
	Reconfigurable assembly systems		X	X	X
	Reconfigurable inspection system				X
	Reconfigurable material handling system				X
	Process monitering system				
Increasing customization level using points of sales system	Measurement devices	X	X		X
Increase level of information integration	Configurators	X	X		X
	Order processing system	X	X		X
Support customers in co-design via human resource	Trained personnel	X	X	X	X

this contention lies in careful determination of the range in which a product or service can be meaningfully customized, and how individuals make options upon this range. Successful MC systems should be able to mix true individualization with high part variety and standardized processes [13].

Gilmore and Pine [14] identified four customization levels based mostly on empirical observation: collaborative (designer dialog with customers), adaptive (standard products can be altered by customers during use), cosmetic (standard products are packaged specially for each customer), and transparent (products are adapted to individual needs). Pine [15] suggests five stages of modular production: customized services (standard products are tailored by people in marketing and delivery before they reach customers), embedded customization (standard products can be altered by customers during use), point-of-delivery customization (additional custom work can be done at the point of sale), providing quick response (short time delivery of products), and modular production (standard components can be configured in a wide variety of products and services). Spira [16] develops a similar framework with four types of customization: customized packaging, customized services, additional custom work, and modular assembly. The combination of these frameworks leads to eight generic levels of MC, ranging from pure customization (individually designed products) to pure standardization; these levels are presented in Table 2.2.

- Level 8 (design) – refers to collaborative project, manufacturing and delivery of products according to individual customer preferences (e.g., residential architecture).
- Level 7 (fabrication) – refers to the manufacturing of customer-tailored products following basic, predetermined designs.
- Level 6 (assembly) – deals with the arranging of modular components into different configurations according to customer orders.
- Levels 5 and 4, MC is achieved by simply adding custom work or services to standard products, often at the point of delivery.
- Level 3, MC is provided by distributing or packaging similar products in different ways using, for example, different box sizes according to specific market segments.
- Level 2, MC occurs only after delivery, through products that can be adapted to different functions or situations.
- Level 1 refers to pure standardization, a strategy that can still be useful in many industrial segments.

TABLE 2.2   Levels of Mass Customization [12]

MC Generic levels	MC Approaches	MC Strategies	Stages of MC	Types of Customization
Design	Collaborative; transparent	Pure customization		
Fabrication		Tailored customization		
Assembly		Customized standardization	Modular production	Assembling standard components into unique configurations
Additional custom work			Point of delivery customization	Performing additional custom work
Additional services			Customized services providing quick response	Providing additional services
Package and distribution	Cosmetic	Segmented standardization		Customizing packaging
Usage	Adaptive		Embedded	
Standardization		Pure standardization		

2.2.3 E-BASED DESIGN SUPPORT AND MASS CUSTOMIZATION SYSTEMS

Manufacturers have to successfully implement strategies to design and develop products in order to satisfy a wide variety of customer requirements. There is need to promote commonality, compatibility, standardization, and modularization among different products or product lines. A modular product would mean that the development time is reduced because once the design is split up into modules, design teams can work in parallel on the different modules. The demand for better products is volatile and challenging to manage, the rapid rate of innovation causes short product lifecycles, thus a short production lead-time associated with modular products enhances a firm's ability to respond in introducing new products. With significantly shortened product life cycles, manufacturers have found that they can no longer capture market share and gain higher profits by producing large volumes of a standard product for a mass market. In this era of globalization and intense competition the long-term health of companies is tied to their ability to innovate successfully and rapidly with the customer in mind thus it will be worthwhile to have tools which aid in that.

A modular clusterization design support system which comprises of software modules which will tap customer requirements, analyze the requirements and interpret them technically, then finally cluster product components according to multi-component relationships is illustrated in case example 1.1. In order to achieve this, there is need for a database model for a research and development (R&D) enterprise with focus on modular product design approach, a software program with the following software modules: Internet based customer's voice gathering, Quality Function Deployment (QFD) matrix construction, particularly the first phase House of Quality (HOQ1) and product module clustering using the Design Structure Matrix (DSM).

CASE EXAMPLE 2.1

Development of a Modular Clusterization Design Support System: Case Study of an Agricultural Implements Manufacturing Company.

Cynthia Mathe

In partial fulfillment of the BEng Hons degree in Industrial and Manufacturing Engineering
Department of Industrial and Manufacturing Engineering
National University of Science and Technology, Zimbabwe

The project serves to illustrate the use of scientifically justifiable means of product development, which are systematic and less rigorous in a Research and Development environment. The ultimate result will be a reduction in product development time, which eventually translates to low costs of production. In this project, the author made use of a Visual Basic program, Mat lab macros as well as a Microsoft access database to come up with an integrated tool, which aids in product development. Software modules were linked in such a way as to obtain customer requirements, analyze data by means of a Quality Function Deployment (QFD), cluster parts according to dependencies and give feedback in terms of summarized reports on the different stages of the analysis. This was achieved through the use of a web based application to gather customer's views for example through a questionnaire that can be filled in online and accessed from the company in-house application system for further analysis. The main thrust being the incorporation of modular clusterization techniques for example the use of design structure matrices.

The software has provisions to store pre designed product modules such that should a customer require a product consisting of existing modules development consists of simply refining and assembling what is there thus reducing costs of development and production. Information can be analyzed quickly and shared across departments through the software.

The following steps explain the system model:

a) Questionnaires about a product design project are posted on a web application online. Customer needs on the product design project are also captured via the application whereby the customer completes the questionnaire and submits it online. The application is linked to a database and data is sent back to the company server.

b) An in-house analysis is then carried out on the customer needs, this subsequently equips the product development team to complete the customer needs (what's) in the Quality Function deployment module.

c) Product components are then clustered by use of a product module-clustering algorithm and the recommended clusters are displayed.

Figure 2.2 shows the system model. The use-case diagram for the system shows two main actors:

a) The customer – Has the ability to read and complete the online questionnaire and then submit it. The information will be served in a database in the company server.

b) The employee – Updates the online application from the company premises, posting questions about the product development project. Also responsible for analyzing the customer requirements and completing a QFD and the finally the using the product module clusterization tool.

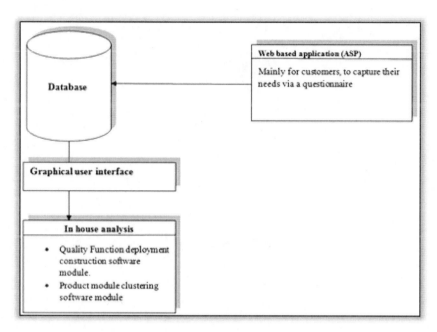

FIGURE 2.2 Modular clusterization system model.

Figure 2.3 shows the use case diagram of the system with the interactions.

2.3 MULTI-LEVEL FUZZY DECISION MAKING IN RECONFIGURABLE MANUFACTURING SYSTEMS

RMS uses Reconfigurable Machine Tools (RMTs) to achieve adjustability, customizability, scalability and diagnosability. RMTs are synthesized using the principles of modular design in order to achieve the required structural design for a particular part to be manufactured. In this chapter, an effective method that uses multi-level fuzzy decisions to create dynamic optimal configurations of machine structures with respect to a given part geometry is presented. A system of modular machine configuration is utilized to arrive at machine configurations considering the fuzzy constraints that are pertinent in this process. With the utilization of fuzzy decisions for the configuration system model, selection of optimal modular tool

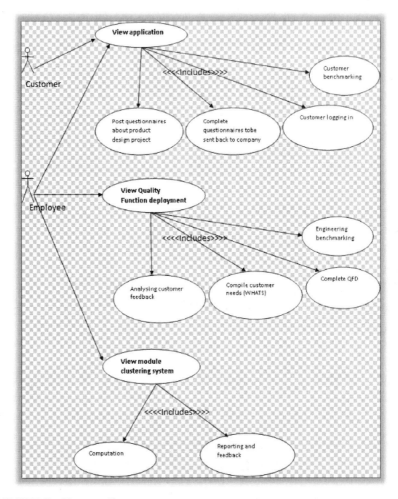

FIGURE 2.3 Use case diagram.

configurations is done. Decisions are made at a particular threshold level so as to verify the appropriateness of such decisions.

In reconfigurable manufacturing systems (RMSs) a critical matter is the conceptualization of the relevant configurations for machining particular part geometries. This is because the manufacturing system is to be designed to be changeable given the variations in the market it is envisioned to penetrate [17]– [20]. The planning of the manufacturing system is assailed with aspects of uncertainty and lack of clarity with respect to, for example the number of degrees of freedom (DOFs) required by the machine tool over a particular interval of time. Zadeh [21], [22] proposed a fuzzy set theory to deal with aspects of uncertainty or inexact events occurring. The fuzziness can be dealt with in the decision making process in reconfigurable manufacturing, giving a tractable, cost effective and low cost solution. The machine required with respect to the number of DOFs is specified with respect to the part geometry, and then the various module combinations are also considered.

2.3.1 CHALLENGES IN RECONFIGURABLE MANUFACTURING SYSTEMS

The critical ambiguity aspects from one generation of manufactured parts to the next that are expected to be encountered in reconfiguration of RMS are as follows [23]:

- a function of time: When parts are going to be machined there will be a varying number of DOFs that will be required to machine particular part geometries. However if the focus is a steady increase or decrease of the number of DOFs, which are required in a particular set of parts then it will be prudent to design such a system that is inclusive of the future needs. This will be advantageous compared with having a machining system that cannot be scaled down or scaled up to the precise number of degrees of freedom so as to minimize on wasted resources.
- alternative combinations of modules: When modules are chosen from the available data-base of modules, a wide range of machine configurations can be found to suffice in machining a certain profile of part geometry. But this has to be done in the context of what modules are readily available to the organization and what the future machining requirements are projected to be.
- lack of exactness: Varying ways of machining will be required for different part geometries. This implies that alternative machine tool configurations will be necessary to carry out the optimal machining process.

2.3.2 RECONFIGURABLE FUZZY MODEL

A domain can be expressed by the following basic function [24]:

$$S_{t+1} = f(S_t, P_t) \tag{1}$$

where S_t and S_{t+1} are the reconfigurable machine tool states at time t and t + 1, respectively. Both will take a value from the machine configuration state set $X = \{X_{11}, X_{21}, X_{31},...,..., I\}$; P_t is the particular part state at time t, and will assume a value from part family set $Y = \{Y_1, Y_2, Y_3,...,..., Y_M\}$. X stands for all the states of the machine configurations to machine a part; Y is the input set to the domain, standing for all feasible parts to be machined; t stands for time, and can be 1, 2, 3,...,..., T; f is the state transformation function which is a fuzzy function in this case. For the multi-decision scenario, a fuzzy decision is calculated at the final interval or final time generation (t = T) with the understanding that:

$$\mu_{BT}(S_T) = [0, 1], (S_T \varepsilon B) \tag{2}$$

where B is the fuzzy decision set that satisfies specific fuzzy goals at time t = T; S_T is a machine configuration state when t = T. It can be thought of as a fuzzy part of B when it represents a machine configuration that can machine a particular part/part family; the valuation set [0,1] is a real type interval–the further the magnitude of $\mu_{BT}(S_T)$ is to 1, the less S_T belongs to B. For particular parts P_t in Eq. (1), there are fuzzy constraints C_t, which are relevant to set Y. This means that for a set of part states $(P_1, P_2, P_3,...,..., P_{T-1})$ there is a fuzzy option group D that is related to Y X Y X Y...,...Y X Y...,... Y X Y. The fuzzy membership function of D then becomes:

$$\mu_D(P_1, P_2, P_3,...,..., P_{T-1}) = \mu_{C1}(P_1) \mu_{C2}(P_2)\mu_{C3}(P_3),...,..., \mu_{C(T-1)}(P_{T-1})\mu_{BT}(S_T) \tag{3}$$

When a decision has to be taken regarding which process is to be implemented, the machine configurations for a particular part-family P_t, will be chosen bearing in mind the features of the part-family. The part-families will all be true to the fuzzy function $S_{t+1} = f(S_t, P_t)$. If the part-families are generally expected to vary in accordance with the following processes needed by the part/part family: drilling, slotting, slabbing, stepping, shaping and reaming, where $Y_1, Y_2,..., Y_6$ are the consecutive machine processes that will be at a particular juncture needing one DOF, two DOFs, three DOFs, four DOFs, five DOFs and six DOFs, respectively. At the onset of decision-making in the establishment of the relevant machine configurations, all alternative parts in the group Y, for the machine configuration St are chosen based on the in-process part features and part-family condition.

The problem of configuration of varying machine structures is a critical one in RMS and has to be solved taking into consideration the constraints that have been identified to affect the system. At a machine level, the arrival at optimal decisions can be achieved through the fuzzy model. Use of an expert system may also make the computation of this model a simpler process.

2.4 RECONFIGURABLE MANUFACTURING MACHINE ADVISOR (RMMA)

In reconfigurable manufacturing the configuration expert makes use of facts and heuristics to arrive at decisions. In some cases the decisions are based on simple rules of thumb, thus trivializing the decision-making activity. However in many engineering configuration problems the decision process is more involving and is not as basic as requiring using only a rule of thumb. Thus a computer program designed to act as an expert to solve a challenge in a specific domain is well known as a knowledge-based expert system (KBES) and is necessary to resolve the configuration problem. This program makes use of the coded information in it and a specified control mechanism to give solutions.

The configuration advisor is constructed as a "classification" expert system. The mechanisms of configuring machine tool architectures are coded as 'if' 'then' rules. Some of the information is kept as facts which are retrieved when the need arises to provide advice. The advice will be used by a manufacturer who is interested in using the machine tool. It may also be used by the machine tool designers who supply manufacturers with the machine modules they need for their production purposes.

2.4.1 SOLICITING THE KNOWLEDGE

The knowhow of the machine configurator is sourced from the available modules from commercial-off-the-shelf organizations. The specifications and the interfaces are supplied with the product manuals. The CAD models may also be available for download from the Internet to use in the creation of machine structures. The information from both the manuals and the Internet is utilized in the conception of the varying architects. Feature knowhow, machine configuration understanding and the respective constraints are discussed in the ensuing sections. This discussion is done in the light that, they are factors that govern the decision process to arrive at a given machine configuration.

The features are essentially the driving factor in the type of machine configuration that is required. The variation in part demands have been indicated in the hypothetical case demonstrated in Figure 2.4. The main features modeled are linear holes, prismatic slot, slab, nonlinear holes and an angled plane. In the configurator these features are described as is for simplicity and clarity to the systems end-user and the technical support team for maintenance of the software.

Case Example 2.2 gives an insight into how the reconfigurable manufacturing machine advisor (RMMA) for appropriate machine configurations can be conceived. The fundamental aspect addressed by this example is the architecture of a knowledge base in the RMMA. The components that make up the expert machine configurator for varying manufacturing demands are discussed. The guidelines regarding how a machine tool user builds a machine structure are also considered. The configuration advisor caters for the right configuration for a specific demand at a specific time within a set of specific constraints.

CASE EXAMPLE 2.2

Development of an Expert System for Reconfigurable Manufacturing Systems.

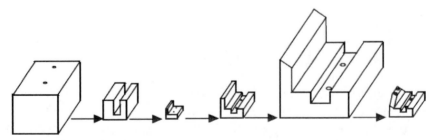

FIGURE 2.4   Hypothetical variations of part/part-family [25].

Khumbulani Mpofu

In partial fulfillment of DTech in Mechanical Engineering
Department of Mechanical Engineering, Tshwane University of Technology, South Africa

C Language Integrated Production System (CLIPS) is utilized in the development of the knowledge based system. The main reason for its use being that it is readily available at no cost to the public. The other advantages that this knowledge-based tool comes with are portability, extensibility (the ability to extend the tool to other parameters not included at the onset), and the range of capabilities it possesses.

Figure 2.5 shows a flow diagram of the how the expert system operates. Its operation is as follows:

1. The user is asked to select the shape type.
2. The response could be a particular shape or that it is not known.
3. The user is then given possible manufacturing configurations if it is known otherwise the user selects their own configuration. The user of the expert system does not necessarily have to have a defined geometry to use the expert system.
4. Then the size of the part is requested from the user.
5. The user is then asked for design parameters if they are there, then these are used to generate the machine configuration.
6. If there are no design parameters then the available information is utilized to give advice.

The configuration advisor stores the knowledge of the manufacturing systems facts and rules in the knowledge-base, the inference mechanism in the CLIPS program determines the manner of operation of the program. The facts are defined as the information that is related to the part shapes, part features, the geometries to be machined (such as hole, slab, slot, etc.). The rules are the specifications of the system requirements with respect to the needs to be addressed for example the DOFs of configuration that will be given to a part-family with given features.

Data may either be inputted at run time or be information that is known the latter being called fixed data and the former being called modifiable data. Figure 2.6 shows a representation of an object-oriented class with the respective attributes and a sample of a CLIPS fact in the advisor inference engine.

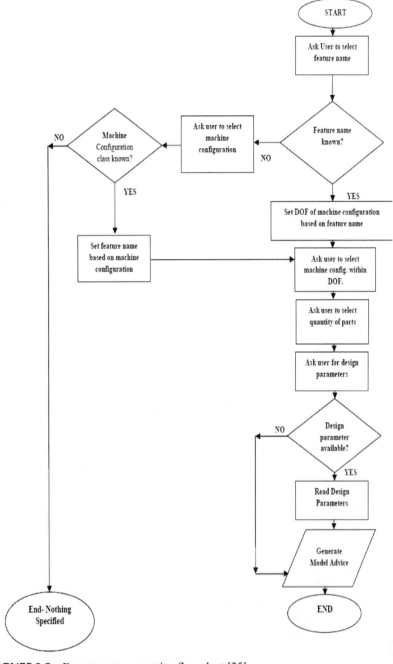

FIGURE 2.5   Expert system operation flow chart [25].

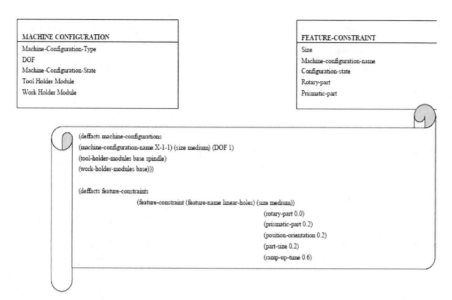

FIGURE 2.6 Templates for machine configuration and feature constraint and a sample of CLIPS facts in the RMMA.

The knowledge-based system generates the configuration suggestions in a textual form. These suggestions contain a summary of the design data that was input during a particular time and the manufacturing advice that has been generated.

2.5 TECHNOLOGY TRANSFER DECISION SUPPORT SYSTEMS (TTDSS)

Technology refers to a set of knowledge contained in technical ideas, information, personal technical skills and expertise, equipment, prototypes, designs or computer codes. It is the useful application of knowledge and expertise into an operation [26, 27]. According to Ramanathan [28], technology transfer (TT) may be defined as a mutually agreed upon, intentional, goal oriented and proactive process by which technology flows from an entity that owns the technology to an entity seeking the technology. Bennett [29] defines technology transfer as a transaction or a process through which technological knowhow is transferred normally between businesses or agencies representing businesses. The transaction takes place when both parties perceive gains and where it can be integrated with other dimensions of the business to improve the competitiveness and performance of the business.

2.5.1 DECISION ISSUES IN TECHNOLOGY TRANSFER

Technology transfer is a complex and difficult process that requires decision making at any stage of its progression [30]. This includes choosing the project, providers and TT method. In all cases decisions have to address the potential benefit that will be realized against the costs and expenditure that will be accrued by the technology acquisition and implementation process [31]. Prominent TT decision issues include technological positioning, communication, market impact, environmental performance and change management.

Technological positioning is an important TT decision factor. The technology taker needs to do a self-check to establish whether they will be able to realize the benefits of the proposed technology. This is described as the absorptive capacity of the recipient, that is the capacity of the recipient to assimilate value and use the knowledge transferred and is linked to various technological positioning indicators within the company [32].

Effective communication is also cited as a determining factor in TT transactions [33]. Any TT process begins with contact between the company seeking the technology and the providers of that technology. Yakhlef [34] points out that interaction is the locus of significant activity in the processes of knowledge transfer, involving various forms of documentation, correspondence, mediation and face-to-face interactions. Top level managers should be involved in the TT process as their experience and personal attributes enable them to connect the elements of a complex heterogeneous network for effective interaction [35].

For manufacturing companies, the major motive for technology transfer is a desire to increase productivity and efficiency hence improve competitiveness and profitability. According to Bozeman [36] this is usually termed the market impact criterion, which measures the commercial success of the new technology. Increase in productivity and efficiency is mostly a direct result of increased technical capacity. The technical capacity may come in the form of electro-mechanical, computerized, designs, code, blueprints, as well as in tacit form. Tacit technology refers to knowledge and expertise possessed by people and organizations through experience and learning [37].

In this age of earthman-ship environmental conservation has become a major factor governing the design and use of technology. The main environmental aspects related to technology transfer are concerned with (a) reducing the unfavorable environmental effects of industry and (b) ensuring that new investment is environmentally sound technology (EST). Any responsible TT process should therefore consider the possible impact, ether positive or negative, of the new technology [29].

The introduction of new technology translates to significant changes to the way people within the company operate. It also affects the company-customer interface due to changes in products or product functionality. As a result the company may face inertial challenges, or resistance to change. Change management becomes vital in these circumstances. It is also suggested that changes that are instigated by customers will be implemented more easily [38].

2.5.2   MODELING A TECHNOLOGY TRANSFER DECISION SUPPORT SYSTEM (TTDSS)

Research questions are designed to capture aspects of the technology transfer process that need decision support at the takers' end together with the nature of decision support required. This is done to ensure congruence of decision support functionality and the actual needs within industry. The following major questions are asked in modeling a TTDSS:

What are the main company objectives when undertaking international technology transfer?
The main drivers to technology transfer are replacement of out-dated machines and the need for high and consistent quality standards. This is followed by reduction of production and maintenance costs which are major issues affecting the competitiveness companies. The objectives are listed in Table 2.3.

What criteria do companies use when selecting technologies?
Table 2.4 summarizes the criteria considered by many companies.

TABLE 2.3   Company Technology Transfer Objectives

Objective
Replacing out-dated machinery
Achieving high and consistent quality standards
Improving manufacturing efficiency
Reducing energy costs
Reducing production costs
Improving response time to changing customer needs
Reducing maintenance problems

TABLE 2.4   Technology Selection Criteria Rating

Criteria
The quality of the provider's technology
Training of personnel by the provider of the technology
The political relationship between the two countries
Guarantees
Provision of spare parts by the provider
After-sale service
Provision of documentation, software, and technical knowledge of design, production and testing
Provision of hardware and special equipment

What functions do respondents expect from a technology transfer decision support system?

The main expectations are listed in Table 2.5 together.

Case Example 2.3 illustrates the development of a TTDSS. The Technology Transfer problem occurs in three domains namely necessity, feasibility and selection. The software system design for a TTDSS includes the modules, database, end user-interface and a web based application that is linked to the in-house TTDSS application.

CASE EXAMPLE 2.3

Development of a Technology Transfer Decision Support System (TTDSS).

TABLE 2.5 Technology Transfer Decision Support System: Respondents' Expectations

Expectation
Allow consideration of multiple options
Outline margins of improvement
Calculate return on investment
Capture acquisition costs
Opportunity costs analysis
Comparison of existing to proposed systems
Report generation for effective presentation
Cash-flow analysis
Resource consumption analysis

Blessing Mwoyongewenyu
In partial fulfillment of BEng Hons Industrial and Manufacturing Engineering
Department of Industrial and Manufacturing Engineering
National University of Science and Technology, Zimbabwe

The core objective of the TTDSS application is to provide decision-making support information for someone anticipating or analyzing a prospective technology transfer project. A technology transfer decision support system should be able to support technology transfer decisions in the three domains namely:

(a) supporting the perceived need for technology transfer (Necessity);
(b) establishing the feasibility of the technology transfer process (Feasibility);
(c) selection of the optimum technology (Selection).

Decisions have to be based on considerations of certain decision support factors. Four main decision factor dimensions namely financial, technical, environment and social dimensions are considered in the case example. The financial dimension deals with quantifiable aspects that relate to the 'monetary' benefits of technology transfer. Those aspects of the decision that pertain to organizational capabilities, interaction, communication and support services make up the technical factors dimension. The impact of technology on the environment and the social fabric of the organization comprise the environment and social factors dimensions.

The financial module has the functions shown in Figure 2.7 that are intended to feed into the three decision support domains.

System Architecture

The TTDSS utilizes the Analytic Hierarchy Process (AHP) as one of the decision support tools especially for technology and provider selection. It also encompasses a web application to facilitate interaction among technology transfer stakeholders. The framework model is depicted in Figure 2.8.

The TTDSS model depicts the relationships that exist among the decision domains and decision dimensions as they lead to the decision support objective. The separation of decision factors into three dimensions facilitates the development of three separate modules that can operate independently with ability to cooperate towards a single decision depending with

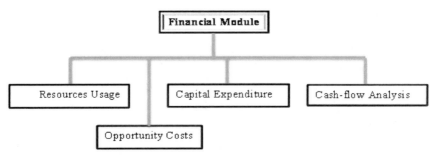

FIGURE 2.7   Financial module components.

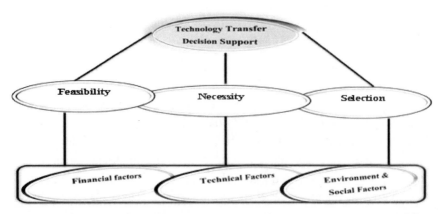

FIGURE 2.8   ETTDSS framework.

requirements and available information. The three modules, that is, financial module, technical factors module and the environmental and social factors module appear in the system illustration of Figure 2.9.

The technology factors module is designed to cater for both the feasibility and selection domains of the technology transfer decision-making process. It therefore has to cover issues of technical performance, operation and maintenance of the technology through its life cycle.

The web application is intended to address the dynamic communication and interaction needs of stakeholders in the industry. The application can be launched with basic functionalities as shown in Figure 2.10.

2.6 CONCLUSION

As the world population continues to grow, production output and product varieties in many companies continue to grow generating a lot of data that should be processed often. Other factors of concern to manufacturers include shorter product life cycles, increasing product variation (mass customization – MC), volatile markets, cost reduction pressures, scarce resources, cleaner production, lack of skilled workforce and aging community. Competition among manufacturers is ever increasing and every player must deliver high quality goods, efficiently and at low cost.

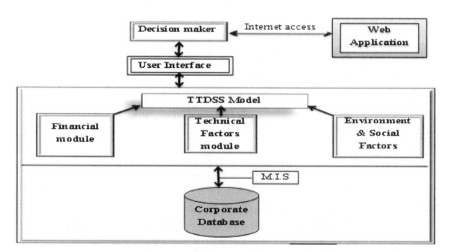

FIGURE 2.9 TTDSS component architecture.

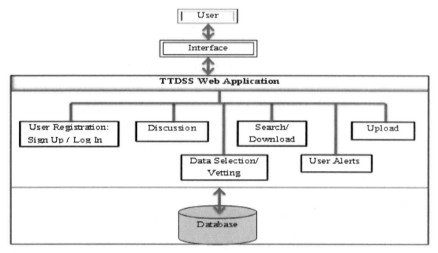

FIGURE 2.10 Web application functions.

Numerous decisions need to be made using the accumulated data so that factories can be more effective in their routine operation.

The chapter discussed Reconfigurable Manufacturing Systems (RMS) and applications of e-decision support systems to solve various problems encountered by manufacturers. In order to realize mass customization, many manufacturers are adopting reconfigurable manufacturing systems (RMS) that can handle a variety of manufactured parts through their adjustable machine operations.

Case examples have been used to illustrate practical applications of decision support in RMS. An effective method that uses multi-level fuzzy decisions to create dynamic optimal configurations of machine structures with respect to a given part geometry has been presented. The method uses a system of modular machine configuration to arrive at machine configurations considering the fuzzy constraints that are pertinent in this process. Another method illustrated uses an expert system (Reconfigurable Manufacturing Machine Advisor- RMMA) as an advising mechanism to help manufacturers to reconfigure machines. The machine modules used in RMS systems are increasingly becoming available as commercial off the shelf (COTS) modules. The end users of the modules are faced with various options that require well-informed decisions to be made before adopting particular technologies. The problem has been described as

Technology Transfer (TT), which is fundamental to the development of new products and the transformation of technology in developed and developing countries. Critical factors, selection criteria and technological positioning indicators that should govern the adoption and implementation of transferred technologies were used to illustrate the conceptualization of a framework for a Technology Transfer Decision Support System (TTDSS) application and the subsequent development of a software system based on the framework. The technology transfer problem occurs in three domains namely necessity, feasibility and selection.

For manufacturing firms not opting for COTS modules but involved in the design and manufacture of modular products, a web-based modular clusterization design support system has been given. The system modules are linked in such a way as to obtain customer requirements, analyze data by means of a quality function deployment (QFD), cluster parts according to dependencies and give feedback in terms of summarized reports on the different stages of the analysis.

KEYWORDS

- e-decision support
- knowledge based expert system (KBES)
- mass customization (MC)
- modular design
- reconfigurable manufacturing machine advisor (RMMA)
- reconfigurable manufacturing system (RMS)
- technology transfer decision support system (TTDSS)

REFERENCES

1. Koren, Y., Heisel, U., Joveane, F., Moriwaki, T., Pritschow, G., Ulsoy, G., Van Brussel, H., Reconfigurable manufacturing systems. *Annals of the CIRP*, 1999, 48(2), 527–540.
2. Malhotra V., Raj T., Arora A. (2009). Reconfigurable manufacturing system: An overview. *International Journal of Machine Intelligence*, ISSN: 0975–2927, Volume 1, Issue 2, 38–46.

3. Chun-Jie Zhang, Xiong-hui Zhou, Cong-Xin, Li. Automatic recognition of intersecting features of freeform sheet metal parts. *Journal of Zhejiang University Science A.* Vol. 10(10), (2009), 1439–1449.

4. Hard, D., et al. (1997). *Next-Generation Manufacturing (NGM) Project, Agility Forum and Leaders for Manufacturing,* Bethlehem, PA.

5. Mailer-Speredelozzi, V., Koren, Y., Hu, S. J. (2003). Convertibility measures for manufacturing systems. *Annals of CIRP*; 52(1), 367–371.

6. Won, Y., Currie k. R. An effective P-median model considering production factors in machine cell-part family formation. *Journal of Manufacturing Systems.* 2006, 25(1), 58–64.

7. Silveira, G. D., Borenstein, D., Fogliatto, S. F. (2001). Mass Customisation: Literature review and research directions. Elsevier Science B. V. Int. J. Production economics, Vol. 72, pp. 1–13.

8. Salvador, F., Forza C. (2004). Configuring products address the customisation responsiveness squeeze: A survey of management issues and opportunities. International Journal of Production economics, Vol. 91, 3rd Edition, pp. 273–291.

9. Koren, Y. (2010). The Global Manufacturing Revolution: Product-Process-Business Integration and Reconfigurable Systems. USA: John Wiley & Sons Inc.

10. Pollard D., S. Chuo, (2008). Strategies for Mass Customization, Journal of Business & Economics Research. Vol. 6, No. 7, pp. 77–85.

11. Xing B., G. Bright, N. S. Tlale, J. Potgieter, (2006). Reconfigurable Manufacturing System for Agile Mass Customization Manufacturing, Proceedings of the 22nd International Conference on CAD/CAM, Robotics and Factories of the Future, pp. 473–482.

12. Hart, L. C. W. (1995). Mass Customisation: conceptual underpinnings, opportunities and limits, International of Journal of Service Industry Management, Vol. 6, No. 2, pp. 36–45.

13. Westbrook R., P. Williamson, (1993). Mass customization: Japan's new frontier, European Management Journal 11(1), Vol. 38, pp. 234–279.

14. Gilmore J., J. Pine, (1997). The four faces of mass customization, Harvard Business Review, Vol. 91, pp. 95–110.

15. Pine J. (1993). Mass customizing products and services, Planning Review 21 (4), Vol. 6.

16. Spira J. (1996). Mass customization through training at Lutron Electronics, Computers in Industry, Vol. 30, pp. 171–174.

17. Wiendahl, H. P., El Mararaghy, H. A., Nyhuis, P., Zah, M. F., Wiendahl, H. H., Duffie, N., et al., (2007). Changeable manufacturing-classification design and operation. CIRP Annals, Manufacturing Technology 56(2):783–809.

18. Bruccoleri Manfedi, Pasek Zbigniew J, Koren Yoram. Operational management in reconfigurable manufacturing systems-reconfiguration for error handling. International Journal of Production Economics 2006, 100(1), 87–100.

19. Mpofu KC, Kumile MN, Tlale S. Design of reconfigurable machine systems: knowledge based approach. Journal of Konbin, Versita, Warsaw volume 2008, 8(1), 135–44.

20. Potgieter J, Bright G, Tlale S. Modular mechatronic control system for Internet manufacturing. In: 18th international conference on CAD/CAM, robotics and factories of the future. 2002, p. 529–36.

21. Zadeh LA. Fuzzy logic and the calculi of fuzzy rules, fuzzy graphs and fuzzy probabilities. Computers & Mathematics with Applications 1999, 37(11–12), 35.

22. Zadeh LA. Fuzzy sets. Information and Control 1965, 8, 338–53.

23. Safaei Nima, Saidi-Mehrabad Mohmmed, Tavakkoli-Moghaddam Reza, Sassani Ferrokh. A fuzzy programming approach for a cell formation problem with dynamic and uncertain conditions. Fuzzy Sets and Systems 2008, 159(2), 215–36.

24. Mpofu, K., and N. S. Tlale. (2012). "Multi-level Decision Making in Reconfigurable Machining Systems Using Fuzzy Logic." Journal of Manufacturing Systems 31, 103–112.

25. Mpofu, K. Expert system design for reconfigurable manufacturing systems, FAIM 2012.

26. Carayannis, E., Rogers, E., Kurihara, K., Albritton, M. (1998). High-Technology Spin-offs from Government R&D Laboratories and Research Universities, *International Journal of Technology Innovation,* 18(1), pp. 1–11.

27. Gee, R. 1993. Technology Transfer Effectiveness in University-Industry Cooperative Research, *International journal of Technology Management,* 8(6), 652–668.

28. Ramanathan, K. (2010). An Overview of Technology Transfer and Technology Transfer Models, 1–24.

29. Bennett, D. (2002). Innovative Technology Transfer Framework Linked to Trade for UNIDO Action, *UNIDO and the World Summit on Sustainable Development,* Johannesburg, 26 August–4 September, 1–54.

30. Hamzei, A. (2011). Decision Support Model in Technology Transfer for Technology Receiver, *International Journal of Natural and Engineering Sciences,* 5(2), 43–48.

31. Ivarsson, M., Gorschek, T. (2009). Technology Transfer Decision Support in Requirements Engineering Research: A Systematic Review of RE Journal, *Requirements Engineering Journal,* 14, 155–175.

32. Carayannis, E., Popescu, D., Sipp, C., Stewart, M. (2006). Technological learning for entrepreneurial development (TL4ED) in the knowledge economy (KE): Case studies and lessons learned, *Technovation,* 26(4), 419–443.

33. Eckl, V. (2012). Creating an Interactive-Recursive Model of Knowledge Transfer, *DRUID Conference,* Copenhagen, 19–21 June.

34. Yakhlef, A. (2007). Knowledge Transfer as the Transformation of Context, *Journal of High Technology Management Research,* 18(1), 43–57.

35. Patriotta, G., Castellano, A., Wright, M. (2012). 'Coordinating Knowledge Transfer: Global Managers as Higher-level Intemediaries,' *Journal of World Business,* Iaries, *Journal of World Business,* http://dx.doi.org/10.1016/j.jwb.2012.09.007.

36. Bozeman, B. (2000). Technology Transfer and Public Policy: A Review of Research and Theory, *Research Policy,* 29, 627–655.

37. Foray, D. (2009). Technology Transfer in the TRIPS Age: the Need for New Types of Partnerships between the Least Developed and Most Advanced Economies, *ICTSD Program on IPRs and Sustainable Development Intellectual Property and Sustainable Development Series,* 23, 1–56.

38. Molina, L., Llorens-Montes, J., Ruiz-Moreno, A. (2007). Relationship between quality management practices, *Journal of Operations Management,* 25, 682–701.

PART 3:

E-GOVERNMENT

CHAPTER 3

SOCIO-DEMOGRAPHIC FACTORS AND CITIZENS' PERCEPTION OF E-GOVERNMENT

P. DEVIKA[1] and N. MATHIYALAGAN[2]

[1]*PhD Research Scholar, Department of Communication, PSG College of Arts and Science, Coimbatore – 641014, Tamil Nadu, India, Tel: +91-9500504674, E-mail: vikram.devika1@gmail.com*

[2]*Associate Professor, Department of Communication, PSG College of Arts and Science, Coimbatore – 641014, Tamil Nadu, India, E-mail: mathiyalagann@yahoo.co.in*

CONTENTS

ABSTRACT

Internet users in developing countries are increasing with emerging economic conditions and Information and Communication Technology (ICT) development. Web portal has become an important means of public administration with increased number of internet users. In India, national, state and local governments aim to deliver public services efficiently to large population through websites that would integrate various citizen services under one roof. Citizen's socio-demographic characteristics and their perception about e-government services, account for the access and reach of the e-government web portals. The study examines the effect of gender, age, education and occupation on citizen's perception towards e-government services. A survey was conducted among 484 citizens in the Coimbatore city, Tamilnadu State, India. The survey intended to analyze citizen's opinion about Coimbatore City Municipal Corporation Website services and identify the role of socio-demographic factors in opinion formation. Coimbatore City Municipal Corporation website serves as a 'one stop' for all Government to Citizen Services of Coimbatore Municipal Corporation (local government system). The results revealed that based on the socio-demographic factors, respondents differ in their perception about the usefulness, ease of use, trust, risk, quality of the website and Intention to Use the e-government web portal services.

3.1   INTRODUCTION

E-government increases efficiency of governments in serving people. Developing countries have implemented various e-government initiatives to provide citizen centric services. "E-government refers to the use of information technologies (such as Wide Area Networks, internet, and mobile computing) by government agencies that have the ability to transform relations with citizens, businesses, and other arms of government." E-government web sites have emerged as a prevalent communication tool between governments and citizens (Bertot & Jaeger, 2006; Bringula & Basa, 2011; Soufi & Maguire, 2008). Web portal Services are called Self-Service Technologies (SSTs), as they enable citizens to get public services without direct involvement of government employees. Online services of web portals reduce dependency on the government officials to avail public services (Chander & Kush, 2012). E-government websites offers e-services to citizens 24 hours a day, 7 days a week, thus act as convenient and cost effective means of communication.

Government of India has allocated huge fund for developing and implementing new technologies to streamline the government process and deliver government services electronically. In India, national, state and local governments have implemented various e-government programs.

An adequate allocation for e-government development was made in the national budget to increase connectivity by enhancing the telecommunication network and to implement Information and Communication Technologies (ICT) in government administration for interaction and service delivery electronically. Indian government supports e-government development through various policies and plans.

Government of India has taken efforts to reform the public administration. During 1970s–1990s, Indian government as a first step towards e-government aimed to connect government departments and used Information Technologies (IT) for data management and for automation of government process. In 1976, National Informatics Centre (NIC) was setup under the Department of Electronics and Information Technology to promote IT infrastructure in the country. The efforts of NIC not only resulted in computerization and connectivity but helped to implement various e-government projects in the country. NIC provides Information and Communication Technologies (ICT) and supports implementation of e-governance projects in the country (National Portal of India; Haque, 2002; Bagga & Gupta, 2009). The country formulated National e-Governance Plan (NeGP) for effective public administration that aims to serve the citizens better.

The aim of National e-Governance Plan (NeGP):*"Make all Government services accessible to the common man in his locality, through common service delivery outlets, and ensure efficiency, transparency, and reliability of such services at affordable costs to realize the basic needs of the common man"* (National Portal of India, Haque, 2002). Development of necessary infrastructure for e-governance is one of the main strategies in NeGP. Under NeGP, digital service delivery infrastructure, portals and data centers were established to provide government services electronically to the citizen. Portals were designed to provide anywhere and anytime access of public services to the citizens. Government portals aim to simplify the process of accessing the public services by the citizen and also avail the services without visiting government department/office. It also would reduce managerial burden for the government and service fulfillment time and costs for the government, businesses, and citizen.

3.1.1 BACKGROUND OF THE STUDY

Local governments play an important role in delivering citizens services. They are responsible for providing services such as education, health, etc. So, e-government implementation at local government level would be more beneficial to the citizens. E-governance in municipalities would improve operations of the local government, quality of citizen services, and interaction between the local government and stakeholders. Also, would enhance transparency and accountability in local government administration and service delivery.

Apart from the state and central governments, the local government has a pivotal role in public administration in India. The local government is the basic level and third level of government in India. In each state, the local government bodies that operate in rural areas are called Panchayats and local government bodies in urban areas are generally called Municipalities. Municipalities render various services to the urban citizens and are responsible for the infrastructure development like roads, public transportation, water supply, sanitation, gardens and maintenance of buildings as well as public safety services like fire and ambulance services.

With increasing urbanization in India, out of the total population of 1210.2 million (in 2011) around 377.1 million live in urban areas. Urban sector (cities and towns) is a one of the major contributor of country's Gross Domestic Product and economic growth. Under 74th Constitutional Amendment Act of 1992, local governments (Urban local bodies) are considered important in service delivery in the urban areas.

Municipalities aim to develop the urban sector by improving the infrastructure and governance. Under National e-Governance Plan, Mission Mode Project for municipalities was developed and implemented to provide various services electronically through e-governance. Bagga and Gupta (2009) stated that National Mission Mode Project for municipalities is to improve the efficiency of municipal services delivered to the citizens. Municipal corporations in many states of India have implemented e-governance through the adoption of appropriate Information and Communication Technologies (ICT). Web portal is an ICT tool that would provide an easy single point of access to public services. Hence, to render services that could be accessed anytime, anywhere by the citizens through single window system, municipal corporations in many states have established their own web portals to provide numerous basic services for citizens in urban areas. The municipal web portals of various states integrate various local government departments to provide the following services:

- Registration and issue of birth and death certificate.
- Payment of property tax, utility bills and management of utilities that come under urban local bodies.
- To register grievances and suggestions.
- Building plan approvals.
- Procurement and monitoring of projects.
  - E-procurement;
  - Project/ward works.
- Health programs.
  - Licenses;
  - Solid waste management.
- Accounting system.
- Personal Information System.
- Grievance handling, including Right to Information Act, Acknowledgement, Resolution monitoring.
- Waste Management Service.

Governments have understood the advantages of the websites to deliver public services and have developed e-government web portals. On the other hand, citizen's expectations of online-public-services offered by the e-government web portal have also increased.

3.1.2 THE STUDY AREA

The study was conducted in Coimbatore city, Tamil Nadu state, India. Tamil Nadu is a south-eastern state in the India, bounded by Karnataka and Andhra Pradesh states on the north, by Kerala state on the west, by Bay of Bengal on the east and by Indian Ocean, on the south. The state is divided into 32 administrative units called districts. Coimbatore district is situated in the extreme west of Tamil Nadu, near the state of Kerala. Coimbatore city is the district head quarter of Coimbatore district. The Coimbatore City Municipal Corporation is the local government administrative body of Coimbatore city. Coimbatore City (Municipal Corporation) is the third highest populated city in the Tamil Nadu state, with highest literacy rate of 91.74% among the highly populated cities in state (2011 Census Data: Census Of India, 2011).

Coimbatore city is an industrial hub and is known as Manchester of South India. It is also emerging as an IT and BPO city. Coimbatore was ranked at 17th among the global outsourcing cities. The city has various small and medium scale business units and is an education center with renowned universities and colleges (Coimbatore City Municipal Corporation, 2013; About Coimbatore: KG Information Systems Private Limited, 2013). Coimbatore city is divided into five administrative zones such as east, west, north, south and central zones and governed by Coimbatore City Municipal Corporation (CCMC). CCMC is one of the largest corporations in the state of Tamil Nadu, India that serves 1,050,721 citizens (2011 Census Data: Census of India Website, 2011). The CCMC has implemented various e-government projects to render public services efficiently. CCMC website is a major e-governance initiative by Coimbatore City Municipal Corporation.

3.1.3 E-GOVERNANCE IN COIMBATORE CITY MUNICIPAL CORPORATION

Initially, computerized data base creation was done for maintenance of e-files. The computerization started in year 1998 under Public Private Partnership (PPP) mode. Under PPP mode, KG Information Systems Limited (KGISL), a private Information Technology (IT) company provided technical support in database creation (KG Information Systems Private Limited, 2014). KGISL also designed a website for Coimbatore City Municipal Corporation to streamline the operation of the departments and enable timely delivery of citizen services. In 2004, e-governance modules for various services of the CCMC were implemented. These modules were designed to improve service delivery and increase public convenience in accessing the services. Municipal departments in all the zones are connected by LAN and WAN. Internet connectivity is provided by Bharat Sanchar Nigam Ltd. (BSNL) one of the largest and leading public sector telecommunication service provider in India (Bharat Sanchar Nigam Limited, 2014). Under Tamil Nadu Urban Development Project, CCMC took various e-governance initiatives for better administration of all departmental actives, efficient service delivery and easy access of all municipal services by the stakeholders (Better Service Through E-Governance, THE HINDU, 2011). The following are the major e-governance initiatives of Coimbatore City Municipal Corporation.

3.1.3.1   Coimbatore City Municipal Corporation (CCMC) Website

The website is a dynamic website with various modules that integrates functions of all municipal departments of Coimbatore City Municipal Corporation such as public health, education, town planning, engineering, water supply, drainage, solid waste management, revenue, accounts and public relation and all zonal offices. The one of the main aim of this integration is to increase accessibility of public service rendered by the Municipal Corporation. CCMC Website provides the following services:

- Issue of birth and death certificates;
- Online submission of building plan for approval;
- Online tax payment for property and water charges;
- Tax calculator;
- Download application forms;
- Status tracking of all applications;
- Status of all development works in real time;
- Grievance redress;
- Publishing tender documents, download tender notice and tender schedule, status tracking of tender;
- Alerts for pending bills and payments;
- Top defaulters list;
- Right to Information;
- Quick links to important government websites;
- Information about the city, transport facilities and time schedules, parks, etc.
- Information about Coimbatore City Municipal Corporation functions and contact details of the officials.

3.1.3.2   Payment of Taxes Through Collection Centers and Online Payment Gateway

Computerized tax collection centers and online payment gateway have eliminated tiresome manual process of tax collection and reduced long queues in the corporation offices to pay tax. There are 29-tax collection centers spread across the city are well connected by WAN or internet. Collection modules were implemented in all these centers for easy and quick tax collection process. Citizens can pay tax in the nearest tax collection centers. Citizens can also pay tax through banks, 20 banks in various locations in the city are also acts as tax collection centers. So, totally 49 centers facilitate tax collection and increase public convenience. Location details of all these centers are given on the CCMC website. 'Anywhere, anytime' is the concept of online payment gateway. The gateway enables round the clock payment of tax from anywhere. Citizen can pay tax anytime through online payment facility in CCMC website using credit/ debit cards (e-Governance: Coimbatore City Municipal Corporation, 2013).

3.1.3.3   Issue of Certificate/ Licenses Through Facilitating Center and Online

Citizens can obtain birth/death certificates, trade licenses from the nearest facilitation centers (Zonal offices of the corporation). These centers are connected by Wide Area Network. The implementation of service module at the facilitation centers resulted in immediate issue of birth/death certificates, trade licenses across the table on receipt of request from citizens. Also, citizens can obtain birth certificates through online. Citizen should summit their request with the required details through the CCMC website and can get birth certificate in PDF format.

3.1.3.4   Online Building Plan Submission and Approval System

Online building plan submission commenced in 2010 with help of Automating Development Control Rules (AUTO DCR) technology. AutoDCR Software was developed and maintained by M/s Softtech Engineers Ltd, Pune.

Coimbatore City Municipal Corporation (CCMC) was chose as pilot run for AUTO DCR and Town Planning department of CCMC introduced AUTO DCR on May 2008. Building plan approval process through Auto DCR software is outsourced to M/s.SoftTech Engineers Private Limited, Pune, in 2010. From 2010 November all manual process were completely stopped and online building plan submission was made mandatory. Building plans can be submitted online through CCMC website. The building plans received online, are verified and processed automatically using Auto DCR software and the approval is given in 3 days (Online plan approval: Coimbatore City Municipal Corporation (CCMC) Website, 2013).

3.1.3.5 Status Tracking of Application

When citizens apply for services either manually or online, they can know the status of their application with the help of "application status" facility in the CCMC website by summiting their application reference number.

3.1.3.6 E-Tendering System

Online tendering is facilitated by Tamil Nadu Government E-Tendering System "TANGETS" developed by National Informatics Center in the state capital Chennai. CCMC website integrates TANGET. Out of 152 Municipalities and 9 Corporations in Tamil Nadu that has implemented E-Tendering System, Coimbatore City Municipal Corporation has recorded the highest number of tenders published through the system (Indian Council for Research on International Economic Relations (ICRIER) website, 2014). E-tendering enables tender creation and publishing online, online bid submission/re-submission or withdrawal of bid, online payment, tender opening online, automatic evaluation of the financial bid, updates of the committee recommendations at each stage of the tender process and award of the contract. E-tendering system reduces the time cycle of the procurement and increases transparency in tendering process. Tender information are on the CCMC website, bidder can access it anytime and need not have to visit the municipal office. Bidders can also easily track the status of the tenders.

3.1.3.7   Biometric Scanning for Attendance Management

Biometric scanning devices and RFID card reader are for attendance management of around 2600 sanitary workers, 700 sanitary staff, 50 sanitary vehicle drivers and staff in 20 Medical centers. CCMC website displays daily attendance reports (Indian Council for Research on International Economic Relations (ICRIER) website, 2014).

3.1.3.8   Waste Truck Monitoring System

Online Waste Truck Monitoring System is linked with Coimbatore City Municipal Corporation (CCMC) website. Complete information about the truck movements and management of solid waste is provided on the website. Online Waste Truck Monitoring is done using Radio Frequency Identification (RFID). RFID system is linked with the Weigh Bridge at the Waste Transfer Stations and the Landfill site. When trucks halts at the Weigh Bridge, the truck number, the ward from which the waste has been collected, name of the driver, the time of entry/exit of the vehicle, weight of the waste collected, etc., is automatically recorded. The system helps to monitor solid waste management activities on a daily basis. Daily reports on exact number of trips made by the trucks and the quantity of the waste transported, processed and disposed at the Landfill site are generated by the Online Waste Truck Monitoring System. The system has resulted in easier and accurate preparation of daily reports about waste management, less dependence on clerical staff to prepare daily weigh reports and generation of accurate transport bills (Indian Council for Research on International Economic Relations (ICRIER) website, 2014).

3.1.3.9   Asset Management Software

Asset Management Software incorporated in CCMC website tracks all the assets of Coimbatore City Municipal Corporation. The system enables automatic record maintenance, removes record duplication, and consolidates assets for financial reporting, provides real time access to information on

CCMC's net worth and presents MIS reports to Top Management (Asset Management System, 2014).

3.1.3.10   Computerized Financial Accounting System

Computerized Financial Accounting System integrated in CCMC website eliminates time consumption manual accounting process such as preparation of reports and budgets and avoids duplication of entries. The system provides integration between the zones and the main office and helps in easy consolidation of financial details. Thus, account management and financial reporting is made easy (Indian Council for Research on International Economic Relations (ICRIER) website, 2014).

3.1.3.11   Communication System

The CCMC Communication System consists of IP (Internet Protocol) PBX (Private branch exchange) – IP PBX, Interactive Voice Response System, Web Portal, Grievance Management System and Call Center, Email, Instant messaging and SMS facility.

CCMC web portal is a major communication tool for interaction between the departments of CCMC, between all the zonal offices and facilitating centers of CCMC and interaction between citizen and CCMC. Interactive Voice Response System is a 24-hour voice response service. Citizen can dial the CCMC's IVRS service number to know their application status, tax payment dues and can also register their grievance. IVR system also has language option for citizen's convenience (Coimbatore City Municipal Corporation, 2013). Web based "Grievance Management system" software is used to redress the citizen's grievances. This system has been integrated with email service of CCMC (grievance@ccmc.gov.in), SMS and CCMC's call center telephone lines. Citizen can register their grievance by calling CCMC's call center numbers, by sending email to grievance@ccmc.gov.in, by SMS to CCMC's registered mobile number or by registering through the grievance registration facility in CCMC web site. CCMC has an email server "mail.ccmc.gov.in" and email IDs are given for all officials and staff. E-mail service improves response time, officials can use 'Email Service' to circulate the documents for approval or comments, to propose agendas and to send meeting minutes, final reports or recommendations. Email Service system also has features such as calendaring for scheduling appointments and meetings and chat services (Indian Council for Research on International Economic Relations (ICRIER) website, 2014). Short Message Service (SMS) is used by the CCMC to send alerts and notifications. There is a facility for citizens to register their mobile number for SMS alerts on Tax Payment Reminders. Also, citizens would receive tax payment confirmation messages to their registered mobile number. Citizens can register their grievance by sending SMS to the mobile number specified by the corporation (Coimbatore City Municipal Corporation, 2013).

Thus, CCMC website is an effective means of communication between the citizen and Coimbatore City Municipal Corporation that integrates various modules for efficient delivery of services, round the clock access of public services through online and simplified the ways of availing the services.

3.1.4 OBJECTIVES OF THE STUDY

The study would identify the role of socio- demographic factors in the citizen's ability to adopt e-government and their perception towards e-government particularly municipal website services. The study puts forward the following objectives to fulfill the purpose of the study.

- To review previous literature for better understanding about adoption of e-government citizen services delivered through web portal.
- To analyze the citizen views regarding the usage of Coimbatore City Municipal Corporation Website (CCMC) web portal services.
- To evaluate the Computer and Internet Skills of citizens and examine the effect of socio- demographic factors on the same.
- To study the influence of socio- demographic factors on psychological factors which shapes citizen's perception towards Coimbatore City Municipal Corporation Website (CCMC) web portal services. The psychological factors examined are perceived usefulness, Perceived Ease of Use, Perceived Risk, trust, perceived website quality and Intention to Use.

3.2   REVIEW OF LITERATURE

Review of previous studies helped the researcher to gain knowledge about e-government, the benefits and the factors that influence e-government adoption. Studies by researcher indicate that socio-demographic characteristic of the individuals have played the role of determinants, predictors and moderators in e-government adoption. Understanding how citizens differ in their perception and level of acceptance of e-government based on socio-demographic factors such as age, gender, education and occupation is important.

E-government system has been adopted by many countries to efficiently deliver citizen centric services to a large population. E-government at national, state and local levels benefits citizens around the world (Huang & Bwoma, 2003; Phang, Sutanto, Li, & Kankanhalli, 2005; Bertot & Jaeger, 2006; Mahadeo & Devi, 2009; Al-Hujran, Al-dalahmeh, & Aloudat, 2011; Zafiropoulos, Karavasilisn & Vrana, 2012). E-government implementation would increase coherence in government organizations and their operations. Also, reduces processing cost by conducting government activities both internal and external electronically. It enhances service delivery quality by reducing bureaucracy, and increasing efficiency of the government in organizing and delivering the services. E-government improves interaction among all stakeholders by exchanging information on an integrated network. Integrated network offers easy to access Government to Citizen Services for individuals and Government to Business Services for business people (El-Sofany, Al-Tourki, Al-Howimel, & Al-Sadoon, 2012).

E-government websites that are dynamic, ubiquitous, interactive, searchable and customizable would render services to wide spread citizens. E-government websites can reduce time and cost involved in direct, physical government and citizen's interaction. These websites should ensure universal access and citizen centric services. To develop websites that would be easily accessible, usable, and functional, a user-centered focus design involving users in the development of the portal and user-centered evaluation methods need to be adopted. Understanding user's needs, demands, and resources available for e-services usage would be useful to increase e-government adoption. E-services could reach large number of people through assistive technologies, particularly for people with disabilities (Bertot & Jaeger, 2006; Bringula & Basa, 2011; Soufi & Maguire, 2008).

Many studies have adopted various technology acceptance models to explain the citizen's adoption of e-government services. Theory of Reasoned Action (TRA), Theory of Planned Behavior, Technology Acceptance Model (TAM), Extension of Technology Acceptance Model (TAM2), Unified Theory of Acceptance and Use of Technology (UTAUT) and Diffusion of Innovations (DOI) were used to study citizen's perception and acceptance towards e-government. The models helped the researchers to identify factors such as perceived usefulness, Perceived Ease of Use, social influence, facilitating conditions, attitude and intension to use that had an impact on citizen's adoption of e-government services. Studies also found that trust in government, trust in internet, Perceived Risk and website quality influenced citizen's Intention to Use e-government services (Carter & Belanger, 2004; Sutanto, Li, & Kankanhalli, 2005; Chen & Dimitrova, 2006; Chee-Wee, Benbasat, & Cenfetelli, 2008; Mahadeo, 2009; Phang, Alomari, Sandhu, & Woods, 2010; Almahamid, Mcadams, Kalaldeh, & Al-Sa'eed, 2010; Vencatachellum & Pudaruth, 2010; Al-Shafi & Weerakkody, 2010; Al-Hujran, Al-dalahmeh, & Aloudat, 2011; Zafiropoulos, Karavasilisn & Vrana, 2012; Alrashidi, 2012; Alzahrani, & Goodwin, 2012; Ahmad, Markkula, & Oivo, 2012). Unified Theory of Acceptance and Use of Technology (UTAUT) by V. Venkatesh, M. G. Morris, G. B. Davis, and F. D. Davis, explained the importance of individual's gender and age, experience and voluntariness of use in the decision making process of technology adoption.

Athmay (2013) found that age, gender, education and employment of the individuals had significant relationship with their perception and satisfaction with e-governance. Around 900 users of e-government services in United Arab Emirates (UAE) were surveyed to understand how gender, age, education, nationality, and employment affect citizen's attitude and satisfaction with e-governance. The findings of the study implied that men were more favorable towards e-governance and e-government portal services compared to women. Age influenced citizen's perception. Older citizens perceived that e-government portals have lead to more open, collaborative and participatory form of governance. Elder respondents expressed that e-government website were beneficial and have eased the process of accessing government services and getting required information from the government. The author explained that older citizens were more satisfied with the e-governance system as they had greater life experience and maturity which resulted in less radical behavior than younger citizens. Respondents with higher level of education exhibited more positive attitude towards e-governance than less educated respondents. The author further stated that respondents with higher educational qualification were more rational in assessing the benefits of e-governance than less educated respondents. With regard to employment, respondents working in public sectors provided favorable views regarding the e-governance and were more satisfied than respondents who work in nonprofit and private sectors.

Verda Canbey Özgüler (2012) also commented that socio-demographic characteristics of the citizens determine their level of acceptance of e-government. The author also established that preference of e-government services also depended on age and education. Preference towards the usage of e-government was more among the citizens with high educational qualification. The author indicated interesting findings such as elder citizens accessed computers and internet more than the young citizens. Also, elders efficiently used e-government services than the young citizens.

A recent study conducted among staff of academic libraries in Oyo state, Nigeria, examined the effect of socio-demographic variable on Perceived Ease of Use and level of accessibility of Information and Communication Technologies (ICT) such as computers, mobile phones and internet. The study revealed that demographic factors such as age, level of education and income significantly correlated with ease of use and accessibility of ICT tools. The study also indicated that age, level of education and income would be important predictors of individual's acceptance of ICT. The study used survey method to collect data from library staff including professional, para-professional and support staff in selected university libraries in Oyo state. The author found that younger age groups were technology savvy and eager to adopt new technology. The author further explained that younger age groups were more ready to face the challenges and complexities of the innovation (Owolabi, 2013). A survey was done by Colesca and Dobrica (2008) in Romania to study citizen's adoption of e-government portal services. The study identified that demographic characteristics of the citizens had a significant effect on perceived usefulness and Perceived Ease of Use the of e-government portal services.

Venkatesh, Morris and Ackerman (2000) analyzed the role of gender in adoption and sustained usage of technology in the workplace. Authors used theory of planned behavior (TPB) to explain technology usage behavior of new software technology application by 355 workers selected from different organization in America. The authors remarked that gender shapes individual decision to adopt technology. According to the theory TPB adopted in the study attitude, subjective norm and perceived behavioral control are the determinants of intension to use the technology. The study results showed that there was a strong relationship between gender and the determinants of intention such as attitude, subjective norm and perceived behavioral control. Men's decision to adopt the new technology was influenced by their attitude toward the technology, whereas, women's decision to adopt the new technology was influence by subjective norm and perceived behavioral control. Women gave more importance to social pressures such as the views and suggestions of the superiors and friends in the work place when compared to men. Women were more concerned about the ease of use of the new technology than men. Men relied more on their attitude to adopt the technology than women. A focus group study in Kuwait showed that a large number of female participants were interested to use e-government services and expressed that e-government services would be advantageous (AlAwadhi & Morris, 2009).

In a survey among users of U.S. federal government services, Morgeson III, Van Amburg and Mithas (2011) found that education and internet use were positive predictors of e-government adoption. Whereas, gender and income were not significant factors. But an online survey conducted among the internet users in the United States revealed that income was positively related to use of government services. Also, the study indicated that age was negatively related to adoption, that is, young people intended to use e-government services more than the elders. Bringula and Basa (2011) stated that young respondents used web portal services more. Also, respondents who used the portal services had high educational qualification and most of the respondents were master degree holders. A study was conducted among internet users in Indonesia where large portion of the respondents were young (21–30) and 80% of respondents were graduates, post graduates, and PhD Holders. They perceived e-government services useful and felt that e-government would ease the process of gathering information from government agencies and interaction with government agencies (Rokhman, 2011).

Occupation is a predictor of e-government adoption. Students, officials and professional were more positive towards e-government web portal (Al-Hujran, Al-dalahmeh &Aloudat, 2011; Mahadeo, 2009; Zafiropoulos, Karavasilis & Vrana, 2012). Pakistani university students adopted a range of e-government services and had proficiency in internet. They spend more than ten hours a week on the internet for information retrieval, transactions, and social networking. Also, they were socially influence to adopt the services and exhibited positive attitude towards e-government services. They perceived the services useful and easy to use (Ahmad, Markkula, & Oivo, 2012). A survey conducted among students, academics, and administrators in Jordanian universities identified that socioeconomic environment in Jordan is favorable for e-government adoption. The author indicated that Jordan has 95% of young people aged between 18 and 24 years and 90% of the adults were literates who would be more willing to accept the e-government services. Also, the author highlighted that the government's e-initiatives such as knowledge stations and e-villages aims to overcome the digital divide between cities and villages in the country. All these were found to be the positive factors for the acceptance of e-government and successful delivery of citizen services through e-government (Elsheikh & Azzeh, 2014).

A quantitative study by Al-eryani and Rashed (2012) examined the impact of gender, age, education, income, occupation level, computer experience and internet experience on e-readiness to adopt e-government services. E-readiness of the citizens refers to the access to the technologies and their ability and willingness to use the technology. E-readiness is an essential element for adoption of e-government. Gender difference existed in the e-readiness. Male respondents were more e-ready and had more confidence in e-government than female respondents. Also, they perceived that necessary technical infrastructure existed in Yemen to adopt e-government. They study also showed that respondents with higher educational qualification, with good computer and internet experience were more e-ready.

Review of past literature suggests that identifying the factors and examining the role of various factors is necessary to know the citizen's intentions towards e-government. Also, analyzing the socio-demographic differences among the citizens is imperative to better understand their level of acceptance of e-government.

3.3   CONSTRUCTS UNDER STUDY

The study had adopted various constructs based on the review of literature. Previous studies suggested that psychological factors and individual's ability to use the technology had an impact on citizen's perception towards e-government services and their level of Intention to Use the same. The studies also revealed that the psychological factors were driven by socio-demographic factors. Psychological factors such as perceived usefulness, Perceived Ease of Use, trust, Perceived Risk, perceived website quality and Intention to Use, individual's skills or ability to use computer and internet and socio-demographic factors such as gender, age, education and occupation were analyzed in the present study.

3.3.1   DEMOGRAPHIC FACTORS

Age, gender, education level and occupation were the demographic factors examined in the study. Previous studies by Al-Shafi and Weerakkody (2010) and Almahamid et al. (2010) revealed that age, gender and education level affected the use of e-government website services. Education was the most significant factor, people with high education level perceived that the government websites provided more information. Age considerably affected the decision to adopt e-government. Younger citizens were more open to the idea of using e-government services than older citizens (Colesca & Dobrica, 2008). Age, gender and education of the individuals have significant relationship with their experience and satisfaction with e-governace (Athmay, 2013). Occupation influences individual's Intention to Use e-governent services (Mahadeo, 2009; Zafiropoulos, Karavasilis, & Vrana, 2012; Ahmad, Markkula, & Oivo, 2012).

3.3.2 PERCEIVED USEFULNESS

"Perceived Usefulness is the degree to which a person believes that using a particular system would enhance his or her job performance" (Davis, 1989). People would prefer e-government services over traditional services for convenience of access, reduction in time, cost and effort and efficient service delivery. E-services delivered by the government web portals must be genuinely useful to the users and should meet their specific needs (AlAwadhi & Morris, 2009).

3.3.3 PERCEIVED EASE OF USE

"Perceived Ease of Use is the degree to which a person believes that using a system would be free of effort" (Davis, 1989). When users feel that they could easily use or could learn to use the e-government website and e-services without much effort, they would intent to adopt the portal. Higher level of Perceived Ease of Use increases the level of Intention to Use e-government services and significantly affected adoption of the same (Almahamid et al., 2010, Suki & Ramayah, 2010).

3.3.4 TRUST

Trust in government is created when citizens believe that the government agencies have the ability to provide reliable e-services and also have technical resources necessary to ensure security and privacy. Trust in the internet, is an important factor in e-government adoption. Citizens must believe that the internet is dependable and information provided through e-government website would be accurate and also the website could be used to conduct secured transactions. Trust is crucial to e-government adoption (Ayyash, Ahmad, & Singh, 2012, AlSaghier et al., 2009).

3.3.4.1 Perceived Risk

Perceived Risk is associated with financial risk as well as data security and privacy. Perceived Risk is a crucial factor in determining adoption of e-government web portal services (AlSaghier et al., 2009; Kumar, Mukerji, Butt, & Persaud, 2007).

3.3.4.2   Website Quality

Individual's opinion about the quality of website has an effect on their Intention to Use the website (Wangpipatwong et al., 2005; Bhattacharya, Gulla, & Gupta, 2007).

3.3.4.3   Computer and Internet Skills

Individual's Computer and Internet Skills would have positive effect on the usage of e-government web portal services (Dimitrova & Chen, 2006; Colesca, 2009; Ahmad, Markkula, & Oivo, 2012, MorgesonIII, VanAmburg, & Mithas, 2011).

3.3.4.4   Intention to Use

Intention is a strong predictor of behavior. Intension to use the web portal services would lead to the actual usage of the same (Venkatesh et al., 2003).

3.4   RESEARCH METHODS

Survey method was used in the present study. The survey was conducted among the students, professionals, executives/officials and business people in Coimbatore city, Tamil Nadu state, India. Self-administered questionnaire were used for the survey. A pilot test was done to pretest the questionnaire. Cronbach's alpha was used to determine the reliability of the questionnaire. After eliminating incomplete and not usable questionnaires, 484 usable responses (questionnaires) were used for the study.

Before the main survey, a pilot survey was conducted with a sample of 45 including students, professionals, officials and business people who would represent the type of respondents who would participate in the study. Pilot study enabled the researcher to understand about the field where the actual survey would be conducted. The data collect through the pilot study was useful to test the reliability of the instrument before using it to conduct the main survey.

3.4.1   RELIABILITY AND VALIDITY OF THE QUESTIONNAIRE

Measurement instrument must possess reliability and validity. Reliable measurement scales of variables can detect relationship between variables and validity of a measuring device ensure that the device measures what it is suppose to measure (Wimmer & Dominick, 2011). So determining the reliability and validity of measurement scales used in the questionnaire is important. The data collected from the pilot study was used to test the reliability of the questionnaire.

Cronbach's Alpha is widely used to ensure reliability of the measurement instrument (Iacobucci & Duhachek, 2003; Singh, 2007; Wimmer & Dominick, 2011). The study used Cronbach's alpha to determine reliability of measurement scales of all constructs in the questionnaire. Alpha more than 0.6 signifies that the measurement is reliable (Rokhman, 2011; Hair et al., 2006; Almahamid et al., 2010; Hinton et al., 2004). Cronbach's alpha coefficients for all constructs were above 0.6, which indicates that the questionnaire used in the study is a reliable measurement instrument (Table 3.1).

Validity of the instrument was ensured as the items were adopted from previous studies, where items measure the constructs under the study. Further, scrutinizing of the items by the panel of judges and modification of the questionnaire based on the suggestions of 15 individuals who would be the type of persons who would participate in the survey increased validity of the instrument.

3.4.2   DEMOGRAPHIC PROFILE OF THE RESPONDENTS OF THE SURVEY

The number of survey respondents was 484. The gender distribution of the survey respondents shows that 63.4 percent of the respondents were males and 36.6 percent of the respondents were females. The large percentage

TABLE 3.1   Cronbach's Alpha Reliability Test

Construct	Cronbach's Alpha	No. of Items
Computer and Internet Skills	0.892	6
Perceived Usefulness	0.801	9
Perceived Ease of Use	0.736	4
Perceived Risk	0.743	7
Trust	0.804	15
Website Quality	0.790	11
Intention to Use	0.882	6

of the respondents was males. In terms of age, 29.5 percent of the survey respondents were in the age group of 20–25, 23.6 percent were in the age group of 26–35, 27.1 percent were in the age group of 36–45 and 19.8 percent were in the age group of Above 45. More number of respondents are in the younger age group of 20–25. Overall, large proposition of the survey respondents were in the age groups of 20–25 and 36–45. In terms of educational qualification, 19.2 percent of the survey respondents were diploma holders, 28.3 percent were graduates and 52.5 percent were post graduates. On the whole, Majority of the respondents (80.8%) were graduates or post graduates. The distribution of sample based on their occupation in the Table 3.2 shows that 15.5 percent of the respondents are students, 26 percent are official/executives, 30 percent are business

TABLE 3.2   Profile of the Survey Respondents N = 484

Demographic factors		Frequency	Percent
Gender	Male	307	63.4
	Female	177	36.6
Age	20–25	143	29.5
	26–35	114	23.6
	36–45	131	27.1
	Above 45	96	19.8
Education	Diploma	93	19.2
	Graduate	137	28.3
	Post Graduate	254	52.5

TABLE 3.2   Continued

Demographic factors		Frequency	Percent
Occupation	Student	75	15.5
	Official/Executives	126	26.0
	Business	145	30.0
	Professional	138	28.5
Monthly Income	Not an Income Earner	75	15.5
	Below 30,000	140	28.9
	30,001–60,000	176	36.4
	Above 60,000	93	19.2

persons and 28 percent are professional. Relatively, small percentage of the respondents was students. In terms of monthly income, 15.5 percent of the respondents were not income earners, 28.9 percent of the respondents belong to Below 30,000 income group, 36.4 percent of the respondents belong to 30,001–60,000 income group and 19.2 percent of the respondents belong to above 60,000 income group. Comparatively large percentage of the respondents belongs to 30,001–60,000.

3.5   DATA ANALYSIS

The quantitative data obtained from the survey was analyzed using appropriate statistical methods. "…Quantitative data are numerical; they are information about the world, in the form of numbers" (Punch, 2000; p. 59). In the study, Mean was used to summarize the data in meaningful way for interpretation. Mean is the average of a set of scores, which is widely used in descriptive statistics.

3.5.1   DESCRIPTIVE STATISTICS

3.5.1.1   Computer and Internet Skills

Computer and Internet Skills of the respondents were measured using 6 items rated on 2 point scale, where Yes = 1 and No = 0. From the Tables 3.3 and 3.4, it is evident that those respondents were proficient in

TABLE 3.3 Score Range for All Constructs

Construct	Low	Medium	High
Computer and Internet Skills	Below 1.5	1.5–4.5	Above 4.5
Facilitating Conditions	Below 11	11–25	Above 25
Awareness	Below 3	3–9	Above 9
Perceived Usefulness	Below 15.75	15.75–22.5	Above 22.5
Perceived Ease of Use	Below 7	7–13	Above 13
Perceived Risk	Below 12.25	12.5–17.5	Above 17.5
Trust	Below 26.25	26.25–48.75	Above 48.75
Website Quality	Below 19.25	19.25–35.75	Above 33.75
Social Influence	Below 7	7–13	Above 13
Intention to Use	Below 9	9–19	Above 19

TABLE 3.4 Mean of All Constructs

Construct	Mean
Computer and Internet Skills	5.576
Perceived Usefulness	26.200
Perceived Ease of Use	11.774
Perceived Risk	24.760
Trust	40.061
Website Quality	30.411
Intention to Use	16.665

using computer and internet (M = 5.5764). They could perform various online activities such as searching information, downloading files/documents, navigating from one website to another, filling online forms and sending and receiving e-mails. Overall, respondents viewed that facilitating conditions were moderate.

3.5.1.2   Perceived Usefulness of the CCMC Website

The perceived usefulness of the CCMC website was measured using 9 items rated on four point scale, where strongly agree = 4, agree = 3, disagree = 2, and strongly disagree = 1. The Tables 3.3 and 3.4 shows that on the whole respondents perceived CCMC website services would be very useful (M = 26.2004). Respondents agreed CCMC website would provides 24/7 service; public could apply for municipal service and pay taxes even if they are not in the city; the website would save traveling time, queuing time and cost; website would facilitate effortless transaction and website would enable them to access all municipal services. Also, they perceived that websites would improve organization and delivery of public services and would result in quick service delivery.

3.5.1.3   Perceived Ease of Use of the CCMC Website

The perceived ease of the CCMC website was measured using four items rated on four point scale, where strongly agree = 4, agree = 3, disagree = 2, and strongly disagree = 1. The Tables 3.3 and 3.4 suggest that on the whole respondents felt that CCMC website would be easy to use (M = 11.7748). They agreed that CCMC website would be easy to use and would be user friendly. They perceived that searching information and making online payment through the website would be easy.

3.5.1.4   Perceived Risk

The Perceived Risk was measured using four items rated on four point scale, where strongly agree = 4, agree = 3, disagree = 2, and strongly disagree = 1. The Tables 3.3 and 3.4 indicate that generally, respondents perceived using internet particularly CCMC website would involve high risk (M = 24.760). They felt that personal information given to CCMC website might be used for other purposes by the corporation/government and CCMC website may allow others to access their personal information without their consent.

3.5.1.5   Trust

Trust was measured using 15 items rated on four point scale, where strongly agree = 4, agree = 3, disagree = 2, and strongly disagree = 1. The scale also consisted of two negative item, so reverse coding was done where strongly agree = 1, agree = 2, disagree = 3, and strongly disagree = 4.

From the Tables 3.3 and 3.4, it is understood that overall, respondents agreed they had trust in government and trust in internet (M = 40.0619). Respondents agreed that e-government websites and its services are for public welfare and e-government services are trustworthy. They perceived CCMC website would be fair in dealing with all their online transactions, increased transparency in the delivery of public services and overall, they trusted CCMC website.

3.5.1.6   Website Quality

Website Quality was measured using 11 items rated on four point scale, where strongly agree = 4, agree = 3, disagree = 2, and strongly disagree = 1. Tables 3.3 and 3.4 imply that generally, respondents considered that the quality of CCMC website would be good (M = 30.4112). Respondents agreed that the information given by the website would be authentic and reliable and the information would be presented in simple and understandable manner. They provided favorable view towards the website for provision of all necessary information. They perceived that CCMC website would be available at all times, contents could be easily accessed and would be easy to navigate.

3.5.1.7   Intention to Use

The Tables 3.3 and 3.4 imply that generally, respondents had Intention to Use CCMC website (M = 16.6652). Respondent's Intention to Use the website was relatively more for paying corporation taxes/bills online and for getting information about Coimbatore Corporation's services and schemes. They also indicated that they would intend to use the website to register complaints or grievances and to avail municipal services.

3.5.2   Z-TEST FOR GENDER

Z-test is done to compare two groups based on their means. In the present study, the respondent's Computer and Internet Skills and their opinion on various constructs such as perceived usefulness, Perceived Ease of Use,

Perceived Risk, trust, website quality, and Intention to Use were compared on the basis of their gender. Hence, Z-test was applied. The following are the hypotheses developed to analyze the significance of gender:

$H_0$1a Both Men and Women have same level of Computer and Internet Skills;

$H_0$1b Both Men and Women have same opinion about usefulness of CCMC web portal services;

$H_0$1c Both Men and Women have same opinion about ease of use of CCMC web portal services;

$H_0$1d Both Men and Women have same opinion about risk in using e-government web portal;

$H_0$1e Both Men and Women have same level of trust in internet and e-government system;

$H_{0_}$1f Both Men and Women have same opinion about of website quality of CCMC web;

$H_{0_}$1g Both Men and Women have same opinion about Intention to Use CCMC website services.

The results in the Table 3.5 suggest the following. $P > 0.05$ for Computer Skills and Internet experience ($Z_0 = 0.780$) ($P = 0.436$), Perceived Usefulness ($Z_0 = 0.991$) ($P = 0.322$), Perceived Ease of Use ($Z_0 = 1.195$) ($P = 0.233$) and Perceived Risk ($Z_0 = 0.612$) ($P = 0.541$). Hence, the Null hypotheses $H_0$1a, $H_0$1b, $H_0$1c and, $H_0$1d, are accepted for these constructs. This indicates that Men and Women respondents on an average had same level of Computer and Internet Skills and had opinion about facilitating condition, perceived usefulness, Perceived Ease of Use and Perceived Risk.

Whereas, $P < 0.05$ for Trust ($Z_0 = 4.084$), Website Quality ($Z_0 = 2.407$), and Intention to Use ($Z_0 = 3.592$). Hence, the Null hypotheses $H_0$1e, $H_0$1f, and $H_0$1g are rejected. This shows that both Men and Women on an average did not have same opinion on trust, website quality, social influence and Intention to Use. Women respondents (M = 40.8983) had more trust than men respondents (M = 39.5798). They (M = 30.8701) perceived that the website quality of CCMC website would be better than Men (M = 30.1466) and also they (M = 9.8644) experienced more social influence than Men (M = 9.4919). They (M = 17.1130) had more Intention to Use CCMC website than Men (M = 16.4072).

TABLE 3.5   Z-test Results for Respondent's Opinion About the Constructs by Gender

Constructs	Gender	Mean	Z_0	Sig.	Remark
Computer and Internet Skills	Male	5.5472	0.780	0.436	≥0.05 Not Significant
	Female	5.6271			
Perceived Usefulness	Male	26.3062	0.991	0.322	≥0.05 Not Significant
	Female	26.0169			
Perceived Ease of Use	Male	11.7166	1.195	0.233	≥0.05 Not Significant
	Female	11.8757			
Perceived Risk	Male	24.7101	0.612	0.541	≥0.05 Not Significant
	Female	24.8475			
Trust	Male	39.5798	4.084	0.000	< 0.05 Significant
	Female	40.8983			
Website Quality	Male	30.1466	2.407	0.016	< 0.05 Significant
	Female	30.8701			
Intention to Use	Male	16.4072	3.592	0.000	< 0.05 Significant
	Female	17.1130			

The level of significance is test at 0.05 level.

3.5.3   ONE WAY ANOVA

Where, three or more number of groups are to be compared on the basis of their mean values, Analysis of Variance (ANOVA) is applied (Singh, 2007). In the study, respondents were classified in to several groups on the basis of their Age, Education, Occupation, and Income. Therefore, to assess or compare their opinion given on the various constructs under study ANOVA technique was used.

3.5.4   ONE WAY ANOVA-AGE

To examine the effect of gender on respondent's Computer and Internet Skills and their perception the following hypotheses were formulated:

$H_0$2a Respondents belonging to various age groups have same level of Computer and Internet Skills

H_02b Respondents belonging to various age groups have same opinion about usefulness of CCMC website services

H_02c Respondents belonging to various age groups have same opinion about ease of use of CCMC website services

H_02d Respondents belonging to various age groups have same opinion about risk in using e-government web portal

H_02e Respondents belonging to various age groups have same level of trust in internet and e-government system

H_02f Respondents belonging to various age groups have same opinion about of Website Quality of CCMC website

H_02g Respondents belonging to various age groups have same opinion about Intention to Use CCMC website services

The Table 3.7 shows that $p \geq 0.05$ for perceived usefulness ($F = 0.521$, $p = 0.668$) and Perceived Ease of Use ($F = 0.546$, $p = 0.651$). Hence, Null hypotheses H_02b and H_02c are accepted for these constructs. This indicate respondents belonging to various age groups had same opinion about perceived usefulness and Perceived Ease of Use of CCMC website.

Also, from the Table 3.7 it is evident that $p < 0.05$ for Computer and Internet Skills ($F = 12.347$, $p = 0.000$), Perceived Risk ($F = 3.588$, $p = 0.014$), trust ($F = 2.669$, $P = 0.047$), website quality ($F = 3.345$, $p = 0.019$), and Intention to Use ($F = 7.417$, $p = 0.000$). Hence, Null hypotheses are H_02a, H_02d, H_02e, H_02f, and H_02g are rejected. This shows that respondents belonging to various age groups did not have on an average same level of Computer and Internet Skills and also did not have same opinion about Perceived Risk, trust, website quality, and Intention to Use. Post-Hoc is applied to find out which age group differs significantly from others in their Computer and Internet Skills and in their opinion on

TABLE 3.6   Table of Means for the Constructs by Age

Age	CI	PU	PEOU	PR	Trust	WQ	IU
20–25	5.867	26.405	11.664	24.552	40.713	31.097	17.279
26–35	5.789	26.175	11.807	25.359	39.956	29.973	16.684
36–45	5.167	25.946	11.778	24.450	39.580	30.122	16.137
Above 45	5.447	26.270	11.895	24.781	39.875	30.302	16.447

CI = Computer and Internet Skills, PU = perceived usefulness, PEOU = Perceived Ease of Use, PR = Perceived Risk, WQ = website quality, IU = Intention to Use.

TABLE 3.7   ANOVA Results for Respondent's Opinion About the Constructs by Age

Constructs		Sum of Squares	df	Mean Square	F	sig.	Remarks
Computer and Internet Skills	Between Groups	40.704	3	13.568	12.347	0.000	<0.05 Significant
	Within Groups	527.468	480	1.099			
	Total	568.171	483				
Perceived Usefulness	Between Groups	15.009	3	5.003	0.521	0.668	≥0.05 Not Significant
	Within Groups	4606.551	480	9.597			
	Total	4621.560	483				
Perceived Ease of Use	Between Groups	3.271	3	1.090	0.546	0.651	≥0.05 Not Significant
	Within Groups	959.181	480	1.998			
	Total	962.452	483				
Perceived Risk	Between Groups	59.754	3	19.918	3.588	0.014	<0.05 Significant
	Within Groups	2664.445	480	5.551			
	Total	2724.198	483				
Trust	Between Groups	95.707	3	31.902	2.669	0.047	<0.05 Significant
	Within Groups	5738.434	480	11.955			
	Total	5834.140	483				
Website Quality	Between Groups	101.344	3	33.781	3.345	0.019	<0.05 Significant
	Within Groups	4847.836	480	10.100			
	Total	4949.180	483				
Intention to Use	Between Groups	95.068	3	31.689	7.417	0.000	<0.05 Significant
	Within Groups	2050.709	480	4.272			
	Total	2145.777	483				

The level of significance is test at 0.05 level.

facilitating condition, Perceived Risk, trust, website quality, social influence and Intention to Use.

Post-Hoc analysis results for age presented in the Table 3.8 reveals that overall, than older respondents, respondents of younger age 20–25 had high level of Computer and Internet Skills and had more trust in e-government and internet. Also, they perceived CCMC website quality would be better. They perceived using online services particularly using CCMC website would be less risky. They intend to use the website more.

3.5.5   ONE WAY ANOVA-EDUCATION

H_03a Respondents belonging to various educational groups have same level of Computer and Internet Skills

H_03b Respondents belonging to various educational groups have same opinion about usefulness of CCMC website services

H_03c Respondents belonging to various educational groups have same opinion about ease of use of CCMC website services

H_03d Respondents belonging to various educational groups have same opinion about risk in using e-government web portal

H_03e Respondents belonging to various educational groups have same level of trust in internet and e-government system

TABLE 3.8   Post-Hoc Results for Respondent's Opinion About the Constructs by Age

Construct	Age		MD	Sig.
Computer and Internet Skills	20–25	36–45	0.699	0.000
		Above 45	0.419	0.014
	26–35	36–45	0.621	0.000
Perceived Risk	26–35	20–25	0.807	0.033
		36–45	0.909	0.014
Trust	20–25	36–45	1.133	0.035
Website Quality	20–25	26–35	1.124	0.026
Intention to Use	20–25	36–45	1.142	0.000
		Above 45	0.831	0.013

TABLE 3.9   Table of Means for Constructs by Educational Qualification

Education	CI	PU	PEOU	PR	Trust	WQ	IU
Diploma	5.580	25.354	11.677	24.591	39.580	29.451	15.698
Graduate	5.708	25.941	11.948	24.678	39.781	30.343	16.394
Post Graduate	5.503	26.649	11.716	24.866	40.389	30.799	17.165

CI = Computer and Internet Skills, PU = perceived usefulness, PEOU = Perceived Ease of Use, PR = Perceived Risk, WQ = website quality, and IU = Intention to Use.

H_0–3f Respondents belonging to various educational groups have same opinion about of website quality of CCMC website

$H_0$3g Respondents belonging to various educational groups have same opinion about Intention to Use CCMC website services

The Table 3.10 shows that $p \geq 0.05$ for Computer and Internet Skills ($F = 1.580$, $p = 0.207$), Perceived Ease of Use ($F = 1.483$, $p = 0.228$), Perceived Risk ($F = 0.567$, $p = 0.568$) and trust ($F = 2.484$, $p = 0.084$). Hence, Null hypotheses $H_0$3a, $H_0$3c, $H_0$3d, and $H_0$3e are accepted for these constructs. This indicate that respondents belonging to various education groups had on an average same level of Computer and Internet Skills and same opinion on Perceived Ease of Use, Perceived Risk and trust. The Table 3.10 also shows that $p < 0.05$ perceived usefulness ($F = 6.791$, $p = 0.001$), Website Quality ($F = 6.207$, $p = 0.002$), and Intention to Use ($F = 19.435$, $p = 0.000$). Hence, Null hypotheses $H_0$3b, $H_0$3f and $H_0$3g are rejected for these constructs. This shows that respondents belonging to various education groups did not have same opinion on perceived usefulness, website quality, and Intention to Use.

Post-Hoc analysis results in Table 3.11 reveals that on the whole, respondents with higher educational qualification perceived CCMC website quality would be better and website would be useful. They also intend to use the website more than the respondents with education qualification of Diploma.

3.5.6   ONE WAY ANOVA-OCCUPATIONAL

$H_0$4a Respondents belonging to various occupational groups have same level of Computer and Internet Skills

TABLE 3.10 ANOVA Results for Respondent's opinion about the constructs by Educational Qualification

Constructs		Sum of Squares	df	Mean Square	F	sig.	Remarks
Computer and Internet Skills	Between Groups	3.709	2	1.855	1.580	0.000	≥0.05 Not Significant
	Within Groups	564.462	481	1.174			
	Total	568.171	483				
Perceived Usefulness	Between Groups	126.922	2	63.461	6.791		<0.05 Significant
	Within Groups	4494.638	481	9.344			
	Total	4621.560	483				
Perceived Ease of Use	Between Groups	5.897	2	2.949	1.483	0.651	≥0.05 Not Significant
	Within Groups	956.555	481	1.989			
	Total	962.452	483				
Perceived Risk	Between Groups	6.408	2	3.204	0.567	0.014	≥0.05 Not Significant
	Within Groups	2717.791	481	5.650			
	Total	2724.198	483				
Trust	Between Groups	59.651	2	29.826	2.484	0.047	≥0.05 Not Significant
	Within Groups	5774.489	481	12.005			
	Total	5834.140	483				
Website Quality	Between Groups	124.512	2	62.256	6.207	0.019	<0.05 Significant
	Within Groups	4824.668	481	10.030			
	Total	4949.180	483				
Intention to Use	Between Groups	160.437	2	80.218	19.435	0.000	<0.05 Significant
	Within Groups	1985.340	481	4.128			
	Total	2145.777	483				

The level of significance is test at 0.05 level.

TABLE 3.11   Post-Hoc Results for Respondent's Opinion About the Constructs by Educational Qualification

Construct	Education		Mean Difference (MD)	Sig.
Perceived Usefulness	Post Graduate	Diploma	1.29477	0.002
Website Quality	Post Graduate	Diploma	1.34760	0.026
Intention to Use	Graduate	Diploma	0.69524	0.030
	Post Graduate	Diploma	1.46643	0.000

$H_0$4b Respondents belonging to various occupational groups have same opinion about usefulness of CCMC website services

$H_0$4c Respondents belonging to various occupational groups have same opinion about ease of use of CCMC website services

$H_0$4d Respondents belonging to various occupational groups have opinion about risk in using e-government web portal

$H_0$4e Respondents belonging to various occupational groups have same level of trust in internet and e-government system

H_0–4f Respondents belonging to various occupational groups have same opinion about of website quality of CCMC website services

$H_0$4g Respondents belonging to various occupational groups have same opinion about Intention to Use CCMC website services

Table 3.13 shows $p \geq 0.05$ Perceived Risk ($F = 2.626$, $p = 0.050$). Hence, Null hypothesis $H_0$4d is accepted. This indicates respondents belonging to various occupational groups had on an average same level of opinion on Perceived Risk. Table 3.13 also shows that $p < 0.05$ Computer Skills and Internet experience ($F = 13.178$, $p = 0.000$), Perceived Usefulness ($F = 8.549$, $p = 0.000$), Perceived Ease of Use ($F = 2.918$, $p = 0.034$), Trust ($F = 11.930$, $p = 0.000$), Website Quality ($F = 6.569$, $p = 0.000$) and Intention to Use ($F = 43.389$, $p = 0.000$). Hence, Null hypotheses $H_0$4a, $H_0$4b, $H_0$4c, $H_0$4e, $H_0$4f and $H_0$4g are rejected. This indicate that respondents belonging to various occupational groups did not have same level of computer skills and internet experience and did not have same level of opinion on perceived usefulness, Perceived Ease of Use, trust, website quality and Intention to Use.

TABLE 3.12   Means for Constructs by Occupation

Occupation	CI	PU	PEOU	PR	Trust	WQ	IU
Student	5.813	26.426	11.453	25.293	39.480	30.346	17.226
Official/ Executives	5.634	26.888	12.031	24.365	40.547	31.103	17.301
Business	5.137	25.165	11.689	24.889	38.917	29.510	15.179
Professional	5.855	26.536	11.804	24.695	41.137	30.760	17.340

Post-Hoc analysis results presented in Table 3.14 reveals that generally, students, officials/executives and professionals had high level of computer skills and internet experience and trust than business people. They also perceived CCMC website would be usefulness, easy to use and website quality would be good. They also had more Intention to Use CCMC website.

3.6   FINDINGS AND DISCUSSION

The study revealed that demographic factors influenced citizen's perception towards e-government services and their Intention to Use the same. Also, demographic factors had an effect on the citizen's Computer and Internet Skills.

3.6.1   INFLUENCE OF DEMOGRAPHIC FACTORS

Gender
Based on the gender, respondents differed in the level of trust and Intention to Use. Also, they differed in their perception about the website quality. Women respondents had more trust than men respondents. They perceived that the website quality of CCMC website would be better and also they had more Intention to Use CCMC website than men.

Age
Age had significant influence on respondent's Computer and Internet Skills, trust, Perceived Risk, website quality and intension to use. Generally than older respondents, respondents of younger age 20–25 had high level of Computer and Internet Skills and trust in e-government and internet more.

TABLE 3.13 ANOVA Results for Respondent's Opinion About the Constructs by Occupation

Constructs		Sum of Squares	df	Mean Square	F	Sig.	Remarks
Computer and Internet Skills		43.236	3	14.412	13.178	0.000	<0.05 Significant
	Within Groups	524.936	480	1.094			
	Total	568.171	483				
Perceived Usefulness	Between Groups	234.422	3	78.141	8.549	0.000	<0.05 Significant
	Within Groups	4387.138	480	9.140			
	Total	4621.560	483				
Perceived Ease of Use		17.241	3	5.747	2.918	0.034	<0.05 Significant
	Within Groups	945.212	480	1.969			
	Total	962.452	483				
Perceived Risk	Between Groups	43.993	3	14.664	2.626	0.050	≥0.05 Not Significant
	Within Groups	2680.205	480	5.584			
	Total	2724.198	483				
Trust	Between Groups	404.815	3	134.938	11.930	0.000	<0.05 Significant
	Within Groups	5429.325	480	11.311			
	Total	5834.140	483				
Website Quality	Between Groups	195.191	3	65.064	6.569	0.000	<0.05 Significant
	Within Groups	4753.989	480	9.904			
	Total	4949.180	483				
Intention to Use	Between Groups	457.760	3	152.587	43.389	0.000	<0.05 Significant
	Within Groups	1688.017	480	3.517			
	Total	2145.777	483				

The level of significance is test at 0.05 level.

TABLE 3.14   Post-Hoc Results Respondent's Opinion About the Constructs by Occupation

Construct	Occupation vs.		Mean Difference (MD)	Sig.
Computer and Internet Skills	Student	Business	0.67540	0.000
	Official/Executive	Business	0.49699	0.001
	Professional	Business	0.71714	0.000
Perceived Usefulness	Student	Business	1.26115	0.018
	Official/Executive	Business	1.72337	0.000
	Professional	Business	1.37071	0.001
Perceived Ease of Use	Official/Executive	Student	0.57841	0.025
Trust	Official/Executive	Business	1.63038	0.000
	Professional	Business	2.22044	0.000
	Professional	Student	1.65768	0.004
Website Quality	Official/Executive	Business	1.59283	0.000
	Professional	Business	1.25052	0.005
Intention to Use	Student	Business	2.04736	0.000
	Official/Executive	Business	2.12228	0.000
	Professional	Business	2.16127	0.000

Also, they perceived CCMC website quality would be better. They perceived using online services particularly using CCMC website would be less risky and they intend to use the website more.

Education

Respondent's educational qualification had an effect on their views regarding the usefulness of the website, website quality and Intention to Use. In general, respondents with higher educational qualification perceived CCMC website quality would be good, website would be useful and they also intend to use the website more than the respondents in the lower education group.

Occupation

Respondent's occupation influenced level of computer skills and internet experience, Intention to Use and also, their perception about the usefulness of the website, ease of use and website quality. On the whole, students, officials/executives and professionals had high level of computer skills and internet experience, trust and Intention to Use CCMC website than business people. They also felt that CCMC website would be useful-ness, easy to use and website quality would be good.

3.6.1.1   Computer and Internet Skills and Demographic Factors

Generally, individuals with Computer and Internet Skills would be feel comfortable to use the e-services. The study showed that respondents were proficient in using computer and internet. They could perform various online activities such as searching information, downloading files/docu-ments, navigating from one website to another, filling online forms and sending and receiving e-mails.

Though, in general respondents had high level of Computer and Internet Skills, their age and occupation had a significant effect. Generally than older respondents, respondents of younger age group 20–25 had high level of Computer and Internet Skills. United Nations E-Government Survey (2012) report also states that individual's age is related their abil-ity to use computer and internet to perform online activities particularly usage of e-government services. In addition, the study indicated that the students, officials/executives and professionals had more computer skills and internet experience, when compared to the business people. Ahmad, Markkula and Oivo (2012) remarked that Pakistani university students who were proficient in internet usage adopted a range of e-government services. Students spend more than 10 hours a week on the Internet for information retrieval, transactions, and social networking.

3.6.1.2   Role of Demographic Factors on Citizen's Perception Towards E-Government Web Portal Services and Their Intention to Use the Same

Athmay (2013) remarked that age, gender, education and employement of the individuals have significant relationship with their perception towards e-governace and e-government portal services. Similarly, the present study also indicated that socio-demographic factors such as age, gender, education and occupation influence citizen's perception and their Intention to Use the e-government website services.

3.6.2 PERCEIVED USEFULNESS AND PERCEIVED EASE OF USE

Respondents perceived CCMC website services would be very useful and easy to use. They agreed that the website provides 24/7 service, the website would save traveling time, queuing time and cost, the website would facilitate effortless transaction and the website would enable them to access municipal services from anywhere. Also, they perceived that the websites would improve organization and delivery of public services and would result in quick service deliver. On the whole, respondents also felt that CCMC website would be easy to use. They agreed that CCMC website would be user friendly and also perceived that searching information and making online payment through the website would be easy.

Respondent's educational qualification influenced the perception about the level of usefulness of CCMC website. Respondents with high educational qualification (post graduated and graduates) perceived CCMC website more useful than the respondents with low educational qualification (diploma holders). In a study conducted in Indonesia, 80% of respondents who were well educated recognized usefulness of e-government services. They felt that e-government services would enhance their efficiency in gathering information from government agencies and make interact with the government agencies easy. Jordanian citizens' level of education had an effect on their perceived usefulness of use of e-government services which was evident from the studies conducted by Almahamid, Mcadams, Kalaldeh, and Al-Sa'eed (2010) and Al-Hujran, Al-dalahmeh, and Aloudat (2011). In both the studies, the respondents with high educational qualification such as post graduation and graduation perceived e-government services useful. Another study was done among the employees of Small and Medium Enterprises (SME) office in United Arab Emirate. The employees who were postgraduate expressed that e-government service were beneficial (Alrashidi, 2012).

Occupation of the respondent had a significant relationship with their perceived usefulness and ease of use of e-government web portal services. This finding of the present study was consistent with previous studies by

Al-Hujran, Al-dalahmeh and Aloudat (2011), Carter and Belanger (2004), Mahadeo (2009) and Zafiropoulos, Karavasilis and Vrana (2012). In the present study, respondents who were students, officials/executives and professionals expressed that CCMC website services would be very useful and would be easy to use. In a study, Jordanian students perceived e-government services usefulness. Study conducted among professional (teaching professionals) in Greece indicated that the respondents opined that e-government web portal services would be advantageous (Zafiropoulos, Karavasilis, & Vrana 2012). Official of 200 organizations surveyed in Mauritius felt that the e-government web portal services useful and easy to use (Mahadeo, 2009).

3.6.3   TRUST

On the whole, respondents agreed they had trust in e-government and trust in internet. They agreed that generally, e-government websites and its services are for public welfare and e-government services are trustworthy. They expressed that they had trust in CCMC website and the website would be fair in dealing with all their online transactions. They believed that the CCMC website would increase transparency in the delivery of public services.

In the present study, gender, age and occupation had a significant influence on trust in e-government and trust in internet. Based on the gender, respondents differed in the level of trust. Women respondents had more trust in internet and in e-government than men respondents. The reasons for high level of trust among the women would be due to their level of usage of new technologies and their familiarity with online activities. Today, women have become active users of internet who do various online activities. This results in more confident in performing tasks using internet. Hence, they would be more confidence in using online services render by the government website. UN report highlighted that social networking sites has increased the usage of internet by women and women spend more time online than men. Also, the report showed that women in Asian countries have become more active users than men (United Nations E-Government Survey, 2012). Younger respondents of this study belonging to the age group of 25–30 had high level of trust in e-government and trust in internet. Studies by Colesca and Dobrica (2008) and Colesca (2009) also pointed out that young people trusted e-government services more than the elders. Occupation of the respondents had an impact on respondent's trust in e-government and trust in internet. Respondents of the study who were students, officials/executives and professionals trusted internet services particularly e-government services. Al-eryani and Rashed (2012) stated that occupation of an individual affected their readiness to accept e-government, where trust determined their readiness.

3.6.4 PERCEIVED RISK

Generally, respondents perceived using internet particularly CCMC website would involve high risk. They opined that CCMC website personal information given to CCMC website might be used for other purposes by the corporation/government and CCMC website may allow others to access their personal information without their consent. Comparatively, younger respondents of age group 25–30 perceived low level of risk. Colesca (2009) stated that younger age group was more open to the idea of using e-government services.

3.6.5 WEBSITE QUALITY

Respondents perceived that quality of the website would be good. Respondents agreed that the information given by the website would be authentic and reliable and the information would be presented in simple and understandable manner. They provided favorable view for the website would provides all necessary information. They perceived that CCMC website would be available at all times, contents could be easily accessed and would be easy to navigate.

Gender, age, education and occupation of the respondents had an effect on their opinion about website quality of CCMC website. Students, officials/executives and professionals perceived that the information and service quality of CCMC was fine. Athmay (2013) found that age, gender and education of the individuals has significant relationship with their experience with e-government services delivered. In a survey, well educated Jordanian citizens expressed positive views regarding the information quality of e-government portal. A study conducted in Indonesia where more than 80% of respondents including graduates, post graduates, and PhD holders felt that e-government web portal would be efficient and would result in easy gathering of information from government agencies and interaction with government agencies. Generally, students, officials and professionals were more positive towards e-government web portal (Al-Hujran, Al-dalahmeh & Aloudat, 2011; Mahadeo, 2009; Zafiropoulos, Karavasilis & Vrana, 2012).

3.6.6 INTENTION TO USE

Respondent's intended to use CCMC website, where Intention to Use the website was relatively more for paying corporation taxes/bills online and for getting information about Coimbatore Corporation's services and schemes than for other mentioned purposes. They also indicated that they would intend to use the website to register grievances and to avail municipal services.

The present study indicates that socio- demographic factors influenced citizen's Intention to Use CCMC website services. This finding confirmed the results of past studies. The factors such as gender, age, education and occupation understudy had a significant influence on Intention to Use CCMC website. VerdaCanbey Özgüler (2012) also identified that socio-demographic characteristics of the citizens determine their level of acceptance of e- government.

Respondent's gender affected Intention to Use the CCMC website services. Women respondent intended to use the CCMC website more than men. AlAwadhi and Morris (2009) and Almahamid, Mcadams, Kalaldeh and Al-Sa'eed (2010) also found that gender played an important role in usage of e-government services. Authors stated that women were more willing to use e-government services than men. Venkatesh, Morris and Ackerman (2000) stated that "Clearly, gender shapes the initial decision process that drives new technology adoption and usage behavior in the short-term, which in turn influences sustained usage…"

Respondents of the study belonging to various age groups differed in their level of Intention to Use CCMC website. Respondents of younger age group 20–25 had high level of Intention to Use the website than the respondents of older age groups 36–45 and Above 45. This indicates that age had an impact on Intention to Use CCMC website. The result is in compliance with the previous researches by Athmay (2013), Ahmad, Markkula and Oivo (2012), Al-Hujran, Al-dalahmeh and Aloudat (2011), Bringula and Basa (2011), Carter and Belanger (2004), Colesca and Dobrica (2008) and Dimitrova and Chen (2006). An online survey conducted by Dimitrova and Chen (2006) among the internet users in the United States revealed that young people were likely to adopt e-services. Studies of Bringula and Basa (2011) and Colesca and Dobrica (2008) also confirmed that young citizens would intented to use e-government web portal services more than citizens of older age groups. The authors suggested that young persons are more likely to trust e-government services than the elders. They perceive e-government web portal services more useful and feel they could use the services easily. In contrary, a recent study in Turkey showed that preference towards the usage of e-government was more among elder citizens and they efficiently used e-government services than the young citizens.

Educational qualification of the respondents influenced their Intention to Use CCMC website. Respondents with high educational qualification (graduates and post graduates) had high level of Intention to Use CCMC website than respondents with low educational qualification (diploma holders). Study by MorgesonIII, VanAmburg and Mithas (2011) substantiated that education is positive predictors of e-government adoption. Another study pointed out that Jordanian citizen's education played a major role in their use of e-government information (Almahamid, Mcadams, Kalaldeh, & Al-Sa'eed, 2010). Occupation of the respondents had a significant relationship with their intention. Students, officials/executives and professional intended to use the CCMC website more than the business people. A survey conducted in Mauritius revealed that officials who participated were positive about government web portal services and intended to use the services (Mahadeo, 2009). Also, results of a study showed that professional (teaching professionals) in Greece intended to use the e-government web portal services. Pakistani university students exhibited affirmative views towards e-government services and also adopted a range of e-government services.

3.7   CONCLUSION

Citizen's acceptance of e-government system and intension to use the e-services provided by the government are crucial to adoption of e-government web portal services. Lack of understanding the reasons for citizen's perception towards e-government web portal services and their Intention to Use the same is a major issue. The study examined the factors that influence citizen's perception and Intention to Use municipal web portal services particularly CCMC website services. The study highlighted the importance of socio-demographic factors in citizen's acceptance of e-government web portal services in a developing country context. The study would enhance the knowledge about impact of citizen's socio-demographic characteristics on psychological factors, which would in turn influence their intentions and level of adoption of municipal government web portal services. The implications of the study would help the e-government project heads to better understanding the citizen's attitude towards the use of government web portals. The findings would guide the web portal designers to develop citizen centric e-government web portals that would reach large population and reduce socio-demographic disparities in accessing the government portal services.

KEYWORDS

- **e-government**
- **intention to use**
- **perceived usefulness and ease of use**
- **socio-demographic factors**
- **trust**

REFERENCES

1. Agrawal, A., Shah, P., Wadhwa, V. (2007). EGOSQ - Users' Assessment of e-Governance Online-Services: A Quality Measurement Instrumentation. In A. Agarwal, & V. V. Ramana (Eds.), Foundations of E-government (pp. 231–244). SIGeGov Publications.

2. Agrawal, V., Mittal, M., Rastogi, L. (2003). Enabling e-Governance: Integrated Citizen Relationship Management Framework – The Indian Perspective. Delhi Business Review , 4 (1), 99–112.

3. Ahmad, M. O., Markkula, J., Oivo, M. (2012). FACTORS INFLUENCING THE ADOPTION OF E-GOVERNMENT SERVICES IN PAKISTAN. European, Mediterranean & Middle Eastern Conference on Information Systems (EMCIS). Munich.

4. Aijaz, R. (2008). Form of Urban Local Government in India. Journal of Asian and African Studies , 43 (2), 131–154.

5. Ajzen, I. (1991). The theory of planned behavior. Organizational Behavior and Human Decision Process , 50, 179–211.

6. Alateyah, S. A., Crowder, R. M., Wills, G. B. (2013). An Exploratory study of proposed factorsto Adopt e-government Services:Saudi Arabia as a case study. International Journal of Advanced Computer Science and Applications (IJACSA) , 4 (11), 57–66.

7. AlAwadhi, S., Morris, A. (2009). Factors Influencing the Adoption of E-government Services. Journal Of Software , 4 (6), 584–590.

8. Al-eryani, A., Rashed, A. (2012). The Impact of The Culture On The E-readiness For E- government in Developing Countries (Yemen). The 13th International Arab Conference on Information Technology ACIT 2012, (pp. 331–339).

9. Al-Hujran, O., Al-dalahmeh, M., Aloudat, A. (2011). The Role of National Culture on Citizen Adoption of eGovernment Services: An Empirical Study. Electronic Journal of e-Government , 9 (2), 93 - 106.

10. Al-Khouri, A. M. (2011). An Innovative Approach for E-Government Transformation. International Journal of Managing Value and Supply Chains (IJMVSC) , 2 (1), 22–43.

11. Almahamid, S., Mcadams, A. C., Kalaldeh, T. A., Al-Sa'eed, M. (2010). The Relationship between Perceived Usefulness, Perceived Ease Of Use, Perceived Information Quality, And Intention to Use E-government. Journal of Theoretical and Applied Information Technology , 11 (1), 30–44.

12. Alomari, M. K., Sandhu, K., Woods, P. (2010). Measuring Social Factors in E-government Adoption in the Hashemite. International Journal of Digital Society (IJDS) , 1 (2), 123–134.

13. Alrashidi, A. (2012). User Acceptance and Motivation of E-Governance Services Based on Employees Levels of Experience in the UAE SME. American Journal of Economics , 2 (6), 132–135.

14. AlSaghier, H., Ford, M., Nguyen, A., Hexel, R. (2009). Conceptualising Citizen's Trust in e-Government: Application of Q Methodology. (F. Bannister, Ed.) Electronic Journal of e-Government , 7 (4), 295–310.

15. Al-Shafi, S., Weerakkody, V. (2010). Factors Affecting E-Government Adoption in the State of Qatar. European and Mediterranean Conference on Information Systems. Abu Dhabi.

16. Alzahrani.M.E, & Goodwin.R.D. (2012). Towards a UTAUT-based Model for the Study of EGovernment Citizen Acceptance in Saudi Arabia. World Academy of Science, Engineering and Technology (64), 8–14.

17. Amritesh, Misra, S. C., Chatterjee, J. (2012). Examining Information Quality for e-Governance Services:Towards a Conceptual Model. International Proceedings of Computer Science and Information Technology. 31, pp. 127–134. Singapore: IAC-SIT Press.

18. Ayyash, M. M., Ahmad, K., Singh, D. (2012). A Questionnaire Approach for User Trust Adoption in Palestinian E-Government Initiative. American Journal of Applied Sciences , 9 (1), 40–46.

19. Azab, N. A., Kamel, S., Dafoulas, G. (2009). A Suggested Framework for Assessing Electronic Government Readiness in Egypt. Electronic Journal of e-Government , 7 (1), 11 - 28.

20. Backus, M. (2001). E-Governance and Developing Countries, Introduction and examples. Netherland: International Institute for Communication and Development (IICD).

21. Bagga, R. K., Gupta, P. (Eds.). (2009). Transforming Government: E-Governance Initiatives In India. Hyderabad, Andhra Pradesh, India: The Icfai University Press.

22. Barnes, S. J., Vidgen, R. (2004). Interactive E-Government: Evaluating the Web Site of the UK Inland Revenue. Journal of Electronic Commerce in Organizations (JECO) , 2 (1), 42–63.

23. Barua, M. (2012). E-Governance Adoption in Government Organization of India. International Journal of Managing Public Sector Information and Communication Technologies (IJMPICT) , 3 (1), 1–20.

24. Belanger, F., Carter, L. (2008). Trust and Risk in eGovernment Adoption. The Journal of Strategic Information Systems , 17 (2), 165–176.

25. (2002). Benchmarking E-Government: A Global Perspective, Assessing the Progress of the UN Member States. New York : United Nations Division for Public Economics and Public Administration (UNDPEPA) .

26. Bertot, J. C., Jaeger, P. T. (2006). User-centered e-government: Challenges and benefits for government Web sites. Government Information Quarterly , 23, 163– 168.

27. Bhatnagar, S. C., Singh, N. (2010). Assessing the Impact of E-Government: A Study of Projects in India. Information Technologies & International Development , 6 (2), 109–128.

28. Bhattacharya, D., Gulla, U., Gupta, M. P. (2007). Assessing Effectiveness of State Government Portals in India. In A. Agarwal, & V. V. Ramana (Eds.), Foundations of E-government (pp. 278–287). SIGeGov Publications .

29. Bringula, R. P., Basa, R. S. (2011). Factors Affecting Faculty Web Portal Usability. Educational Technology & Society , 14 (4), 253–265.

30. Burke, M. (2012). A Decade of e-Government Research in Africa. The African Journal of Information and Communication 2012 (12), 1–25.

31. Carter, L., Belanger, F. (2004). The Influence of Perceived Characteristics of Innovating on Egovernment Adoption. Electronic Journal of E-Government , 2 (1), 11–20.

32. Chakravarthi, B., M.Venugopal. (2008). "Citizen Centric Service Delivery through e-Governance Portal - Present Scenario in India". Hyderabad: National Institute for Smart Government.
33. Chander, S., Kush, A. (2012). E-Governance Web Portals Assessment of Two States. International Journal of Advanced Research in Computer Science and Software Engineering , 2 (2).
34. Chander, S., Kush, A. (2012). Performance Analysis using metrics of two egovernment Portal services. The International Journal of Computer Science & Applications , 1 (4), 32–40.
35. Chander, S., Kush, A. (2012). Web Portal Analysis of Asian Region Countries. International Journal of Information Engineering and Electronic Business , 4 (4), 25–32.
36. Chatzopoulos, K.-C., Economides, A. A. (2009). A holistic evaluation of Greek municipalities' websites. Electronic Government , 6 (2), 193–212.
37. Chee-Wee, T., Benbasat, I., Cenfetelli, R. (2008). Building Citizen Trust towards E-Government Services: Do High Quality Websites Matter? Proceedings of the 41st Hawaii International Conference on System Sciences (HICSS'08) (p. 217). Big Island, Hawaii: IEEE Computer Society Washington, DC, USA.
38. Chen, Y.-C., Dimitrova, D. V. (2006). Electronic Government and Online Engagement: Citizen Interaction with Government via Web Portals. International Journal of Electronic Government Research , 2 (1), 54–76.
39. Choudrie, J., Ghinea, G., Weerakkody, V. (2004). Evaluating Global e-Government Sites: A View using Web Diagnostic Tools. Electronic Journal of e-Government , 2 (2), 105–114.
40. Coimbatore City Municipal Corporation. (2014). Retrieved from Coimbatore City Municipal Corporation Website: https://www.ccmc.gov.in
41. Colesca, S. E. (2009). Increasing E-Trust: A Solution to Minimize Risk in E-Government Adoption. Journal of Applied Quantitative Methods , 4 (1), 31–44.
42. Colesca, S. E., Dobrica, L. (2008). Adoption And Use Of E-Government Services: The Case Of Romania. Journal of Applied Research and Technology , 6 (3), 204–217.
43. Conklin, W. A. (2007). Barriers to Adoption of e-Government. 40th Annual Hawaii International Conference on System Sciences (HICSS'07) (p. 98a). Big Island, Hawaii : IEEE Computer Society.
44. Datar, M. (2006). Determining Priorities of E-Government: A Model Building Approach. In A. Agarwal, & V. V. Ramana (Eds.), Foundations of E-government (pp. 76–85). SIGeGov Publications .
45. Detlor, B., Hupfer, M. E., Ruhi, U. (2010). Internal factors affecting the adoption and use of government websites. Electronic Government, An International Journal , 7 (2), 120–136.
46. Dimitrova, D. V., Chen, Y.-C. (2006). Profiling the Adopters of E-Government Information and Services: The Influence of Psychological Characteristics, Civic Mindedness, and Information Channels. Social Science Computer Review , 24 (2), 172–188.
47. Dominic, P., Jati, H., Sellappan, P., Nee, G. K. (2011). A comparison of Asian e-government websites quality: using a non-parametric test. Int. J. Business Information Systems, , 7 (2), 220- 246.

48. E- governance. (2012). Retrieved May 19, 2013, from Department of Electronics and Information Technologies: http://deity.gov.in/content/e-governance-infrastructure

49. Ebrahim, Z., Irani, Z. (2005). E-government adoption:architecture and barriers. Business Process Management Journal , 11 (5), 589–611.

50. e-Governance: Coimbatore City Municipal Corporation. (2013). Retrieved July 9, 2014, from Coimbatore City Municipal Corporation Website: https://www.ccmc.gov.in

51. E-government: The World Bank. (2011). Retrieved January 22, 2014, from The World Bank: http://web.worldbank.org

52. Elsheikh, Y., Azzeh, M. (2014). What Facilitates the Delivery of Citizen-Centric E-Government Services in Developing Countries:Model Development and Validation Through Structural Equation Modeling. International Journal of Computer Science & Information Technology (IJCSIT) , 6 (1), 77- 98.

53. El-Sofany, H. F., Al-Tourki, T., Al-Howimel, H., Al-Sadoon, A. (2012). E-government in Saudi Arabia: Barriers, Challenges and its Role of Development. International Journal of Computer Applications , 48 (5), 16–22.

54. Eynon, R., Margetts, H. (2007). European Journal of ePractice. 1, 1–13.

55. Goings, D. A., Young, D., Hendry, S. H. (2003). Critical Factors in the Delivery of e-Government Services: Perceptions of Technology Executives. Communications of the International Information Management Association , 3 (3), 1–15.

56. Government of Tamil Nadu . (2013). Retrieved November 7, 2013, from Government of Tamil Nadu website: http://www.tn.gov.in

57. Haque, M. S. (2002). E-governance in India: its impacts on relations among citizens, politicians and public servants. International Review of Administrative Sciences , 68, 231–250.

58. Hilbert, M. (2005). Development Trends and Challenges For Local e-Governments: Evidence From Municipalities in Chile and Peru. Santiago of Chile: United Nations Publication.

59. Holzer, M., Manoharan, A. (2007). Global Trends in Municipal E-Government: An Online Assessment of Worldwide Municipal Web Portals. In A. Agarwal, & V. V. Ramana (Eds.), Foundations of E-government (pp. 178–188).

60. Holzer, M., Manoharan, A., Ryzin, G. V. (2010). Global Cities on The Web: An Empirical Typology of Municipal Websites. International Public Management Review , 11 (3), 104–121.

61. Huang, Z. (2006). E-Government Practices At Local Levels:An Analysis Of U.S. Counties' Websites. Issues in Information Systems , 5 (2), 165–170.

62. Jaeger, P. T., Thompson, K. M. (2003). E-government around the world: Lessons, challenges, and future directions. Government Information Quarterly , 20, 389–394.

63. Kaisara, G. a. (2009). e-Government in South Africa: e-Service Quality Access and Adoption Factors. Cape Peninsula University of Technology. Informatics & Design Papers and Reports,Digital Knowledge.

64. Kamal, M. M., Hackney, R. (2012). Inhibiting Factors For E-Government Adoption: The Pakistan Context. Pacific Asia Conference on Information Systems (PACIS) 2012 Proceedings. Paper 112.

65. Kamarulzaman, Y., Azmi, A. A. (2010). Tax E-filing Adoption in Malaysia: A Conceptual Model. Journal of E-Government Studies and Best Practices , volume 2010, 1–6.

66. Kumar, V., Mukerji, B., Butt, I., Persaud, A. (2007). Factors for Successful e-Government Adoption: a Conceptual Framework. Electronic Journal of e-Government , 5 (1), 63 - 76.

67. Lai, C. S., Pires, G. (2010). Testing of a Model Evaluating e Government Portal Acceptance and Satisfaction. (E. Frisk, & K. Grunden, Eds.) Electronic Journal of Information Systems Evaluation , 13 (1), 35 46 .

68. Mahadeo, J. D. (2009). Towards an Understanding of the Factors Influencing theAcceptance and Diffusion of e-Government Services. Electronic Journal of e-Government , 7 (4), 391–402.

69. Mann, I. J., Kumar, V., Mann, H., Kumar, U. (2008). Scope of City E-Government Initiative. In A. Ojha (Ed.), E Governance in Practice (pp. 173–184). Hyderabad, Andhra Pradesh, India: CSI SIGeGOV .

70. Mathews, K. (2012, december 10). E-Government in the United States: Steps to Advance its Success. Bloomington, Indiana, United States.

71. Misra, D. C. (2006). Defining e-government: a citizen-centric criteria based approach. 10th National Conference on e-Governance, (pp. 1–9). Bhopal.

72. Miyata, M. (2011). Measuring impacts of e-government support in least developed countries:a case study of the vehicle registration service in Bhutan. Information Technology for Development , 17 (2), 133–152.

73. Mofleh, S. I., Wanous, M. (2008). Understanding Factors Influencing Citizens' Adoption of e-Government Services in the Developing World: Jordan as a Case Study. (H. A. Costa, Ed.) INFOCOMP Journal of Computer Science , 7 (2), 1–11.

74. Moon, M. J. (2002). The Evolution of E-Government among Municipalities: Rhetoric or Reality? Public Administration Review , 62 (4), 424.

75. Mpinganjira, M. (2012). Factors affecting adoption of e-government services: A conceptual model. African Journal of Business Management , 6 (11), 4245–4249.

76. National Portal of India. (2005). Retrieved May 17, 2013, from National Portal of India Website: http://www.india.gov.in/india-glance/profile

77. Nkwe, N. (2012). E-Government: Challenges and Opportunities in Botswana. International Journal of Humanities and Social Science , 2 (17), 39–48.

78. Nurdin, N., Stockdale, R., Scheepers, H. (2011). Understanding Organizational Barriers Influencing Local Electronic Government Adoption and Implementation: The Electronic Government Implementation Framework. Journal of Theoretical and Applied Electronic Commerce Research , 6 (3), 13–27.

79. Online plan approval: Coimbatore City Municipal Corporation (CCMC) Website. (2013). Retrieved January 10, 2014, from Coimbatore City Municipal Corporation Website (CCMC): https://www.ccmc.gov.in

80. Owolabi, E. S. (2013). Socio-Demographic Factors as Determinants of Access and Use of ICT by Staff of University Libraries in Oyo State. Library Philosophy and Practice (e-journal) , Paper 947. http://digitalcommons.unl.edu/libphilprac/947.

81. Pandey, M. R., Kapil, M., Garg, S. (2012). Beginning of an Effective E-Governance in India by using Informative and Communicative Mechanism. International Journal of Soft Computing and Engineering (IJSCE) , 2 (2), 107–109.

82. Pandey, N., Geetika. (2008). Strategic Marketing of E-Government for Technology Adoption Facilitation. In J. Bhattacharya (Ed.), Critical Thinking in E-Governance (pp. 51–60). Andhra Pradesh: SIGeGov Publications.

83. Parajuli, J. (2007). A Content Analysis of Selected Government Web Sites: a Case Study of Nepal. Electronic Journal of e-Government , 5 (1), 87–94.

84. Patel, H., Jacobson, D. (2008). Factors Influencing Citizen Adoption of E-Government: A Review and Critical Assessment. 16th European Conference on Information Systems (ECIS), (pp. 1058–1069.). Galway (Ireland).

85. Phang, C. W., Sutanto, J., Li, Y., Kankanhalli, A. (2005). Senior Citizens' Adoption of E-Government:In Quest of the Antecedents of Perceived Usefulness. Proceedings of the Thirty-Eighth Annual Hawaii International Conference on System Sciences. Hawaii, United States.

86. Prakash, G., Singh, A. (2008). A New Public Management Perspective in Indian E-Governance Initiatives. In J. Bhattacharya (Ed.), Critical Thinking in E-Governance (pp. 71–80). AndhraPradesh: SIGeGov Publications .

87. Prima, S., Ibrahim, R. b. (2011). Citizen Awareness to E-Government Services for Information Personalization. International Journal of Innovative Computing , 1 (1).

88. Ray, S., Rao, V. V. (2004). Evaluating Government Service: A Customers' Perspective of e-Government. Proceedings of the 4th European Conference on e-Government 'Towards Innovative Transformation inthe Public Sector' – ECEG 2004 (pp. 627–638). Dublin: Department of the Taoiseach of the Republic of Ireland.

89. Roadmap for E-government in the Developing World. (2010). Retrieved December 28, 2013, from United Nations Public Administration Network (UNPAN): http://unpan1.un.org/intradoc/groups/public/documents/apcity/unpan005030.pdf

90. Rokhman, A. (2011). E-Government Adoption in Developing Countries; the Case of Indonesia. Journal of Emerging Trends in Computing and Information Sciences , 2 (5), 228–236.

91. Rupanagunta, K. (2006). E-Governance in Public Financial Management:. IIMB Management Review , 18 (4), 403–413.

92. Scott, J. K. (2006). " E " the People: Do U.S. Municipal Government Web Sites Support Public Involvement? Public Administration Review , 341–353.

93. Shah, M. (2007). E-Governance in India: Dream or reality? International Journal of Education and Development using Information and Communication Technology (IJEDICT) , 3 (2), 125–137.

94. Shajari, M., Ismail, Z. (2011). Key Factors Influencing the Adoption of E-government in Iran. ICIC '11 Proceedings of the 2011 Fourth International Conference on Information and Computing (pp. 457–460). Washington,DC: IEEE Computer Society.

95. Shajari, M., Ismail, Z. (2012). Trustworthiness: A key Factor for Adoption Models of e-Government Services in Developing Countries. International Proceedings of Economics Development and Research (IPEDR) , 30, 22- 26.

96. Sharma, P., Mishra, A., Mishara, P. (2011). E-Governance in India is the Effectual and Challenging Approach to Governance. nt.J.Buss.Mgt.Eco.Res. , 2 (5), 297–304.

97. Singh, M., Sarkar, P., Dissanayake, D., Pittachayawa, S. (2008). Diffusion of E-Government Services in Australia: Citizens' Perspectives. European Conference on Information Systems (ECIS) 2008 Proceedings.

98. Soufi, B., Maguire, M. (2008). Usability and Accessibility in E-commerce Web Sites. In R. K. Ching, J. Tommila, B.-J. Choi, & G. Lee (Ed.), Proceedings of the Eighth

International Conference on Electronic Business (pp. 103–112). Hawai'i, USA: International Consortium for Electronic Business.

99. Sousa, J. M., Lopez, V. W. (2007). Analyzing The Development of Municipal E-Governmnet in Peruvian Cities. Proceedings of the 9th International Conference on Social Implications of Computers in Developing Countries. São Paulo, Brazil.

100. Suki, N. M., Ramayah, T. (2010). User Acceptance of the E-Government Services in Malaysia: Structural Equation Modelling Approach. (Z. Kovacic, Ed.) Interdisciplinary Journal of Information, Knowledge, and Management , 5, 395–413.

101. THE CONSTITUTION OF INDIA. (n.d.). Retrieved May 17, 2013, from National Portal of India: http://www.india.gov.in/

102. Tripathi, R., Gupta, M. P., Bhattacharya, J. (2007). Selected Aspects of Interoperability in One-stop Government Portal of India. In J. Bhattacharya (Ed.), Towards Next Generation E-government (pp. 1–10). SIGeGov Publications

103. (2012). United Nations E-Government Survey 2012. Department of Economic and Social Affairs. New York: United Nations Public Administration Network (UNPAN).

104. Urban Growth. (2011). Retrieved December 10, 2013, from Ministry of Urban Development website: http://moud.gov.in/

105. Vasavi, S., Kishore, S. (2011). Need for Semantic Interoperability of E-Government webservices within one stop web portals: A Case Study. International Journal of Computer Science & Technology , 2 (Sp 1), 136–140.

106. Vencatachellum, I., Pudaruth, S. (2010). Investigating E-Government Services Uptake in Mauritius: A User's Perspective. International Research Symposium in Service Management. Mauritius.

107. Venkatesh, V., Chan, F. K., Thong, J. Y. (2012). Designing e-government services: Key service attributes and citizens' preference structures. Journal of Operations Management , 30 (1–2), 116–133.

108. Venkatesh, V., Morris, M. G., Ackerman, P. L. (2000). A Longitudinal Field Investigation of Gender Differences in Individual Technology Adoption Decision-Making Processes. Organizational Behavior and Human Decision Processes , 83 (1), 33–60.

109. VerdaCanbeyÖzgüler. (2012). Enjoyment of E-Government Services by Different Socio-demographic Groups: The Case Of Eskisehir/Turkey. Journal of Business, Economics & Finance , 1 (4), 95–115.

110. Verma, N., Mishra, A. (2010). India's Approach in Constructing One-Stop Solution Towards e-Government. Journal of E-Governance , 33, 144–156.

111. Wangpipatwong, S., Chutimaskul, W., Papasratorn, B. (2005). Factors Influencing the Adoption of Thai eGovernment Websites:Information Quality and System Quality Approach. Special Issue of the International Journal of the Computer, the Internet and Management , 13 (3), 14.1–7.

112. Wangpipatwong, S., Chutimaskul, W., Papasratorn, B. (2008). Understanding Citizen's Continuance Intention to Use e-Government Website: a Composite View of TechnologyAcceptance Model and Computer Self-Efficacy. Electronic Journal of e-Government , 6 (1), 55–64.

113. West, D. M. (2008). State and Federal Electronic Government in the United States. Washington, DC: Brookings Institution.

114. Zafiropoulos, K., Karavasilis, I., Vrana, V. (2012). Assessing the Adoption of e-Government Services by Teachers in Greece. Future Internet , 4, 528–544.

115. Zhang, N., Guan, X., Meng, Q. (2011). Exploring Different Roles between Service Expectation and Technology Expectation In Citizen's E-Government Continuance Adoption: An Extended Expectation-Confirmation Model. PACIS 2011 Proceedings. Paper 224. Brisbane, Australia: Queensland University of Technology.

116. (2007, March 12–13). Measuring and Evaluating E-Government in Arab Countries . Dubai, United Arab Emirates: OECD and Dubai School of Government.

117. 2011 Census Data: Census of India Website. (2011). Retrieved September 17, 2014, from Census of India Website: http://www.censusindia.gov.in

CHAPTER 4

ELECTRONIC COURT MANAGEMENT SYSTEM IN MALAYSIA: THE LEGAL AND FUNCTIONAL REQUIREMENTS FOR COURT RECORDS

WAN SATIRAH WAN MOHD SAMAN[1] and NURUSSOBAH HUSSIN[2]

[1]*School of Information Management, Universiti Teknologi MARA (UiTM), Malaysia*

[2]*Faculty of Information Management, Universiti Teknologi MARA (UiTM), Malaysia*

CONTENTS

4.1 TECHNOLOGIES IN COURTS

ICT development is one of the best allies of justice when used wisely. Courtroom technology is increasing in scope and complexity, indeed, those opinions stating that computerization in the courtroom constitutes a threat are fading. The standard of living in many countries is high and this has been brought about by the Information and Communication Technologies (ICT). Court leaders and judges believe that ICT plays a key role in improving the current justice systems. In fact it is the catalyst to radical change in court performance. Literature reviews show the shift from the traditional case management system to applications developed to support electronic filing, electronic case management, electronic data interchange and e-justice. Today most countries around the world have already embarked on e-justice, introducing various types of electronic court case management applications.

4.2 MALAYSIAN LEGAL SYSTEM

In Malaysia, the Civil Court stands side-by-side with the Shariah Court, with different jurisdictions, by virtue of Article 121(1A) of the Federal

Constitution. The jurisdiction of the Shariah Court is relatively small, since it only applies to Muslims and solely on personal law matters, for example marriage, inheritance, and apostasy [8]. Other than that, all cases fall under the Civil Court's jurisdiction.

Under the existing legal system the Malaysian Federal Constitution defines the separation of power between the Federal and State Governments. Under Schedule 9, List 1, the constitution provides that all matters of civil and criminal law and legal administration fall under the Federal list. Conversely, List 2 of the same schedule states that Shariah or Islamic law matters are vested to state governments. As a result, the Civil and Shariah Courts became separate independent entities having their own specific jurisdiction as provided by Article 121(1A) of the constitution [1, 6, 21].

Today, both the Civil and Shariah Malaysian judiciary systems' administration have moved forward, especially over the past few years. Traditionally in the Malaysian judiciary, whether civil or Shariah, cases have taken a long time to be decided. Previously, the backlog of Shariah cases was the subject of much public criticism [15]. Indeed, it took years for a case to be settled for a number of reasons including: a limited number of judges and Shariah Court officials, a high volume of Shariah cases, poor infrastructure in place, limited budget allocated for Shariah departments and so on. However, the most significant reason was the unavailability of complete information relating to the cases as and when required [11]. Some cases lingered on for more than a year just because of the fact that information available to the judges was incomplete, and therefore they could not arrive at a conclusive decision. According to the Malaysian Federal Constitution, the federal government of Malaysia does not have any direct control over the administration and functioning of Shariah Courts. On the other hand, state governments are dependent on the Malaysian federal government for budget allocations; they find it hard to maintain a standardized policy for the smooth functioning of the Shariah Courts. This is why, traditionally, there has been significant discrepancy and mismatch between the way Shariah Courts function throughout the country. This discrepancy, however, is restricted to the workflow of the court rather than in the decisions carried out by the courts.

4.3   MALAYSIAN COURTS CASE ADJUDICATION

Case adjudication is the principle business of the court. It involves case processing. In the administration of Civil Court cases, they are divided into criminal and civil cases, respectively. The Working Procedure Manual for Civil and Criminal cases published by the Chief Registrar's Office of the Federal Court of Malaysia lists types of cases and coded them as code 1 to 89, with the specifications shown in Table 4.1.

The workflow procedure for all the above cases differs from matter to matter, according to the case flow requirements. A typical case in a civil court starts with case registration by the petitioner or plaintiff, case preparation by a court clerk, data key in in EFS system, verification and case management via CMS system. While the case is being managed the QMS system is also implemented. When the case is ready for hearing, it is scheduled in the CMS. During the hearing the CRT system is used for reporting and transcription. Finally, the court decision is recorded in the CMS and the court order is implemented. Data of the case is properly stored and shared with other courts for the purpose of appeal, records management and implementation of order [5].

In the Shariah Courts, originally the types of cases and flow of case management were diverse due to individual state administration and their own particular practices. When the Malaysia Shariah Judiciary Department (JKSM) took the initiative to standardize the procedural steps between states, the case flow become parallel in every state. JKSM

TABLE 4.1   Types of Cases in Civil Court

Code	Types of cases
1–4	Civil Appeal to Federal Court and Court of Appeal
5–8	Criminal Appeal to Federal Court and Court of Appeal
11–38	High Court Civil cases
41–45	High Court Criminal cases
51–58	Sessions Court Civil cases
61–64	Sessions Court Criminal cases
71–78	Magistrate Court Civil cases
81–89	Magistrate Court Criminal cases

created a new classification number scheme to be adopted by state courts through Shariah Court Practice Direction No. 1 Year 2000, which consists of State Codes, Court Hierarchy Codes, Hierarchical Court, District and Courtroom Codes and Case Codes. The case code numbers for Shariah civil cases are 001–099 and 101–299 for Shariah criminal cases

4.3.1   E-GOVERNANCE

E-governance is defined as digitization of government information and online transactions to improve public service delivery [11, 13, 17]. ICTs are an ideal enabler for a one-stop solution of court services to the general public. Various types of transactions such as: online case registration and filing, online submission of legal documents, case status reviews, court calendar searches, reference material searches, payment of court fees, legal information searching, updated cause lists, case judgments can all be available to the public via the court portal. Provision of court information through other online services such as: enquiries through email, *faraid* calculator, important links, form download, court directories, court latest news and court procedures can all be accessible to current and future court clients. ICTs also offer courts the provision of e-governance. E-governance has not only the ability to handle momentum and complexity, but also to underpin the regulatory reform [3].

E-governance allows data integration makes it possible for the best decision be made on certain issues. With different types of information coming from different departments/sources, it gives more meaning, significance and integrity to the information. Record integrity refers to the completeness of the record, which depends very much on the three key attributes of a record, for example, their content, context and structure [9]. For a court of law, comprehensive data concerning an accused person as a result of data integration between various departments will provide accurate information for the crucial decision to serve impartial justice to the parties involved. When information systems are properly integrated, no duplication of work is required and no redundancy of information is created in different departments. An example could be a court officer only having to use a smart card reader to retrieve all the personal particular of the parties involved in a case, since the information is readily available in the National ID card produced by the National Registration Department. This will save him having to key the data into the court's system. This is a kind of smart partnership than can save a great deal of a court's time and resources [14]. In Malaysia, three out of six Public Sector ICT policies, known as "no wrong door policy" targets are focusing on integration and information-sharing between departments:

a. Inculcating information sharing interoperability;
b. Cross agency collaboration towards seamless service;
c. Government shared services [14].

Integrated data allows the government to provide services to the public more easily than before through the e-government one stop solution. The electronic government concept embraced by most existing authorities allows the citizen to resort to one single point of reference for all government services. This 'no wrong door' policy can eliminate the red tape that Third World citizens have suffered for decades. This one-stop solution allows citizens to have access to a reliable justice system. Information dissemination on legal rights and responsibilities can also be made available online, providing more legal awareness through the use of IT. More importantly, any online transactions made via the e-government channel are secured, speedy and economical, because they are implemented with a high level of legal, public policy and IT compliance. The services to the public can be wireless, thus supporting unique applications anywhere.

4.4   THE NEED FOR LEGAL FRAMEWORK AND FUNCTIONAL REQUIREMENTS FOR ELECTRONIC COURT RECORDS

The court system in Malaysia has been frequently criticized because of its dilatoriness in resolving disputes resulting backlog of cases. Accordingly, a new system called e-court system has been introduced in 2011 [10]. However, according to Aliza [28] the trustworthiness and integrity of records in the electronic form are always questioned due to the lost or missing documentation in recordkeeping procedures. The controls of managing records in paper form fail to be regulated in the electronic environment, as information systems created by non-records professionals fail to capture the necessary information needed in providing the evidence in the organization.

Various models and framework such as Business-Driven Recordkeeping Model [29], Model of High Level Functional Requirements for ERM [30], Trust Model of Record's Life Cycle (http://www.interpares.org/display_file.cfm?doc=ip2_longrec_external_description_2006.pdf) suggests that the systematic recordkeeping functional requirements is necessary to be complied by all organizations in order to manage the electronic records effectively. This view is supported by several studies, which revealed that one of the significant approaches for guaranteeing the trustworthiness of electronic records in the organization is by embedding recordkeeping requirements [4, 22]. However, based on the investigation, there is lack of policies or procedures in managing electronic court records in the Malaysian Judiciary. Until the time of writing, the only available policies and procedures that are relevant but not comprehensive enough to ensure the authenticity and integrity of electronic court records throughout their entire life cycle are: (i) Retention Schedule for Court Records (www2.arkib.gov.my/english/index.html); (ii) ICT Security Policy [12]; (iii) ICT Security Policy (www.kehakiman.gov.my) and (iv) Classification Codes (www.kehakiman.gov.my).

Furthermore, in the context of court environment, the scope and application of a piece of legislation refer to as its *jurisdiction*. If the jurisdiction of a particular policy or act is not clear, this can create difficulties for record managers and archivists. For instance, in Australia there has been uncertainty about the jurisdiction of archival legislation over the court records. This is because the legislation does not explicitly cover case files and transcripts of the courts. These documents therefore, may not be affected by the legislation. By contrast, in Namibia, the Archives Act 1992 explicitly states the legal records to which it applies [2]. This view is supported by Saman and Haider [20] recommending that clear policy of court records management, paper and electronic should spell out requirements on every phase of court records' life cycle encompassing records processes from the creation until the disposition of court and case files regardless of the forms. In addition, records retention schedule and long-term preservation training should also be included in the policy.

On this basis, the specific functional requirements for the management of electronic court records need to be developed in the Malaysian Court of Appeal through a comprehensive research study. The functional requirements should serve as the basis for designing systems that facilitate the recordkeeping process and the benchmark for measuring the performance of the existing system. This is vital for the long-term preservation of those court records as evidence of transactions and accountability that the records held for their current use.

4.5   MALAYSIAN LEGAL FRAMEWORK: STATUTORY AND QUALITY COMPLIANCE

Achievement in technology implementation/assimilation in court is not valid, unless it is legally compliant, that is, it is consistent with all the requirements set up by legal rules and standards. For the sake of quality management, technology implementation/assimilation needs to conform to ISO 15489 (Record management Standard) and ISO 2700 – Information Security Management Standard (ISMS) standards which require certain criteria and procedures for which appropriate paper and electronic records are maintained. This section tackles the legal requirements and compliance issues associated with technology adoption in court. The first part deals with the existing legal requirement in Malaysia and the second part discusses arising issues related to it faced in the course of establishing technology institutionalization in court. The available policies and requirements are divided into four categories: legislation, international standards, national policies, and organizational policies (circular, practice directions). Figure 4.1 shows the four quadrant of compliance framework.

Legal compliance aspect is discussed in the following section reflecting the legal compliance issues in four dimensions – ISO standards, written legislations, national and organizational policies.

4.5.1   INTERNATIONAL STANDARD ISO

A number of international standards (ISO) are appropriately related to electronic court management. There is a drive to have ISO-certified material existing in court due to mimetic and normative pressure from other government agencies in Malaysia. However, due to the voluminous workload of court staff, acquiring ISO becomes secondary. In the Kuala Lumpur court complex, the only court, which is ISO-certified is the Criminal Section of the Lower Court, as confirmed by one of the interviewees:

FIGURE 4.1   Statutory and Quality Compliance Framework.

It is always up to the Managing Deputy Registrar whether we want to go with ISO or not. In this court complex, only the Lower Court (Criminal Section) took up ISO. Other courts are not able to do it because we could not manage to have time for that matter.

—(Deputy Registrar)

This is possible since criminal cases normally take less time to settle and are less complicated. The said certification is ISO 9001:2008 for quality management system in the public sector. A more related international standard is the ISO 15489 (International Standard for Records Management) and Information Security Management Standard (ISMS), the ISO 27001:2005 for information and data security. In fact, ISO 15489 was adopted by Malaysia as Malaysian Standard ISO MS 2223 in 2009. However, when asked about the application or acknowledgement of these standards in courts, none of the respondents gave a positive answer to that effect. Most of them are not familiar at all with ISO 15489/MS 2223. However, the interview with the IT developer reveals that they do acknowledge the ISMS framework.

In the Shariah legal environment, three international standards which are appropriately related to electronic court management include: the ISO 9001:2008 for quality management system in the public sector, the ISO 15489 for Electronic Records Management and the ISMS ISO 27001:2005 for information and data security. The Shariah Courts are quite advanced in the applications of these standards; in fact, more so than the civil courts.

In fact, JKSM is among the first government departments to employ the ISO standards in its business operation. Both the KL Shariah Court as well as JKSM in Putrajaya are certified with ISO9001:2008 qualifications. In addition, they observed the ISO 15489. It is not the isomorphic pressures that have forced them to move forward, but rather their own determinism. An interview with the Head of Records Unit in JKSM shows that the department really observes the guideline provided by ISO 15489 in its administration of court cases.

> *Although ISO 15486 has no certification system, we do internal audits based on it, as well as arranging training. In my opinion, this ISO really helps JKSM and all Shariah Courts to have awareness of the importance of proper record keeping (both manual and electronic). On top of that, the ISO 9001:2008, which almost all Shariah Courts in Malaysia are certified with, really gives a kick off to the importance of records because ISO 9001:2008 itself encompasses records components (Clause 424).*
>
> *—(Head of Records Unit)*

Since the development of E-Shariah, there has been close monitoring by MAMPU in the beginning and the following steps were entirely taken over by MAMPU. Accordingly, ISMS standards such as ISO 9001:2008 and other standards are observed and in fact are assimilated into MAMPU's ICT policies.

4.5.2   LEGISLATIONS AND STATUTES

The primary legislations for all court actions are the Rules of Court 2012, the Court of Judicature Act 1964 and the Subordinate Court Rules Act 1955. Other legislations are the High Court Rules 1980 and the Subordinate Court Rules 1980. An interviewer with a High Court Deputy Registrar confirmed this when saying:

> *The administration of this court and court records/filing is governed by Rules of High Court 1980. We do not have ISO. Here in the High Court, we based our conduct on the administrative Practice Directions issued from time to time by the Palace of Justice. Based on that directive, the Managing Deputy Registrars decided how to take care of the case files in their own courts. So, there is always a minor difference from one court to another.*

The recently-enacted Rules of Court 2012 are applicable to both superior and subordinate courts and incorporate the use of technologies in courts. Before this Act, both types of courts were governed under separate legislations. However, this new Act does not repeal the existing acts and rules, but rather complements them in the way that it defines the new way of implementing justice using technology. The fundamental principles (substantive and procedural) of justice as spelt out in existing acts like the Civil Procedure Act and the Criminal Procedure Code are still maintained. Another legislation directly related to E-Court implementation is the Electronic Government Activities Act 2007.

Court records are not classified as confidential, but they are public documents. Like other government agencies, courts are subject to all public department rules and regulations, including those pertaining to records management such as the National Archives Act, the Security Act and the Information Technology Security Act. In addition, the court has its own policies, rules and directives, usually issued by the Office of the Chief Registrar of the Federal Court (the apex court in the Malaysian courts hierarchy). At its own level, the court issued, among others, Rules and Procedures in ICT and an ICT strategic plan. Other related acts and policies pertaining to electronic records are as follows:

a. Electronic Government Activities Act 2007
b. Digital Signature Act
c. Computer Crimes Act
d. Copyright Act 1987 and Copyright (amendment) Act 1997
e. Personal Data protection Act/Bill
f. National Archive's Service Direction Chapter 3
g. National Archive's Service Direction Chapter 5
h. General Circular No 1/2003
i. National Archive's electronic records policies (eSPARK).

According to the Director of IT Department of BHEUU, other laws/regulations involved are specifically:

a. Public Sector Data Dictionary (Data Dictionary Sektor Awam)
b. ICT Security Policy-MAMPU (Dasar Keselamatan ICT-MAMPU
c. Biometric User Guidelines for Public Sector Agencies (Garis Panduan Penggunaan Biometrik Bagi Agensi Sektor Awam)
d. Information Technology Directive (Arahan Teknologi Maklumat)

e. Malaysian Public Sector ICT Security Management Guidelines (Garispanduan Pengurusan Keselamatan ICT Sektor Awan Malaysia) (MyMIS)

f. (Pekeliling Am Bilangan 3 Tahun 2000 (Rangka Dasar Keselamatan Teknologi Maklumatdan Komunikasi Kerajaan)

g. The Malaysian Government Interoperability Framework for Open Source Software (My GIFOSS).

At the beginning of E-Court implementation, a serious issue arose regarding the use of technology in the Criminal court, to the effect that it is now a legal infringement to use the Court Recording and Transcribing (CRT) system. It was considered to be in contradiction of Sections 265–272 of the Criminal Procedure Code regarding the mode of taking and recording of evidence in inquiries and trial of criminal cases. Chapter XXV of the code consists of the following sections:

a. Section 264 – Evidence to be taken in presence of accused

b. Section 265 – Manner of recording evidence

c. Section 266 – Recording evidence in summons cases

d. Section 267 – Recording evidence in other cases

e. Section 268 – Record to be in narrative form

f. Section 269 – Reading over evidence and correction

g. Section 270 – Interpretation of evidence to accused

h. Section 271 – Remarks as to demeanor of witness

i. Section 272 – Judge to take notes of evidence

j. Section 272A – Other persons may be authorized to take down notes of evidence.

For a trial in the Magistrates Court, Section 266 deals with the mode of taking notes in summons cases while Section 267 deals with the mode of recording evidence in other cases. Section 266 (1) states: "In summons cases tried before a Magistrate, the Magistrate shall, as the examination of each witness proceeds, make a note of the substance of what the witness deposes, and such note shall be written by the Magistrate with his own hand in legible handwriting and shall form part of the record." Section 267 states: "In all other trials before a Magistrate's Court, and in all inquiries…, the evidence of each witness shall be taken down in legible handwriting by the presiding Magistrate and shall form part of the record." For High

Court criminal case trials, the High Court Judge is required to take notes of evidence in handwriting by virtue of Section 272 of Criminal Procedure Code. It provides: "In all criminal cases tried before the High Court the Judge shall take down in writing notes of the evidence adduced." Section 272A allows a judge, besides having his own notes, to instruct any other person to record verbatim notes of what each witness deposes. This section does not mention any other modes of recording evidence in court. Hence, an audio or video recording is not legally acknowledged as forming part of the trial record. Thus the provisions clearly require all notes of evidence in criminal cases be taken in the judge's handwriting, in narrative form.

This was the big hurdle for the implementation of the Court Recording and Transcribing system when it was first introduced. It forced an amendment to be made to the statute. Finally, the Criminal Procedure Code (Amendment) Act 2009 (Act A1350) was passed by the Parliament in April 2009, by inserting a new chapter, for example, Chapter XXVA after Chapter XXV in the Criminal Procedure Code (Act 539). The new Chapter XXVA mandates the recording of proceedings by mechanical means. Section 272C explicitly permits any mechanical means to be employed for the recording of any proceedings before all courts in Malaysia. It provides "Notwithstanding the provisions...dealing with the mode of taking and recording of evidence, any mechanical means may be employed for the recording of any proceedings...and where mechanical means are employed the provisions of this Chapter shall apply." Mechanical means is defined as being any equipment, device, apparatus or medium operated digitally, electronically, magnetically or mechanically (272D (b)). "Proceedings" includes, specifically: any trial, inquiry, appeal or revision, or any part of it, as well as any application, judgment, decision, ruling, direction, address, submission and any other matter done or said by or before a Court, including matters relating to procedure (272D(c). The term "electronic record" means any digitally, electronically, magnetically or mechanically-produced records stored in any equipment, device, apparatus or medium or any other form of storage such as disc, tape, film, sound track, and which includes a replication of such recording to a separate storage equipment, device, apparatus or medium or any other form of storage (272D(a). Section 272E further explained that proceedings may be recorded by mechanical means or a combination of mechanical means and other methods. The rest of the provisions in this new Act deal with the transcription of electronic records (272F), safe custody of electronic record and transcripts (272G), transcript to form part of records or notes of proceedings or evidence (272I), electronic filing, lodgement, submission and transmission of document (272J) and issuance of practice direction of the court relating to the use of mechanical means and any matter related to it (272K). So now, with this amendment, the legal issue regarding the authenticity of electronic records in criminal court has been resolved.

For civil cases, there is no explicit requirement or prohibition regarding any mode of taking notes of evidence. From among the Civil Procedure Code, the High Court Rules, the Subordinate Court Rules and the Court of Judicature Act, none of them mention that the note-taking must be hand-written or otherwise. In this instance, according to one of the senior High Court Judges in Kuala Lumpur, electronic recording is deemed legal and permissible. He quoted: '...there are no specific laws in respect to electronic court records for civil matters. But, there is no prohibition. It is just a procedural matter. We do not need any form of laws for procedural matters. But, maybe in the future the High Court Rules and other related legislations will be amended to incorporate the e-filing and so forth.'

In Shariah Courts, the case administration of the Shariah Courts is governed by State legislations. This means that each of the 14 states has its own acts, and is different from one another (see Table 4.1). The Kuala Lumpur Shariah Court's administration of justice is governed by, namely: the Administration of Islamic Law (Federal Territories) Act 1993, the Shariah Civil Procedure (Federal Territory) Act 1998 and the Shariah Criminal (Federal Territory) Act 1997. For specific matters such as family matters, it has its own act, for example, the Islamic Family Law (Federal Territory) Act 1984.

For the Shariah Court system, legal issues arise when Islamic matters are placed under state jurisdiction by virtue of Schedule A of the Federal Constitution. This ultimately means that Shariah laws are enforced differently in the 14 different states in Malaysia. Each state has its own sets of laws and enactments. They also have different sets of procedural laws and manage their records in the way they deem best. In an effort to standardize and manage all the courts, a special department was established under the Prime Minister's Department. This was named as the Shariah Judiciary

Department of Malaysia in the year 1998. JKSM faced various problems holding the responsibility for standardizing the policies and procedures of all courts that fall under 14 different states' jurisdictions. It was a real struggle to bring together all 14 bodies that had been used to their own style of legacy in managing cases. All 14 states are tied with their own statutes and different procedural codes as reflected in Table 4.2.

To resolve the issue of diverse sets of laws definitively and without having to amend or change any of the existing law of the states, JKSM conducted a series of meetings and discussions attended by all Shariah Court policy makers from all states. All end results of the discussions were

TABLE 4.2 State Laws Regulating Shariah Courts in Each of the States in Malaysia

No	State	Statute
1	Federal Territories	Administration of Islamic Law (Federal Territories) Act 1993 (AIL (FT) Act 1993) (Act 505)
2	Selangor	Administration of Islamic Law Enactment 1989 (Selangor) (No. 2 of 1989)
3	Johor	Administration of Muslim Law Enactment 1978 (Johore) (No. 14 of 1978)
4	Kelantan	Administration of Shariah Courts Enactment 1982 (Kelantan) (No. 3 of 1982)
5	Melaka	Administration of the Shariah Courts Enactment 1985 (Melaka) (No. 6 of 1985)
6	N. Sembilan	Administration of Islamic Law Enactment 1991 (Negeri Sembilan) (No. 1 of 1991)
7	Pahang	Administration of Islamic Law Enactment 1991 (Pahang) (No. 3 of 1991)
8	Perak	Administration of Muslim Law Enactment 1992 (Perak) (No. 2 of 1992)
9	Perlis	Shariah Courts Enactment 1992 (Perlis) (No. 5 of 1992)
10	Sabah	Shariah Courts Enactment 1992 (Sabah) (No. 14 of 1992)
11	Penang	Administration of Islamic Religious Affairs Enactment 1993 (Penang) (No. 7 of 1993)
12	Kedah	Shariah Courts Enactment 1993 (Kedah) (No. 4 of 1994)
13	Terengganu	Shariah Courts (Terengganu) Enactment 2001 (No. 3 of 2001)
14	Sarawak	Shariah Courts Ordinance 2001 (Sarawak) (Ord. 4/2001)

codified in a series of Practice Directions, with the endorsement of JKSM's Director General/Shariah Chief Judge. They were then distributed to all Shariah Courts nationwide, as a guideline that needed to be adhered to. This effort was successful in bringing about uniformity of work process in Shariah Courts in Malaysia. For legal compliance with E-Shariah, the laws and regulations, which need to be adhered are as follows:

a. Electronic Government Activities Act 2007
b. Digital Signature Act
c. Computer Crimes Act
d. Copyright (amendment) Act 1997.

4.5.3   NATIONAL POLICIES

National policies in this context refer to the policies issued by the Federal Government to all ministries and organizations. The legal-based national policies can come from the Prime Minister's Department and its units, such as the Attorney-General's Department and the MAMPU. Most of the directives on E-Government are applicable to courts, especially concerning the electronic services and management by MAMPU and electronic records by the National Archives. The national policies applicable to Shariah Courts are mostly the same as those applicable to the civil courts, as well as those Practice Directions and circulars issued by the JKSM. These are as listed below:

a. Shariah Court Retention Schedule
b. Public Sector Data Dictionary 2002 [12]
c. ICT Security Policy Version 5.3 (Malaysian Administrative Modernization and Management Planning Unit 2010) [23]
d. Biometric User Guidelines for Public Sector Agencies [12]
e. Information Technology Instructions (Malaysian Administrative Modernization and Management Planning Unit 2007) [24]
f. Malaysian Public Sector ICT Security Management Guidelines (Malaysian Administrative Modernization and Management Planning Unit 2002a) [25]
g. Government General Circular No 3 Year 2000: Information and Communication Security Policy Framework (Malaysian Administrative Modernization and Management Planning Unit 2000) [26]

h. The Malaysian Government Interoperability Framework for Open Source Software (My GIFOSS) (Malaysian Administrative Modernization and Management Planning Unit 2006) [27]
i. JKSM Practice Directions (Year 2008–2013)

Most practice directions are issued as a result of mutual agreement among states and JKSM concluded in their series of meetings.

4.5.4 ORGANIZATIONAL POLICIES

Organizational policies for civil courts are normally issued by the Federal Court's Chief Registrar's Office and the Chief Justice. Some policies are written as a book and need to be followed strictly by courts, for example, the Civil Court Work Procedure Manual. Other policies are normally issued from time to time through a court's Practice Directions, which are binding upon all civil courts in Malaysia. They are distributed to an individual court through the Managing Registrar/Deputy Registrar who will distribute them to the affected individuals only. It means every court staff member will receive only those Practice Directions/policies directly related to him/her. When asked whether these Practice Directions are compiled or not, a Deputy Registrar replied:

We do not normally compile the Practice Directions. They come in at different times, with different issues and instructions. The managing Deputy Registrar will copy and distribute them to the DR/ SAR concerned with the matters only.

Organizational policies also refer to the internal policies issued by the individual court to its staff. However the matters are usually administrative in nature. For example, the way the case files are stored for easy retrieval is different from one court to another depending on local circumstances like the scattered or centralized building locations, etc.

Organizational policies for the Shariah Court are policies issued by the JKSM as well as State Shariah Courts. The policies are issued through JKSM Practice Directions, which are binding upon all Shariah Courts in Malaysia. They are distributed to State courts, and the State courts then distribute them to District Shariah Courts in their own states. Every year, all the Practice Directions are compiled and printed properly for reference. The past year's Practice Directions and circulars are also compiled according to year and then reprinted and distributed to Practice Directions. This action is carried out through the Managing Registrar/Deputy Registrar who will distribute them to the State Courts. This practice is very different from and much more systematic than that of the civil court. An interviewee says:

We compile the practice directions on a yearly basis and give the cop-
ies to all Shariah Courts. This situation allows everybody to be aware
of important policies set for the court. There are so many important
things in the Practice Directions and circulars that everyone should
know and it ensures standardization of practice in all Shariah Courts.
—(Head of Department)

Organizational policies for Shariah Courts also include internal policies issued by the individual court to its staff, with these matters usually being administrative in nature.

4.6   THE FUNCTIONAL REQUIREMENTS FOR ELECTRONIC COURT RECORDS

Functional requirements mean what the system should be able to do and the functions it should perform in order to generate the desired result. They are known as behavioral requirements as they address what the system does which include inputs, outputs, calculations, external interfaces, communications, and special management information needs. Besides, there are other several definitions of functional requirements depending on the context. Below are some existing definitions of functional requirements within the context of electronic records management.

a)   Functional requirements define the functions that a computer system needs to perform in order to meet the organization's needs.
b)   Functional requirement is defined as the tasks a computer application must perform to carry out a process satisfactorily, or the conditions or performance standards that a computer system should meet in order to support the business of the organization.
c)   Functional requirement is a description of the requirements of a system for it to fulfill its responsibilities and to support users in performing tasks relating to those responsibilities.
d)   Functional requirement is a description of an organization's computer processing needs to carry out its programs and satisfy its mission.

Understanding the recordkeeping functional requirements will help to identify what data comprises a record and what does not. To manage electronic records effectively, organizations must incorporate a certain number of requirements into the electronic records management system in order to facilitate the correct management of electronic records.

For the purpose of this study, the most relevant definition is by the International Council on Archives (ICA), for example, the functional requirements in the perspective of recordkeeping processes are divided into create, maintain, disseminate and administer. The functional requirements focused on the outcomes required to ensure records are managed appropriately. They do not specify particular processes, as it is recognized that the techniques and strategies to achieve the outcomes will depend on the organization and electronic records management system being used. Thus, the underlying tasks of the functional requirements are to identify the controls needed to ensure a software system will complete the tasks it is intended to complete. In other words, in order to manage electronic records effectively, it is necessary to identify what are the types of functionality a recordkeeping system should possess.

Recordkeeping functional requirements describe what the system should be able to do and the functions it should perform to ensure records are managed properly. This specification is based on the analysis of seventeen international and national best practices of functional requirements for electronic records management and court's management; and in-depth interviews with records practitioners and IT Personnel in the Malaysian Judiciary. The purposes of this specification are:

(i) to define the court records management functionality to be included in a design specification when building, purchasing or upgrading electronic court records management software;
(ii) to provide basic guidance to create, maintain, disseminate and administer electronic court records.
(iii) to assist in recordkeeping audit or review of the records management functionality or assess the capability of available software packages that is currently in place.

A list of contents and description of the functional requirement of electronic court records are shown below. The subsequent sections explains in detail about the requirements:

A. Creation
- (i) Capture,
- (ii) Identification,
- (iii) Classification and indexing,

B. Maintenance
- (i) Managing the authenticity, reliability and integrity of court records
- (ii) Retention and disposal

C. Dissemination
- (i) Search and retrieve
- (ii) Render

D. Administration
- (i) Administrator
- (ii) Metadata administration
- (iii) Reporting

4.6.1 CREATION: CAPTURE

Capture is defined as the process of fixing and initiate the content, structure and context of a record to ensure that it is a reliable and authentic representation of the activities in which it was created. The functional requirements for 'Capture' phase are:

a) The system must ensure that electronic court records that have been created or received can be captured, registered and stored along with associated metadata.

b) The system must support mechanisms for capturing electronic records received by the system that are either fully automated (electronic or imaged) or a combination of automated and manual.

c) Records may also comprise more than one component. Therefore, when capturing an electronic record that has more than one component, the system must maintain a relationship between all components and associated metadata for it to be managed as a single record and retain the structural integrity of the record.

d) The system must be able to capture in bulk records imported from other systems.

e) The requirements must support Malaysian e-court system. The Malaysian e-court system consists of e-filing, case management system, queue management system and court recording and transcription system. However, these requirements strongly support e-filing and case management system of the e-court due to the fact that these two systems are focusing on managing cases, documents and records.

f) The system must *alert* a user to any failures to successfully capture records.

g) The system must be able, where possible and appropriate, to provide a warning if an attempt is made to capture a record that is incomplete or inconsistent.

4.6.2 CREATION: IDENTIFICATION

Identification is defined as the process of constantly linking court record with a unique identifier. The functional requirements for identification phase are:

a) The system must ensure each electronic record is uniquely identifiable.

b) The system must be able to register, generate and store unique identifiers automatically, and prevent users from inputting the unique identifier manually and from subsequently modifying it.

4.6.3 CREATION: CLASSIFICATION AND INDEXING

Classification and indexing is defined as the systematic process of organizing court records in groups or categories according to methods, procedures or conventions represented in a classification scheme and establishing access points to facilitate retrieval of court records. The index is created at case initiation and maintained throughout the life of a case. The functional requirements for classification and indexing phase are:

a) The system must support and be compatible with the court classification scheme.

b) The system must support close linkage and interaction between a record's classification and other records management processes, such as capture, access and security, disposition, searching and retrieval, and reporting.

c) The system must allocate a unique identifier to each term defined within a records classification scheme.

The computerized system must be able to recognize all cases according to subject codes, as already been used in its conventional system, as presented in Appendix A for Civil Courts and Appendix B for Shariah Courts. Taking the Shariah Court as an example in an offense of false declaration for marriage, the case unique identifier of the case following the sequence of "case code number-court number-running number-year" come to this effect: "149-010-0130-2009."

4.6.4 MAINTENANCE: MANAGING THE AUTHENTICITY, RELIABILITY AND INTEGRITY OF COURT RECORDS

It is referred to the activities associated with controlling, storing, tracking, appraising, preserving and disposing of manual, electronic and imaged records in order to maintain the trustworthiness of court records. The requirements include:

a) The system must maintain authenticity, reliability and integrity of court records and metadata at all times, regardless of maintenance activities and other user actions or failure of system components.

b) The system must enable the Courts to demonstrate the authenticity, reliability and integrity of records against media deterioration, following transmission (from one system to another), and over time.

c) The system must be able to maintain and display the audit trail of file additions, modifications and deletions.

d) The system must enable the courts to set multiple levels of security assign to system users.

e) The system must be able to authenticate, by whom and the means of authentication by using the existing PKI and System Login used by the Malaysian Judiciary.

f) Only authorized user who is successfully identified and verified are allowed to carry out any action in the system as follows:

    i. The system must allow authorized users to save court records directly to the appropriate file location.

    ii. The application must allow users to send court records via the agreed communication device.

    iii. The system must allow users to search for court records or file locations in a number of formats.

g) The system must ensure electronic court records cannot be modified without authorized person notification.

h) Allow access and similar privileges based on authorizations defined, maintained, and controlled by the user.

i) Restrict local and remote access and permissible operations (i.e., view, add, change, delete; combinations of view, add, change, delete; and output) on case types, case categories, files, parts of files and system functions, device (e.g., personal computers) locations, users and groups of users.

j) Provide adequate security if public access is allowed.

k) Provide secure password for the user, containing strong combinations of alphanumeric, numbers and symbols. The length of the passwords must be at least 12 characters.

l) The system must be able to protect the audit trail against modifications by any user, including a System Administrator.

m) Provide for disaster recovery (e.g., reconstruct the status of the system and its case processing and, financial functions and data such as permitting access authorization tables).

n) Management of hybrid court records. A hybrid system is defined as an environment consisting of both electronic and non-electronic records. The management of hybrid court records includes functionality for managing court records in both non-electronic and electronic format. Electronic court records can be linked to non-electronic court records through a secure bound metadata relationship to form a hybrid record.

o) The system must be able to accept the importation of non-electronic court records in the electronic records in accordance with the statutes and rules. Under certain circumstances, the system must allow both kinds of record to be managed in an integrated manner.

p) The system must allow non-electronic court records that are associated as a hybrid with an electronic aggregation (class, file, sub-file, volume) to use the same title and numerical reference code but with an added indicator that it is a hybrid non-electronic records.

q) The system must allow different records management metadata set to be configured for non-electronic and electronic aggregations. Non-electronic aggregation records management metadata must include information on the physical location of the non-electronic aggregation.

r) The system must ensure that retrieval of non-electronic court records display the records management metadata for both electronic and non-electronic records associated with it.

s) The system must include features to control and records access to non-electronic aggregations, including controls based on the security category, which is comparable with the features for electronic aggregations.

t) The system must support tracking of non-electronic aggregations by the provision of request, check out and check in facilities that reflect the current location of the item concerned.

u) Scanning and imaging.

According to InterPARES3 Glossary of TEAM Malaysia (http://www.interpares.org/display_file.cfm?doc=ip3_malaysia_%20cs02_final_report.pdf), reliability is trusted representation of the transactions, activities to which they indicate and can be depended upon throughout subsequent transactions or activities. Authenticity means that the record is what it claims to be and has been created by the organization with which it is identified. While record integrity refers to the completeness of the record, which highly depends on the three key attributes of a record, for example, their content, context and structure.

As regard to the requirement for level of security assigned to users, the division of levels are suggested as follow:

a) Level 1 – For court users (e.g., clerk's office staff) who individually have different privileges on the system but collectively can enter data and records, access most data and records, and change some data and records. The system, data and records must be protected from unauthorized access.

b) Level 2 – For official users outside the court who frequently submit filings and need information from the system (e.g., attorneys of record), there must be protection from access to unauthorized sections of the system, from submission of incorrect data and records, and from direct entry of data and records (i.e., only Level 1, users would be permitted to enter data and records directly into the system).

c) Level 3 – For unofficial users (e.g., the public), there must be protection from any access that goes beyond viewing limited sections of the system's data and records.

The activities associated with controlling over access and security support the maintenance of authenticity, reliability, integrity and usability during normal operations and after a system failure of the outage. Under the imaging functionalities, the system must have the ability to scan images in standard formats and allow users to capture scanned images as records. Scanning feature must be capable of automatically sending scanned images to a queue after scanning (e.g., indexing, quality assurance). It must have the ability to attach documents to a case (i.e., minute entries), ensure appropriate security (access) to case images and documents from within the case management system, having ability for the authorized users to search and retrieve scanned images and documents and capable to hide specific information like demographic info and personal identifiers. Scanning and imaging requirements are essential and need to be included in this specification since Malaysian Judiciary managed hybrid records (both non-electronic records and electronic records).

4.6.5   MAINTENANCE: RETENTION AND DISPOSAL

Disposal according to the Malaysian National Archives Act 2003 means the manner of managing the segregation of records with a view to destroy, transfer or otherwise. Courts must ensure that all court records have to be disposed upon maturity of their retention periods as specified in the Retention Schedule for Malaysian Judiciary 2010. The system therefore must be able to control the retention and disposal of records in accordance with disposal authorization and preservation strategies. The requirements for retention and disposal are:

a) The system must support the controlled disposal of the Court of Appeal records legally authorized for disposal, in accordance with approved disposal authorities issued by the National Archives of Malaysia or in accordance with the Court of Appeal requirements. In order to be able to identify court records to be legally destroyed or retained permanently, Malaysian Judiciary must refer to the Retention Schedule for Court Records developed by the National Archives of Malaysia (2010).

b) The system must allocate a unique identifier and unique record series to each retention and disposition schedule when it is created.

c) The system must display disposition type (i.e., type of judgment) including those involving entire cases, individual parties, cross complaints etc.

d) The system must automatically track retention period (refer to the Retention Schedule for Court Records developed by the National Archives of Malaysia) that have been allocated to all entities.

e) The system must maintain an unalterable history of changes and deletions (audit trail) that are made by users including the date change or deletion.Deletions or changes of retention and disposition schedules must be controlled closely to reduce the risk of records being destroyed inappropriately.

f) The system must ensure that any amendment to a retention and disposition schedule immediately applies to all entities to which the retention and disposition schedule is allocated.

g) When the retention period applicable to some record(s) because of a retention and disposition schedule reaches its end, the system must automatically prompt user regarding appropriate action (e.g.: prepare a notice of motion to dismiss, extend dates) or initiate the processing of the disposition decision.

h) Each retention and disposition schedule must include either:

    i. a retention period or a disposition date
    ii. a disposition action or a reason
    iii. a description or a mandate

i) Preservation strategies

    i. Whenever the system transfers or exports any records, it must transfer or export the contents, structures and all its components and must preserve the correct relationships between them.

    ii. Whenever the system transfers or exports any class, file, sub-file or volume, the transfer or export must include:

- (for classes) all files and records in the class;
- (for files) all volumes and sub-files in the file;
- all records in all these files, sub-files or volumes;
- all or selected metadata associated with all of the above;
- all or selected audit trails for all of the above.

iii. Be able to include a copy of the entire metadata and retain copies of all electronic records that have been transferred, at least until such time as a successful transfer is confirmed in their active/original environment.

iv. Whenever the system exports or transfers information, it should be able to produce on request a report listing the records exported or transferred according to their security categories.

v. The system must ensure that destruction results in complete destruction so that they cannot be restored through specialized data recovery facilities.

vi. The system must support migration processes.

4.6.6 DISSEMINATION: SEARCH AND RETRIEVE

The system must be able to retrieve electronic court records by a variety of search methods. The requirements are:

a) The system must allow users to search for and retrieve:

  i. complete electronic court records and all its contents and contextual metadata.

  ii. every level of aggregation of records (class, file, sub-file, volume) and their associated metadata at any level of the classification system.

  iii. a single aggregation or more than one aggregation.

  iv. records directly through the use of unique identifier.

b) The system must never allow a search retrieval function to reveal to a user any information (records management metadata or record content) that the access and security settings are intended to hide from the user.

c) The system must present faultless functionality when searching and reporting non-electronic and hybrid aggregation.

d) The system must allow users to refine searches (e.g., a user should be able to start with the result list from a search and then initiate a further search within that list).

4.6.7 DISSEMINATION: RENDER

This means the system must be able to render the results. The rendering process includes displaying records on screen, printing or outputting to appropriate media records, which cannot be printed (e.g., audio files). The requirements for render are:

a) Rendering: Displaying records:

i. The system must never present information to any user who is not entitled to access it.
ii. The system must be able to render all types of electronic court records specified by the court in a manner that preserves the information in the records and which renders all components of an electronic record in their original relationship.
iii. The system must be able to render its contents and its metadata (subject to access controls) by a single click.

b) Rendering: Printing

i. The system must be able to print all or specified contents and metadata for any class, file, sub-file, volume or record. However, the system must also be able to block printing function for contents and metadata that are prohibited for printing.
ii. The system must be able to print out the summary list of selected records (e.g., the content of the case).
iii. The system must allow authorized user to print the results list.
iv. Printing must preserve the layout produced by generating application package and include all (printable) components of the electronic records.

c) Rendering: Other

i. The system must include features for presenting and outputting to appropriate media records which cannot be printed (e.g., Court Recording and Transcription System).

4.6.8   ADMINISTRATION: ADMINISTRATOR

The requirements associated for managing system parameters, back-up and restore, system management and user administration. Those include:

a) Configure the system in accordance with Malaysian Judiciary procedures and rules.

b) Manage, retrieve, display and re-configure system parameters and to re-allocate users and functions between a user's role.

c) Make changes (e.g., alter the metadata of a record within the system and reporting conflicts, in summary for individual resolution) and to allow finalization/correction of the record profile. When changes have been made, the system must automatically log this in the audit trail.

d) Provide back-up facilities so that records and their records management metadata can be recreated using a combination of restored backup and metadata. Backups must be performed daily, weekly, monthly and annually. Frequency depends on the levels of backing up critical information.

e) Provide recovery, update error and must notify the administrator of the results (e.g., allow administrators to 'undo' a series of transactions until a status of assuring the database integrity is reached).

f) Communicate errors encountered in storing data and monitor available storage space and notify the administrator when action is needed because available space is at a low level, or because it needs another administrative attention.

g) The system must allow administrative roles, and only administrative roles can create and maintain retention and disposition schedules.

h) Print all aggregations, metadata, records disposal schedules, classification scheme and audit trail.

i) The system must provide the appropriate administrators to have control of the assignment and revocation of security levels and privileges.

4.6.9   ADMINISTRATION: METADATA ADMINISTRATION

Metadata schemas have to be administered, including the creation, addition, deletion or alteration of metadata elements, and the semantic and syntactical rules and obligation status applied to those elements. The functional requirements of metadata are as follows:

a) Allow the administrator to create, define, modify and delete metadata elements, including custom fields.
b) Allow the administrator to configure the system to restrict the viewing or modifying of metadata elements by group, functional role or user.
c) Document all metadata administration activities.
d) Ensure all records management metadata and metadata data are handled correctly and completely at all times.

4.6.10 ADMINISTRATION: REPORTING

The system must be able to report the caseload, case flow, and workload statistics and other court financial operations and staff management information, either by the system itself or by an integrated or interfaced external records management mechanism, during a specified period of time (daily/monthly/annually) The requirement for reporting includes:

a) Provide flexible reporting facilities for the administrator. They must include, at a minimum, the ability to report the following:

   i. numbers of aggregations, volumes and records;
  ii. transaction statistics for aggregations, volumes and records; and activity reports for individual users.

b) Allow the administrator to report on metadata based on selected:

    i. aggregations, volumes and record objects;
   ii. users and user groups;
  iii. security categories;
  iv. other records management metadata.
   v. time periods and time retention;
  vi. file formats and instances of each format.

c) Be able to produce a report, listing the details and outcome of all records functions process.
d) Allow the administrator to request and find daily, monthly and yearly reports.

e) Allow the administrator to restrict users' access to selected reports.

f) Provider statistical report function, for example, provide access to statistical reports such as case load, closed cases, the number of cases assigned per judge, case flow reports, retention period, number of pending cases at each proceeding stage, average time intervals between proceeding stages and workload statistics.

g) Provide management report function, for example, provide access to management reports, providing information on cases (e.g., compliance with rules of court and state statutes; summary information on case parties, filings and events; detailed information on case aspects, such as no action) and enable trend analysis. (e.g., changing file formats in use).The system must be able to send monthly reports to the Chief of Justice.

4.7   CONCLUSION

This chapter discusses the legal and functional requirements for electronic court record, with special reference to Malaysian judicial environment. It presents the legal compliance framework and the functional requirements, which should be fulfilled in electronic court records management. Since the framework is developed based on Malaysian legal environment, this framework and the functional requirements are suggested to be tested in other jurisdictions in the future research endeavors.

KEYWORDS

- **electronic court**
- **ICT**
- **judicial environment**
- **legal requirements**
- **Malaysian Legal System**
- **Technologies in Courts**

REFERENCES

1. Bari, A. A. (2007). 'British Westminster System in Asia – the Malaysian variation,' *U.S.-China Law Review*, vol. 4, p. 1.
2. IRMT (2002). *Case study-legal and judicial records and information systems in Argentina: Evidence-based governance in the electronic age*, International Records Management Trust, London.
3. Deka, G. C., Zain, J. M., Mahanti, P. (2012). 'ICT's role in e-governance in India and Malaysia: a review,' viewed 7 March (2013). http://arxiv.org/abs/1206.0681.
4. Duranti, L. (2010). 'Concepts and principles for the management of electronic records, or records management theory is archival diplomatics,' *Records Management Journal*, vol. 20, no. 1, pp. 78–95.
5. Federal Court of Malaysia (1997). *Working procedure manual for civil and criminal cases in courts*, Federal Court of Malaysia, Kuala Lumpur.
6. Fernando, J. M. (2006). 'The position of Islam in the Constitution of Malaysia,' *Journal of Southeast Asian Studies*, vol. 37, Cambridge Journals Online, no. 2, pp. 249–266.
7. Galves, F. (2000). 'Where the not-so-wild things are: computers in the courtroom, the federal rules of evidence, and the need for institutional reform and more judicial acceptance,' *Harvard Journal of Law & Technology*, vol. 13, no. 2., pp. 165–178.
8. Hamzah, A., Bulan, R. (2003). *An introduction to the Malaysian Legal System*, Oxford Fajar, Kuala Lumpur.
9. Ismail, A., Jamaludin, A. (2009). 'Towards establishing a framework for managing trusted records in the electronic environment,' *Records Management Journal*, vol. 19, no. 2, pp. 135–146.
10. Kamal Halili Hassan, Maizatul Farisah Mokhtar, (2011). 'The E-Court System in Malaysia,' 2nd International Conference on Education and Management Technology, IPEDR vol. 13 (2011) IACSIT Press, Singapore
11. Kaushik, P. (2002). *E-governance: Government Initiatives in India, Bangladesh, Nepal and Sri Lanka*, International Policy Fellowship.
12. Malaysian Administrative Modernization and Management Planning Unit (2009). *Garis panduan penggunaan biometrik bagi agensi-agensi sektor awam*, Malaysian Administrative Modernization and Management Planning Unit, viewed 21 November (2012). http://www.mampu.gov.my/c/document_library/get_file?uuid=f6ed752d-1c80-414f-ac9b-cc9627877603&groupId = 10136.
13. Mathur, D., Gupta, P., Sridevi, A. (2009). 'e-Governance approach in India – The National e-Governance Plan (NeGP),' in *Transforming Government – eGovernment Initiatives in India*, The ICFAI University Press, India, viewed 29 March (2012). http://www.nisg.org/.
14. Memorandum on Managing Government Records, Administration of Barack Obama (2011). Federal Register Library – HeinOnline.org.
15. Rafie, R. (2011). 'Mahkamah Syariah perlu pembaharuan elak persepsi buruk,' *Berita Harian*, 23 March, Kuala Lumpur.
16. Rahmah (2008). *Records management reform in Malaysian court of appeal*, Briefing, Federal Court of Malaysia, Putrajaya.
17. Rahman, T., Khan, N. A. (2012). 'Reckoning electronic government progress in Bangladesh,' *International Journal of Public Administration*, vol. 35, no. 2, pp. 112–121.

18. Law Review Board (2006). *Federal Constitution*, International Law Books Service, Petaling Jaya.

19. Saman, W. S., Haider, A. (2011). 'The implementation of electronic records management system: a case study in Malaysian judiciary,' paper no. 170, *AMCIS 2011 Proceedings*, Association of information System, USA.

20. Saman, W. S., Haider, A. (2012). 'Electronic court records management: a case study,' *Journal of e-Government Studies and Best Practices*, vol. (2012). pp. 1–11.

21. Sherif, S. M. (2010). 'The contempt power: a sword or a shield,' Thesis, Durham University, United Kingdom.

22. Yusof, Z. M., Chell, R. W. (1999). 'The eluding definitions of records and records management: is a universally acceptable definition possible? Part 2: Defining records management,' *Records Management Journal*, vol. 9, no. 1, pp. 9–20.

23. MAMPU: Malaysian Administrative Modernisation and Management Planning Units. (2010). Profile. Retrieved November 10, 2010, from http://www.mampu.gov.my/web/bi_mampu/profile

24. MAMPU: Malaysian Administrative Modernisation and Management Planning Units. (2007). Profile. Retrieved November 10, 2007, from http://www.mampu.gov.my/web/bi_mampu/profile

25. MAMPU: Malaysian Administrative Modernisation and Management Planning Units. (2002a). Profile. Retrieved November 15, 2010, from http://www.mampu.gov.my/web/bi_mampu/profile

26. MAMPU: Malaysian Administrative Modernisation and Management Planning Units. (2000). Profile. Retrieved November 10, 2014, from http://www.mampu.gov.my/web/bi_mampu/profile

27. MAMPU: Malaysian Administrative Modernisation and Management Planning Units. (2006). Profile. Retrieved November 10, 2006, from http://www.mampu.gov.my/web/bi_mampu/profile

28. Ismail, Aliza (2010). Assessing the practice of trusted electronic records management in Malaysian government-controlled companies / Aliza Ismail. PhD thesis, Universiti Teknologi MARA.

29. InterPARES (2002). Authenticity task force report. http://www.interpares.org/book/interpares_book_d_part1.pdf. Accessed 11 January 2015.

30. ICA (2008). Principles and Functional Requirements for Records in Electronic Office Environments. Module 1. Overview and Statement of Principles (Paris: ICA).

APPENDIX A

UNIQUE CODE IDENTIFIER ACCORDING TO TYPES OF CASES IN CIVIL COURT SYSTEM

Code	Type of Cases in Civil Court System	Type of Court/ Jurisdiction
1–4	Civil appeal	Federal Court
5	Criminal appeal	
8	Application for leave of appeal	

APPENDIX A   Continued

Code	Type of Cases in Civil Court System	Type of Court/ Jurisdiction
1–4	Civil appeal from High Court	Court of Appeal
5	Criminal appeal from High Court	
7	Application for leave of criminal appeal	
8	Application for leave of civil appeal	
11	Civil appeal from Magistrate Court	High Court
12	Civil appeal from Sessions Court	
13	Civil review	
14	Tax appeal	
15	Land reference	
16	Appeal from administrative tribunal	
17	Disciplinary proceeding and "Short call"	
18	Petition for lawyer's admission to High Court	
21	Civil suits against government	
22	General civil suits	
23	Civil suits (Tort cases)	
24	Originating summons	
26	Petition	
27	Admiralty-in-rem	High Court
28	Company winding-up	
29	Bankruptcy	
31	Letters of administration/power of attorney	
32	Probate	
33	Divorce	
34	Adoption of child	
36	Application of execution (prohibition order)	
37	Application of execution (moveable property)	
38	Application of execution (immoveable property)	
41	Criminal appeal from Magistrate Court	
42	Criminal appeal from Sessions Court	
43	Criminal review	
44	Criminal application	
45	Criminal trial	

APPENDIX A   Continued

Code	Type of Cases in Civil Court System	Type of Court/ Jurisdiction
51	Proceeding by or against Malaysian Government in Sessions Court	Sessions Court Civil Jurisdiction
52	Civil Cases management	
53	Tort	
54	Originating summon	
56	Registration of enforcement application, writ of seizure and sale, public auction, interpleader, land sale, ownership writ, writ of transmission delivery, distress writ, judgment debtor's summon, judgment notice	
57	Nationality investigation case management	
58	Petition of adoption	
61	Public servant arrest cases	Sessions Court Criminal Jurisdiction
62	Arrest cases (general)	
63	Summons Cases	
64	Application under Section 32(2) and 32(3) Dangerous Drug Act (property forfeiture) 1988	
71	Proceeding by or against Malaysian Government in Magistrate Court	Magistrate Court Civil Jurisdiction
72	Civil Cases management	
73	Tort	
74	Originating summon	
76	Writ of seizure and sale, public auction, judgment debtor's summon, judgment notice	
77	Small claims	
78	Cases under Married Women and Children (Maintenance) Act 1950	
81	Committal cases to High Court	Magistrate Court Criminal Jurisdiction
82	Public servant arrest cases	
83	Arrest cases (general)	
84	Juvenile cases	
85	Criminal summons	
86	Traffic cases	
87	Public agencies' summons	
88	Death investigation	
89	Miscellaneous criminal applications	

APPENDIX B

UNIQUE CODE IDENTIFIER ACCORDING TO TYPES OF CASES IN SHARIAH COURTS SYSTEM

Code	Type of Shariah civil cases
001	Appeal
002	Application of leave for appeal
003	Review
004	Application for prohibition order/Injunction from bringing a child out of Malaysia
005	Application for prohibition order/Injunction from interference
006	Application for validity of child's status
007	*Waqaf* validity application
008	*Nazar* validity application
009	Betrothal compensation claim
010	Application for marriage validity/order
011	Application for polygamy permission
012	Application for polygamous marriage registration
013	Application for declaration of nullity of marriage due to change of religion
014	*Fasakh* application
015	Application for assumption of death
016	*Muta'ah* claim
017	Matrimonial property claim
018	Wife's alimony claim
019	Claim of alimony for disabled person
020	Alimony security claims
021	*Eddah* alimony claim
022	Application to vary order of alimony
023	Claim of deferred alimony
024	Claim of child support
025	Application to vary order of child custody rights or alimony
026	Application to vary agreement of child custody rights or alimony
027	Application to annul alimony orders
028	Application for custody of child
029	Application for dismissal of custody rights
030	Application for prohibition relating to property of a minor
031	Application for custody of an orphan

APPENDIX B Continued

Code	Type of Shariah civil cases
032	Application for freezing of property transactions
033	Application for enforcement of maintenance orders
034	Application for execution of court orders
035	Garnishment claims
036	Application for contempt of court order
037	Judgment debtor claims
038	Application for an interim order
039	Probate confirmation application
040	Application for *faraid* declaration certificate/authority
041	Interlocutory application
042	Application for *faraq* of a wedlock
043	Application of apostasy
044	Application for validation of *hibah*
045	Application for validation of *hibah* made during sick before dying
046	Application for verification of beneficiaries
047	Application to sue or defend as a poor person
048	Interpleader application
049	Claims for marital compensations
050	Marriage authorization application of a minor
051	Marriage authorization application using *wali* hakim
052	Application for guardian refusal
053	Marriage authorization application for widow
054	Application for verification of pronouncement of divorce
055	Divorce claims
056	*Khulu'* application
057	Application for verification of *ta'lik* divorce
058	Application for *nusyuz* conviction
059	Application for right of abode
060	Application of order for husband to stay together
061	Application for confirmation of *ruju'*
099	Others
100	Application of order for wife to stay together
101	Appeal
102	Application for leave of appeal

APPENDIX B   Continued

Code	Type of Shariah civil cases
103	Revision
104	Deviant worship
105	Accusing/charging a non-Muslim
106	*Takfir* (apostasy accusation)
107	False doctrine
108	False claim
109	Spreading opinion contrary to fatwa
110	Disobeying court order
111	Publication contrary to S*yarak'*/Islamic teachings
112	Incest
113	Matchmaking
114	Preparation of intercourse out of wedlock
115	Same sex intercourse
116	Intercourse beyond the natural way
117	Enticing a married woman
118	Instigating husband and wife to divorce
119	Selling/giving a child to non-Muslim
120	*Qazaf*
121	Sodomy
122	*Muhaqashah*
123	Building a mosque without permission
124	Enticing a woman
125	Breach of trust
126	Leaks out a secret
127	Officer disobeying rules to harm anybody
128	Not presenting document to religious authority
129	Declining to take oath
130	Declining to answer questions posed by religious authority
131	Insulting officers on duty in court proceeding
132	Encouraging immoral conduct
133	Mocking Al-Quran/Islam
134	Humiliating Islam
135	Disgracing places of worship

APPENDIX B   Continued

Code	Type of Shariah civil cases
136	Humiliating religious authority
137	Preaching without credentials given by authority
138	Gambling
139	Drinking liquor
140	Not paying zakat
141	Brothel
142	Preparation of sexual intercourse out of wedlock
143	Seclusion
144	Prohibiting husband and wife from living together
145	Colleting zakat without authority
146	Misuse of halal sign
147	Abetment
148	Interference into a marriage
149	False declaration for marriage
150	Being unreasonably unfair to wife/wives
151	Apostasy in order to annul marriage
152	Instigating to neglect religious duties
153	Attempt to make offenses specified in statute
154	Offences with unspecified sentences in the statutes
155	Contempt of court
156	Pleading for various crime
157	Disrespecting the month of *Ramadhan*
158	Not performing *Jumaat* prayer
159	Being a transgender woman
160	Behaving indecently in public
161	Being absent before court
162	Illegal marriage certificate
163	Solemnizing a marriage without authority
164	Offences relating to solemnization
165	Getting married in contrary to legal provision
166	Polygamous marriage without court's permission
167	Divorce without court's permission
168	Not lodging a report regarding family law case

APPENDIX B   Continued

Code	Type of Shariah civil cases
169	Leaving a wife purposely
170	Wife mistreatment
171	*Nusyuz*
172	Intercourse between divorced persons
173	Purposive carelessness in disobeying rules provided by Family law
174	Reconciliation without wife's agreement
175	Not registering reconciliation
176	Intercourse within reconciliatory divorce period
177	Collecting charitable fund without permission
178	Giving false declaration to a *Muallaf* registrar
179	Trespassing into *waqf* property
180	Getting pregnant out of wedlock
181	Selling intoxicated drinks
182	Not abiding rules set for starting of *Ramadhan* and *Eid*
183	Acting falsely as imam in *Jumaat* prayer
184	Preaching in a mosque without permission
185	Offences regarding *Baitulmal*
186	Refuse to make statement
187	Practicing as Shariah lawyer without credential
199	Others

CHAPTER 5

E-GOVERNMENT FORMATION CHALLENGES AND SOLUTION PERSPECTIVES

RASIM ALGULIYEV[1] and FARHAD YUSIFOV[2]

[1]*Professor, Director of the Institute of Information Technology of ANAS, Baku, Azerbaijan, Tel: +994125390167; E-mail: rasim@ science.az*

[2]*PhD, Department of Information Society, Institute of Information Technology of ANAS, Baku, Azerbaijan, Tel: +994125101260, E-mail: farhadyusifov@gmail.com*

CONTENTS

5.1   INTRODUCTION

Nowadays, the wide implication of information technologies in developed countries is affecting their social-economic development. The number of citizens, centers, organizations, institutes having access to and using internet for satisfying their needs is being rapidly increased. In this situation, there is an increasing need for more mobility and interactivity in transparency principles of state services and neutrality principles from political point of view. It is worth to mention that the implication opportunities of political and social technologies in administration are being widened nowadays.

In some sources, "electronic government" term is used during translation and different definition, especially in the developing countries. As "electronic government" is currently under construction especially in the developing countries, it has not been fully formed as a definition. As shown in relevant documents, it does not only include the central executive authority, it also includes the three branches of government – executive, legislative and judicial [1–5].

The conducted research shows that the definition of implication of electronic government (e-government) is expanded not only as application of information technologies, but also as a tool of administration of public services in the world. Sometimes, as a key of success of e-government, customer relationship management (CRM) systems are indicated [6, 7].

The forming of national e-governments in post-industrial countries is carried out based on reform of all public administration system. The main objective here is the compliance of public administration with Information society. New public administration contains its substantial clarity, transparency, competition environment, responsibility for the outcomes of its actions, increase of the role of ethical requirements, and active mutual relationship with civil society. It is essential that, during the use of information-communication technologies (ICT) in public administration, also other factors affecting the character of socio-political, economic, cultural, mental, and government-society relationships are considered.

Nowadays, the governments invest big amounts of money to realization of e-government projects for further upgrade of services supplied by the government to citizens and reduction of costs in whole world. The governments can increase the efficiency of actions and carry out administrative operations more easily by using ICT. By considering this important fact, the specification of researches in direction of e-government establishment and also the most successfully applied models and their research are remarkably necessary.

5.2 THE CONCEPTUAL AND ARCHITECTURAL PRINCIPLES OF E-GOVERNMENT

There are many definitions for "electronic government" term, each of which accents attention on certain functions [8–12]. Analyzing this definition, we cannote that, "electronic government" – means new opportunities for public administration, application of ICT, innovations in activity of government authorities for improvement of interrelation with business-sector, non-governmental organizations and citizens of the country. In other words, the main goal of e-government is – to make public administration systems such, that in majority, they would consider interests of citizens and organizations, and would give them the opportunity to participate in government activities, as well as improve the interaction between them.

It is known that e-government has started forming in the cross of two centuries. It is known from history that, each transition to new quality has been accompanied by several complications, sometimes by serious crisis. Following this practice, the government can prevent the possible social-economic crises by modernizing the administrative mechanisms. In this regard, the government was required to conduct some reforms for the transformation to a new phase in public administration. The transition to a new phase necessitates the conduction of important scientific-research works.

Let's define the functions of electronic government as mechanism of public administration in different aspects: information technologies in public administration; organization of public administration based on electronic means of processing, transfer and dissemination of information, rendering of services by government authorities, provision of citizens with information about activities of public authorities; automated government services, use of innovative technologies by public administration authorities, use of ICT in the Internet network as an instrument, allowing to achieve a more effective administration [11–14].

Particularly, international organizations such as UN, ITU, OECD, World Bank, CID etc. play a major role in research of e-government. Also, researches of companies such as Gartner, Microsoft, CISCO, HP etc. in the field of implementation of electronic government is a good sign, as given organizations and companies possess a sufficient material base for conduction of qualitative and important scientific-practical researches. In its turn, it allows to attract significant financial resources to the field of research of electronic government. It also must be noted that, given organizations and consulting companies are interested in certain coverage of electronic government problem by social-political goals.

According to "E-government Act" of USA dated 2002, this term was accepted as an expansion of access to government information of agencies and government structures by means of Internet and other information technologies, also as an implication of information technologies and the use for the increase of efficiency [15, 16]. In official documents, electronic government (e-government) is comprehended as a mutual relationship system with information character of local public authorities and the society by using ICT. Gartner company reckons that e-government – is the concept of administration by incessant optimization of services process, participation of citizens in political processes, also by changing of internal and external relations with the help of technical tools, Internet and modern mass media [4, 17].

In some research works, "electronic government" term is defined as Internet-technologies providing the informative mutual relationship of public authorities with population and civil society institutions. E-government is specified as an integral, socially responsible enterprise having regular counter-relation and open to information.

In general, e-government is specified as a mutual relationship between specialized complex system of public authorities and citizens, civil society and business structures by means of Internet. As following steps of a mutual relationship, C2B (customer-to-business) – between citizen and business;

B2B (business-to-business) – between private companies; G2C (government-to-citizen) – between government services (on government, departments and regions level) and citizens; G2B (government-to-business) – between government and business sectors; G2G (government-to-government) – between public authorities can be shown [1–5, 18].

It is worth to mention that the conceptual model of e-government is based on government structure existing in countries of democratic society and market economy (Figure 5.1b). If to approach this problem conceptually, the reforms conducted in public administration in the beginning of 90-s showed the larger share of government in forming of e-government (Figure 5.1a).

Alongside, it must be considered that, the society not only obtains the access to information, but also gets the opportunity to affect the decision-making process of government and participate interactively in the process of preparation of decisions, as a result, the transparency of public sector performance increases.

In general, e-government creates new opportunities for development of democracy. It provides the mutual information relationship between population and civil society institution and public authorities by means of ICT. In other words, e-government comprises the mutual relations system of citizens, civil society and business-structures, and executive government structures by means of Internet. Implication of ICT in government performance, transparency and accessibility of government information, feedback principle between population and public authorities, government

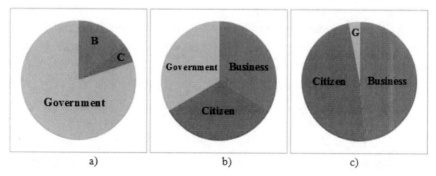

FIGURE 5.1 Evolution model of E-government.

responsibility for the decisions made and etc. issues in different countries are the main characteristics specifying e-government.

It is essential that the transformation to Information society, e-government strategy based on democratic values necessitates the gradual change of government model, the increase of share of civil and business structures, minimization of government share (Figure 5.1c).

According to definition of European Committee, e-government – is the use of ICT in public structures and improvement of performance of government employees and public authorities in the back ground of realization of organizational reforms and forming of skills directed to the increase of level of services provided by them [19].

According to the concept of e-government, the whole system of public authorities performs as an integral service organization for the provision of services to citizens. The performance of e-government must be clear, transparent and accessible in terms of information for citizens. The specific attention is drawn to establishment of feedback mechanism, efficiency of services provision and execution period by using the centralized systems. These all enable to increase either the quality of provision of services provided by the government to citizens, or the performance efficiency of government.

5.2.1   TRANSFORMATION OF PUBLIC ADMINISTRATION

Based on international experience, we cannote that, e-government requires transformation of order of interrelations of government and society [10–13, 20, 21]. Transformation process requires a consequent strategy, starting from monitoring of current condition of the country and the ability of its citizens to use new technologies. In order to improve the transformation process, government must developed a model of interrelations with main persons of interest and consider opinions of the country citizens.

Based on international experience, we cannote that, as there is no unique general strategy on implementation of e-government, it can emphasize following foundation issues of transformation process:

- Legalization of application of ICT in the activity of government authorities and interrelations of government authorities with business and citizens must be the main constituent of the e-government;

- Creation of new processes and new interrelations between controlled and controlling ones is required for successful implementation of e-government. For this reason, if correctly planned, real revolutionary changes can be introduced in the process of public management itself;
- During the process of passage to e-government, there must be employees on all levels of government system that comprehend the importance of the new technology, understand the goals of government policy and are determined to actively conduct the innovation in general;
- During the process of passage to e-government, developing countries must thoroughly select projects in order to optimize their expenses of time and resources. Projects must be characterized by the increased the level of government participation in processes of public administration. For this reason, it is necessary to define standards and target directions;
- Public authorities must test a new order of interrelations among government institutions, as well as within the framework of partnership with private sector and non-governmental organizations for provision of quality and accessibility of the e-government;
- Considering the fact that concept of the e-government is oriented to the interests of citizens, it represents an initiative directed at improving the life of citizens. For this reason, efforts to stimulate citizen participation in government actions form a very important task.

It is important to note that, during the process of transformation of public administration, following can be emphasized among difficult tasks:

- Problems related to Internet network access. For example, if developed countries generally have an unlimited Internet access, developing countries use Internet mainly at work or at Internet-clubs. On the other hand, generally developing countries do not have the financial resources to be invested in implementation of e-government. The problem is, planning of expenses, ICT related expenses or expenses spent on internet-infrastructure of government authorities for effective achievement of practical results are used purposelessly.
- Problems of connection of interrelation of public authorities with the society consist of conformability of systems, acceptance of unique standards, possibility of exchange of information, unified search in government information resources etc. E-government must be

integrated into the process of reformation of public administration and activities on development of information society as a mechanism.

• Problems related to access of citizens to government information and government services through Internet and use of possibilities of Internet in an incomplete volume. For elimination of this problem, it is necessary to stimulate the citizens of the country, for example, to reduce the Internet access fees and application of new technologies for improvement of possibility of access the government information resources.

5.3   THE ISSUES OF ASSESSMENT OF FORMING PROCESSES AND MANAGEMENT OF E-GOVERNMENT

One of the up-to-date issues regarding e-government is the assessment and monitoring of its forming processes. It can be justified with such fact that e-government is an online environment with the quite complicated structure. On the other hand, e-government is the sum of vertically and horizontally interrelated corporative information spaces.

The issue of establishment of complex indicators system for the monitoring of the efficiency of state governance and the use of ICT in different areas, the methodology of practical implication has been started to forming at the end of 1990s [3, 4, 22, 23]. Nowadays, the existing practical experience and methodical potential for the assessment of electronic readiness, the monitoring, potential analysis and comparative analysis of governments are present. As such methodologies, ICT development index (IDI) of International Telecommunication Union regarding the assessment of Information society, e-government development index of UN regarding the forming and the use of e-government, networked readiness index of World Economic Forum, digital opportunity index, indicators system for the assessment of development level of e-government of European Union countries (Capgemini company) can be shown [24–27].

The implication of international indexes for the development of methods of the assessment and monitoring of e-government forming processes can be considered as an important factor. Also, it can be mentioned that the position of the country in international ratings has a great importance in terms of the position of the country attained in the region. These indicators exhibit the carrying out of development strategy of Information society of the country.

It is worth to mention that, alongside with the assessment and monitoring of digital differences either at national, corporative or at enterprise level; these are the important information for carrying out of expedient management of the forming process of aimed electronic environment (e-environment).

Taking into consideration the necessity of realization of five sequential phases (communicatization, computerization, networking, informatization, and virtualization) of e-environment forming, the balanced relation must be provided among the separate phases of it. Management centers are the intellectual systems enabling the efficient decision-making bases on the indicators characterizing the progress of the process (Figure 5.2).

The following can be shown as the indicators characterizing the virtualization, socialization phase in e-environment:

• The indicators characterizing the social networks created in considered e-environment;
• The indicators characterizing the classification and activeness (age, specialty, gender, space, time, etc.) of users;

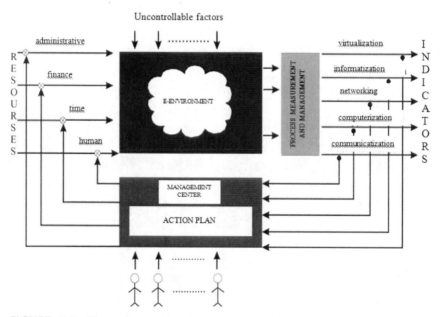

FIGURE 5.2 The assessment and management scheme of forming process of e-environment.

- Classification and rating indicators of used contents;
- The indicators characterizing the transparency, accessibility and sequences of information in e-environment;
- The indicators characterizing the virtual relations established and contents turning over in e-environment, etc.

It is essential that the solution of several problems (technological, normative-legal base, cadres' education, scientific, etc.) is an important condition. By taking the leading practice into consideration, e-government must be formed based on both horizontal and vertical management principles. From this point of view, each institution included in e-government must have an action plan, the indicators characterizing its plan must be specified the management of this process must be executed and the continuing (or on specific cycles) monitoring must be carried out.

It must be mentioned that, nowadays monitoring and assessment issues has a timely importance for the realization of e-government projects, programs specifically. From this point of view, there is a need for establishment of complex system of assessment and monitoring. The indicators accepted at international level and parameters meeting the local needs must be included in the system itself.

5.4   INTELLECTUAL ANALYSIS ISSUES OF WEB-RESOURCES PERFORMING IN E-GOVERNMENT ENVIRONMENT

While considering the e-government programs carried out in different countries of the world, it becomes clear that e-government will be more accessible and efficient under the "single window" principle in the near future. This is mainly related to rapid development of content mining methods, web technologies, social networks [28–31]. From this point of view, the analysis of web-resources and development of management mechanisms is of great importance in carrying out the e-government projects.

The implication of web content mining, Web analytics and social networks are strong tools in improvement of e-government management effectiveness and establishment of feedback mechanism. If to consider that one of the main issues in realization of e-government programs are the analysis of web-resources and establishment of the effective management policy, then the implication of innovations, new technologies widen the communication capacity significantly. This, in turn, enables to achieve new integration forms between business sector and citizens.

It is clear that web-resource of each institution performing in government sector is created. As a result, web-infrastructure of e-government is formed. Established web-resources creates online interactive social communication environment between public authorities and citizens. More information is gathered in this environment eventually. Thus, new opportunities are created for the intellectual analysis of web-infrastructure and more efficient management of the society.

The internal structure of electronic communities existing inside each online environment in disguise can be revealed by applying the social networks theory. The analysis of e-government web-infrastructures, web mining technology can be implied for obtaining the following information [28–30]:

- Which issues are mostly discussed by the citizens;
- The monitoring whether the discussions are related to public sector;
- The classification based on different criteria's (space, time, age, specialty, activeness, etc.) of citizens applying mostly to which institutions;
- The online monitoring of realization status of requirements claimed against web-resources;
- The classification based on different criteria's (countries, institutions, issues, time, etc.) of inquiries to web-resources from foreign countries performing in e-government environment, etc.

An effective web analytics of sites, portals and also, sites providing online services to citizens is revealing existing program, technical, content related errors and adjustment to requests of citizens, and users. By using Web analytics, the reasons for leaving the site by users, their actions and behaviors at website regarding a site or particular service can be revealed. It is clear that Web analytics is not limited with particular statistics and enables to obtain more detailed information for analysis.

Also we cannote that, there are no globally accepted definitions within web analytics as the industry bodies have been trying to agree definitions that are useful and definitive for some time. The main bodies who have had input in this area have been JICWEBS (The Joint Industry Committee for Web Standards in the UK and Ireland), ABCe (Audit Bureau of Circulations electronic, UK and Europe), The WAA (Web Analytics Association, USA) and to a lesser extent the IAB (Interactive Advertising Bureau) [32–35]. Both the WAA and the ABCe provide more definitive lists for those who are declaring their statistics using the metrics defined by either [33–34].

The analysis log-files gathered in servers, information gathered in e-mails play a prominent role in effective decision-making by e-government parties in the process of establishment of online relations between citizens and public authorities [28–31]. This, in turn, enables the development of feedback mechanisms for e-government management.

For the provision of citizen access to public authorities countrywide, the forming of social Internet centers and points infrastructure in settlements is one of the important issues in e-government building. So that, e-government must perform sustainably, be reliable and secured against threats. Namely, information, energy and etc. security of e-government must be provided and be prepared for information war.

With the advent of the automation tools, texts in the electronic form, effective 60s of past century an initial development of the content mining of the information of large volumes was obtained – the databases and interactive media-sources. The traditional political use of content mining contemporary technologies was augmented by the unlimited list of headings and subjects that cover production and social spheres, business and finances, culture and science. This process was, in turn, accompanied by the large number of different program systems [36–41].

The interesting feature of content mining is the fact that until recently this methodology was attributed to a certain sphere of human activity (policy and sociology). Nevertheless, today content mining is increasingly used in many fields of political and economic life and this contributes to the larger applied relevance to use sociology and policy in the methodology of the content mining. Based on the received information conclusions are made and recommendations are formed at the stage of the data mining regarding making that very managerial decision which everything started with. After the preparation of report about the carried out estimation and "feedback" according to the results of estimation the final stage comes – making a decision considering the received information.

The distribution of the methodology of content mining to two branches is considered as universally recognized: qualitative and quantitative. The basis of quantitative content mining is frequency of occurrence of certain content characteristics in the documents. The method of qualitative content mining is based on the very fact of presence or absence of one or several characteristics of the content in the text.

The method of qualitative content mining is based on the fact that at any phase of quantitative content mining an expert can be involved for the evaluations of results. Thus, this method is intended to provide the expert with the necessary tools for conclusions and additional results. The expert with the aid of such tools can reveal specific properties of the part of information and verify them against the general text flow and extend general properties of the text flow to its specific subject part.

It should be noted that content mining is first of all a quantitative method that assumes the numerical estimation of some components of the text that can be supplemented also with different qualitative classifications and revelation of various structural regularities. Therefore, the most successful definition of content mining can be considered the one provided in the relatively recent literature – it is a systematic numerical treatment, estimation and interpretation of form and content of the information source.

From the point of view of ICT specialists the content mining is a typical example of the applied information text analysis that contributes to the extraction of some special components that are interesting to a researcher from the entire variety of existing information in it and their presentation in a form convenient for the perception and subsequent analysis. The numerous specific versions of content mining are distinguished depending on what these components are and what is precisely understood under the text [42, 43].

Thus, the idea of content-monitoring can be formulated in the simplest form as a constant implementation of content mining of continuous information traffics that is narrowly outlined by its problems. Let us emphasize that it is continuous reproduction of input data processing in time is the very characteristic of content-monitoring. Actually, content mining comes out here as composite, and content-monitoring has got its own problems and methods to solve applied problems.

Discovering of knowledge can be defined as finding and analysis of useful information. Use of methods of content mining gives rise to the need of creation of server and client intellectual systems, which can effectively analyze the content of Web-resources.

5.5   INFORMATION SECURITY PROVISION ISSUES OF E-GOVERNMENT

The modern world is characterized with the transformation of industrial society into information society. The major objectives of establishing information society include issues, such as the formation of legal basis for the IS, development of human resources, rights of the citizens for obtaining and using information, formation of electronic government and electronic trade, in particular, strengthening the intellectual potential of the country, establishment of information and knowledge-based economy, development of modern information and communication infrastructure, formation of national electronic information space and provision of information security. As a development stage of the civilization, information society is characterized by the increase of the role of information and knowledge in the society, growth in the share of information communications, products and services in the flow of domestic goods, establishment of the global information space ensuring effective exchange of information and access into the global information resources of the people. Currently, the wide application of ICT in various spheres of the society, including economy, energy, ecology, etc. emerges the problem of information security. In general, the more dependence of all sectors of society and the people on ICT is stronger, the more obvious the importance of information security becomes. As one of the major duties of the information society establishment, information security is becoming one of the main directions of providing security of the government, society and person.

As the society is becoming informatized, the people became more dependent on information. Non-provision of information security can cause major consequences for the society. The priorities of information security in a particular country are specified based on the balanced ratio of government, society and citizens interests. As one of the main components of the safety of society, the duties of information security are the confidentiality of information, information integrity, information accessibility and the fight with harmful computers. A number of programs, projects, techniques have been developed in the international practice with the purpose of ensuring the global information security. Resolution on the Global Information Security Culture adopted by the UN in 2002, the Global Information Security program adopted by the International Telecommunication Union in 2008 can be examples [44, 45].

The several issues with technical and administrative-legal characters must be solved for carrying out the e-government program. The preparation of mutual relationship regalements, the creation of government services classification, also integral technical architecture, realization of program platform and the provision of information security can be indicated among them.

For the provision of normal performance of e-government, it is necessary to provide the security of each level constituting e-government (Figure 5.3).

In general, the up-to-date issues in the framework of provision of information security of e-government can be classified as following:

- Development of conceptual-architectural models for provision and management of e-government information security and sustainable performance;
- Development of models for the analysis of information security risks and management;

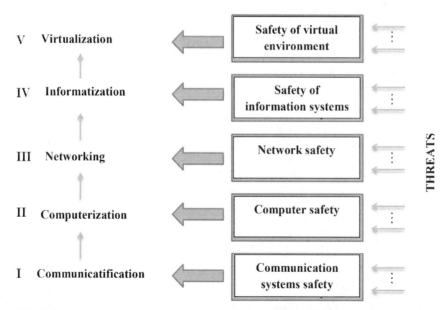

FIGURE 5.3   The conceptual model of information security provision of e-government.

- Development of cybercrime defenses technologies;
- Revealing the disguised criminal social networks creating threats for e-government environment and development of analysis methods;
- Development of intellectual monitoring system of corporative network environment;
- Development of spam busters methods and algorithms by means of data intellectual analysis technologies;
- Protection of individual information in e-government environment and development of user-oriented security mechanisms;
- The creation of Computer Emergency Response Team (CERT) in e-government;
- Investigation of information war, information attack and information attack defenses technologies and development of new methods and algorithms.

It is worth to mention that complex and systematic approach is required to information security provision issues of e-government. With the development of information society, the necessity of establishment of integral and multilevel nation-wide information security system appears in the process of e-government forming.

In general, the forming of Information society perplexes the provision of information security of countries, the increase of immunity, the sole fight against threats of different nature and scale and thus, the forming of global information security environment must be of interest of all countries, civil societies, companies and people.

5.6   ELECTRONIC DEMOCRACY AND DIGITAL CITIZEN ISSUES

Electronic democracy (e-democracy) is considered such organization form of citizens' social-political activity that the wide use of ICT provides the establishment of more effective relations at new level either among citizens, or between citizens and public authorities, civil society and business sector [46, 47]. In other words, for the strengthening of democratic institutions, the expansion of participation of citizens in political activity, the use of ICT constitutes the essence of e-democracy. E-democracy term means the consideration of citizens' thoughts and the engagement of citizens and organizations to political relations and processes. In this phase, the issue of how close the citizens are engaged in social-political processes is characterized with digital citizen problems.

Starting from initial phases of awakening of e-democracy, the provision of access opportunities to socially important information of public authorities by the citizens was constrained by creating of voting opportunities regarding particular decisions of the government [46–48]. The further development has widened the opportunities of both sides, the citizen and the government and close participation of citizens in social-political processes was provided. This meant the establishment of the opportunity of expressing the thoughts by citizens in any level of decision-making and the not worthy increase of transparency.

The following are related to e-democracy mechanisms [46, 47]:

- Electronic voting (voting with mobile phone, Internet-elections, etc.);
- Online collective discussion mechanisms of subjects with social-political content and socially important issues;
- The forming mechanisms of online communities, groups, social networks;
- The mechanisms of realization of citizens' incentives;
- Citizens' control mechanisms on public authorities performance, etc.

While the organization of election process in agricultural, industrial societies encounters the major organizational, financial and other problems, there was no alternative of democracy based on quantity for the reasons of information distribution, the difficulty of operative interactive relations between electorate and candidates. In the democracy based on the quantity, the age and the living address of elector was mainly considered, that is, his intellect, the parameters characterizing his way of thinking were not considered. In other words, the votes of electors were not distinguished.

In knowledge society the result of election can be calculated by considering the general intellect (IQ) of electorate and candidates. Hence, the application of privilege of votes of elector with higher intellect coefficient over the elector with lower intellect coefficient is a possible alternative.

It is worth to mention that the government is a bottom-up, multistage superstructure consisting of legal relations system among people (vertical relations system). The civil society a relations system unlimitedly builds among people (horizontal relations system). On the assumption of this principle, the internal relations system of the society is described in Figure 5.4.

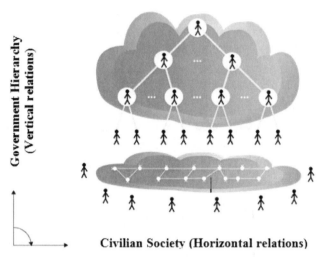

FIGURE 5.4   Internal relationships system of the society.

Figuratively speaking, the main currency of the democracy is information and communication. In the presence of those, the citizens are self-organized, start to govern themselves and digital citizen is formed. Social networks, blogs and etc. play a prominent role in forming of civil society. As the civil society, that is, horizontal relations system is being shaped, self-governance opportunities (municipalities, NGO's, etc.) are created, that is, it takes some functions of the government. Other functions are carried out by business sector.

Alongside with mentioned above, the transition to Information process does not requires only an automatization of existing processes in government management, but also their re-building based on particular interests of citizens and a group of interests of the society. Considering those principles, nowadays, the direct e-democracy principles are not sufficiently supported by business sector. E-democracy concept has several inconsistencies and is reasonably criticized. Hence, recently-government concept is dominant in socio-political and scientific literature, which is the basis for carrying out the reforms in government management sphere by means of ICT.

5.7   EFFECTIVE E-GOVERNMENT MANAGEMENT MECHANISMS: CONCEPTUAL APPROACHES

Application of web content mining, Web analytics and social networks is a strong asset for improvement of e-government management effectiveness and creation of feedback mechanism. Considering that formation of effective administration policy is one of the important issues for implementation of e-government programs, then application of especially new technologies significantly broadens the political communication capabilities. This allows achieving new integration forms among government, business sector and citizens.

5.7.1 WEB ANALYTICS AS A FEEDBACK MECHANISM

Increasing role of information and information technologies leads to development of e-democracy, information economy, e-government, e-governance, digital markets etc., for example, concept of information society implicates method of conduction of information aspects of government actions. Realization of e-government concept implicates integration of many functional elements, including local administration institutions.

Within the framework of growing informatization of government administration, provision of accessible information to population and business is one of the most important functions of local administration institutions. Provision of information services is carried out through development and competent support of website of local administration institution [25, 28, 34, 49, 50]. Multiple problems of sphere of local administration institutions' websites stipulate relevance of scientific research in this field. Web analytics – is monitoring of web site operation, based on which audience of visitors is determined and behavior of web- audience is studied for making decisions on development and expansion of functional capabilities of web-resource [28]. Application of Web analytics methods is necessary for many aspects of site development, such as [28, 34]:

- Development of functionality of the site based on tendencies in behavior of visitors;
- Evaluation of effectiveness of internet-marketing;
- Detection of problematic points in information architecture, design and content of site.
- Formulation of recommendations for improvement of resource and relevant optimization.

Web analytics includes following methods: website traffic analysis; usability analysis, visitor behavior analysis; benchmarking; expert evaluation.

Effective web analytics – is detailed analysis of behaviors and navigations of users on a website and portal. It is impossible to make measures regarding optimization of a site without a precise knowledge of what users do on the site [28, 37, 51, 41]. Optimization and administration of web resources is an important condition for adaptation of a site in accordance with demands of users and citizens. Effective web analytics of sites and portals of government institutions, as well as web sites rendering online service to citizens, is discovering of existing program, technical, content errors and their adaptation to requirements of citizens, users [30, 52, 53]. It's possible to determine the reason users leave the web site, generally behavior of users related to site or websites of a certain service using web analytics. It is clear that, web are not limited to certain statistics, and allow to obtaining more detailed information for analysis [41, 53–55]. Otherwise, web usage mining allows understanding following [28, 30, 37, 53–56]:

- Quantity of browsed web-pages,
- Keywords and phrases which visitors use to find the site in search systems,
- Geography of visitors,
- Time, spent on the web-page by visitor,
- Passages between web-pages,
- Audience of the site (accidental, permanent visitors etc.),
- Site navigation convenience for visitors, etc.

Also, it must be noted that, processing of technologies of decision making in online environment, studying public opinions, conductions of referendums and elections, as well as creation of situation center for each institution included in e-government are one of the quite important issues. Considering that such situation centers are created based on computers, development of mutual connection among them in accordance with safety policy requirements can be considered as one of the most important duties. Analysis of log files collected from servers and information collected from e-mails during development of on-line connections of citizens with administration institutions plays an important role while making effective decisions by subjects of e-government as a result of use of Web analytics. In its turn, this allows operation of feedback mechanisms for management of e-government.

5.7.2 SOCIAL NETWORKS AS A GOVERNMENT ADMINISTRATION MECHANISM

Social networks (SN) form an integral part of our lives, uniting people with common interests in internet-environment. Currently virtual networks supporting technology such as blogs and Wiki compose the majority of websites. Let's note that, term SN is currently often used in sociology. Network modeling of social processes represents private philosophy for data analysis. In its turn, this allows uniting different mathematical approaches, statistics, system analysis, imitation models in single organism for solution of social tasks. Network analysis facilitates modeling of structural interrelation among social individuals, people, staffs or organization [56, 57]. In relation to this, let's note that, SN analysis if currently widely used in economics, management, medicine, criminology and other fields. Currently, development and sufficient popularization of SN among users promotes their use in different field. From this standpoint, application of SN during the process of creation and management of e-government is one of the relevant issues of our day. Application of SN in processes of creation of information society, as well as transformation and development of government administration mechanisms allows developing effective solution methods.

Position of users is determined during the process of web analysis, which has a special value for determination of their role in the network. Several principal stages can be demonstrated during the process of web analysis. Each stage solves its specific methodological task. Firstly, indicators characterizing specifics of network connections are developed. Based on that, a survey is created for network users. Collected data is analyzed and the network is established. Methods measuring Indicators, reflecting the structural characteristics of the network are developed on the next stage. In such manner, SN is considered as the key element for consecutive development of the society, and shortly they will become an inseparable part of any activity sphere. Multiple general-director or niche social services occur every year, and active users of Internet typically have several profiles in different social networks. Discovering profiles, belonging to the same person, in several social networks, allows to obtain more complete social graph, which can be useful in many tasks, such as information search, effective administration, recommendation systems etc. The most significant work for user identification is Veldman [58], which represents many heuristics using information on profiles, as well as connections among them and similar researches are described [31, 57, 59, 60]. In works Motoyama et al. [59] compared a couple of profiles between the "Facebook" network and "MySpace," in its turn Gaewon et al. [57] solved the analogical task for "Twitter" and "EntityCube" [57, 59]. In works Raad et al. [60] in their research generated accidental social graphs with accidentally formed profiles and applied multiple complex heuristics to them, in order not to miss a single potentially useful information source, accessible in social network [60]. In the work of Vosecky et al. [31], profiles from "Facebook" and "StudiVZ" are presented as vectors of factors, to which subsequently operations of precise, partial and fuzzy comparison were applied to, and identification was carried out in accordance with their results [31]. Application of SN in conducted researches is considered as a new direction for effective e-government management, which is based on results on analysis of massive e-documents and monitoring of information environment. Proposals for adoption of administrative solutions are formed on data analysis level and based on obtain information; and conclusion is provided. Decision making in accordance with obtained information is the conclusive stage after report preparation based on feedback as a result of conduction of evaluation measures.

5.8   CONCLUSION

During review of practice of leading countries it is revealed that, existing e-government projects have different objectives and different models, conceptual approaches are suggested by executive institutions, organizations for the development of e-government. By considering this fact, the inspection of research conducted in direction of e-government establishment in international practice is remarkably necessary.

It is clear that, having covered all spheres of the society, the scientific-research works on e-government are timely for several scientific spheres. From this point of view, development of scientific-theoretical principles of forming of e-government is of great importance. By considering the international practice in research, some up-to-date scientific-theoretical problems of forming of e-government have been researched. Some conceptual and architectural principles of e-government forming were inspected and some suggestions were made. The assessment and monitoring of forming of e-government, intellectual analysis of web-resources and the provision of information security, e-democracy and digital citizen issues were researched, conceptual approaches and solutions were suggested. Important research directions were specified by considering the main principles of e-government theory.

ACKNOWLEDGMENT

This work was supported by the Science Development Foundation under the President of the Republic of Azerbaijan – Grant № EİF/ GAM-2-2013-2(8)-25/03/1.

KEYWORDS

- **content mining**
- **digital citizen**
- **electronic democracy**
- **electronic government**
- **information security**
- **information society**
- **intellectual analysis**
- **public administration**
- **social networks**
- **transformation**
- **web analytics**

REFERENCES

1. Alhomod, S.M. and Shafi, M.M. Best Practices in E-government: A review of Some Innovative Models Proposed in Different Countries. International Journal of Electrical and Computer Sciences. 2012, vol. 12, No 01, pp. 1–6.

2. Definition of E-Government. World Bank. 2002, www.worldbank.org (accessed September, 2014).

3. Yildiz, M. E-government research: Reviewing the literature, limitations, and ways forward. Government Information Quarterly, 2007, No 24, pp. 646–665.

4. Fang, Z. E-Government in Digital Era: Concept, Practice, and Development. International Journal of The Computer, The Internet and Management, 2002, vol. 10, No.2, pp. 1–22.

5. Heeks, R., Bailur, S. Analyzing e-government research: Perspectives, philosophies, theories, methods, and practice. Government Information Quarterly, 2007, No 24, pp. 243–265.

6. Vulić, M., Dadić, J., Simić, K. et al., CRM e-government services in the cloud. www.fos.unm.si (accessed September, 2014).

7. Lowery, L. M. Developing a Successful E-Government Strategy. http://unpan1.un.org (accessed September, 2014).

8. Godse, V., Garg, A. From E-Government to E-Governance. ICEG'2007, 2007, pp. 13–20.

9. Jain Palvia, S. C., Sharma, S. S. E-Government and E-Governance: Definitions/Domain Framework and Status around the World. ICEG'2007, 2007, pp. 1–12.

10. Kovalev, M. Development of electronic government considering the international experience. Transactions Banking, Cherven, 2006, pp. 16–25.

11. Zouridis, S., Thaens, M. E-Government: Towards a Public management Approach. Asian journal of public administration, 2003, vol. 25, No 2, pp. 159–183.

12. Alguliev, R. M., Yusifov, F. F. Electronic governance as transformation technology of public management. Proceedings of the 3rd International Conference on Application of Information and Communication Technologies (AICT'2009), Baku, 2009, pp. 1–5

13. Gordon, T. F. e-Governance and its Value for Public Administration, www.tfgordon.de/publications/Gordon2004a.pdf (accessed September, 2014).

14. Joseph, R. C., Kitlan, D. P. Key Issues in E-Government and Public Administration, https://irma-international.org (accessed September, 2014).

15. Snead J. T., Wright E. E-government research in the United States, Government Information Quarterly, 2014, No 31, pp. 129–136.

16. E-Government Act of 2002, U.S.A., www.gpo.gov (accessed September, 2014).

17. Gartner company, www.gartner.com (accessed September, 2014).

18. Nograšek, J. Change Management as a Critical Success Factor in e-Government Implementation. Business Systems Research Journal, 2011, vol.2, No.2, pp. 1–56.

19. ICT for Government and Public Services, European Commission. http://ec.europa.eu (accessed September, 2014).

20. Pardo, T. A., Jiang Y. Electronic Governance and Organizational Transformation. ICEGOV'2007, 2007, pp. 99–107.

21. Traunmüller, R., Leitner, C. e-Government: State and Perspectives. ICEGOV'2008, 2008, pp. 4–7.

22. Potnis, D. D. Measuring e-Governance as an innovation in the public sector. Government Information Quarterly, 2010, No 27, pp. 41–48.

23. Koh, C. E., Prybutok, V. R., Zhang, X. Measuring e-government readiness. Information and Management, 2008, No 45, pp. 540–546.

24. The United Nations E-Government Survey 2014: E-Government for the Future We Want, www.unpan.org (accessed September, 2014).

25. Global Information Technology Report 2014. www.weforum.org (accessed September, 2014).
26. Measuring the Information Society 2012. www.itu.int (accessed September, 2014).
27. eGovernment Benchmark Framework 2012–2015. http://ec.europa.eu (accessed September, 2014).
28. Kaushik, A. Web Analytics 2.0 – The Art of Online Accountability and Science of Customer Centricity. Wiley Publishing, Inc. 2010, 447 p.
29. Alguliyev, R. M., Aliguliyev, R. M., Yusifov, F. F. Automatic Identification of the Interests of Web Users, Automatic Control and Computer Sciences, 2007, vol. 41, No 6, pp. 320–331.
30. Liu, H., Keselj, V. Combined mining of web server logs and web contents for classifying user navigation patterns and predicting users' future requests. In: Data and Knowledge Engineering, 2007, vol. 61, No 2, p. 304–330.
31. Vosecky, J., Dan Hong, Shen, V. Y. User identification across multiple social networks. Proceedings of First International Conference on Networked Digital Technologies. 2009, pp. 360–365.
32. JICWEBS (The Joint Industry Committee for Web Standards in the UK and Ireland). www.jicwebs.org (accessed September, 2014).
33. ABCe (Audit Bureau of Circulations electronic, UK and Europe). www.abc.org.uk (accessed September, 2014).
34. Web Analytics Association, www.webanalyticsassociation.org (accessed September, 2014).
35. Interactive Advertising Bureau, www.iab.net/about_the_iab (accessed September, 2014).
36. Stumme, G. Hotho, A., Berendt, B. Usage mining for on the Semantic Web. 2004, www.kde.cs.uni-kassel.de/stumme/papers/2004/berendt04usage.pdf (accessed September, 2014).
37. Cooley, R. The Use of Web Structure and Content to Identify Subjectively Interesting Web Usage Patterns. ACM Transactions on Internet Technology, 2003, vol. 3, No. 2, pp. 93–116.
38. Meo, R., Lanzi, P. L., Matera, M. Integrating Web Conceptual Modeling and Web Usage Mining. Proc. of International Workshop on Web Mining and Web Usage Analysis, ACM Press. U.S.A., 2004. http://maya.cs.depaul.edu (accessed September, 2014).
39. Spiliopoulou, M., Pohle, C., Teltzrow, M. Modeling and Mining Web Site Usage Strategies. In Proceedings of the Multi-Konferenz Wirtschaftsinformatik, Nurnberg, Germany, 2002. http://www.iw.uni-karlsruhe.de (accessed September, 2014).
40. Abraham, A. Business Intelligence from Web Usage Mining. Journal of Information & Knowledge Management, 2003, vol. 2, No. 4, http://arxiv.org/pdf/cs.AI/0405030. (accessed September, 2014).
41. Balog, K., Hofgesang, P. I., Kowalczyk, W. Modeling Navigation Patterns of Visitors of Unstructured Websites. The Twenty-fifth SGAI International Conference on Innovative Techniques and Applications of Artificial Intelligence (Springer Verlag). 2005, www.cs.vu.nl/ci/DataMine/DIANA/pages_eng/publications.html (accessed September, 2014).
42. Shadbolt, N., Berners-Lee T., Hall, W. The Semantic Web Revisited. IEEE Intelligent Systems. 2006, 21(3), pp. 96–101, http://eprints.ecs.soton.ac.uk/12614/01/Semantic_Web_Revisted.pdf (accessed September, 2014).

43. Norguet, J.P., Zimanyi, E. and Steinberger, R. Semantic Analysis of Web Site Audience. Proceedings of the 2006, ACM Symposium on Applied Computing. 2006. http://code.ulb.ac.be/dbfiles/media260.pdf (accessed September, 2014).

44. Creation of a Global Culture of Cyber Security. 2002. www.un.org (accessed September, 2014).

45. Global Cybersecurity Agenda. 2008. www.itu.int (accessed September, 2014).

46. Anttiroiko, A.-V. Building Strong E-Democracy – The Role of Technology in Developing Democracy for the Information Age. Communications of the ACM vol. 2003, 46, No. 9, pp. 121–128.

47. Meier, A. eDemocracy and eGovernment. Springer-Verlag. Berlin, Heidelberg, 2012

48. Hilbert, M. The Maturing Concept of E-Democracy: From E-Voting and Online Consultations to Democratic Value Out of Jumbled Online Chatter. In: Journal of Information Technology & Politics, 2009, vol. 6, pp. 87–110.

49. Wenhua, X., Jian, Y. E-government and the change of government management mode. In: International Conference on E-Business and E-Government, 2010, pp. 675–678

50. Fong, S., Meng, H.S. Web-based Performance Monitoring System fore-Government Services. In: ICEGOV2009, Bogota, Colombia, 2009.

51. Baglioni, M., Ferrara, U., Romei1, A., Ruggieri, S., and Turini, F. Preprocessing and Mining Web Log Data for Web Personalization. In: Proccedings of the 8th Congress of the Italian Association for Artificial Intelligence. Pisa, Italy, 2003, vol. 2829, pp. 237–249.

52. Everts, J., Bulacu, M. Assignment: Clustering of Web Users. Groningen University. Netherlands. 2005, http://www.ai.rug.nl/ki2/assignments/ki2-assig03.pdf (accessed September, 2014).

53. Wanga, X., Abraham, A., Smitha, K. A. Intelligent web traffic mining and analysis. In: Journal of Network and Computer Applications, 2005, vol. 28, p. 147–165.

54. Smith, K., Ng, A. Web page clustering using a self-organizing map of user navigation patterns. In: Decision Support Systems. 2003, vol.35, No 2, p. 245–256.

55. Qiu, F., Cho, J. Automatic identification of user interest for personalized search. In: Proccedings of the 15th WWW Conference (WWW15). Edinburgh, Scotland, 2006, p. 727–736.

56. Bouquet, P., Bortoli, S. Entity-centric Social Profile Integration. In: Proceedings of the International Workshop on Linking of User Profiles and Applications in the Social Semantic Web (LUPAS 2010). 2010, pp. 52–57.

57. Gae-Won, Y., Seung-Won, H., Zaiqing, N., Ji-Rong, W. Social Search: Enhancing Entity Search with Social Network Matching. EDBT, 2011, http://ids.postech.ac.kr/papers/socialsearch.pdf (accessed September, 2014).

58. Veldman, I. Matching Profiles from Social Network Sites. Master's thesis, University of Twente, essay.utwente.nl/59436/1/scriptie_I_Veldman.pdf (accessed September, 2014).

59. Motoyama, M., Varghese, G. I Seek You – Searching and Matching Individuals. In: Social Networks. WIDM '09: Proceeding of the eleventh international workshop on Web information and data management. http://cseweb.ucsd.edu/~mmotoyam/widm09-iseekyou.pdf (accessed September, 2014).

60. Raad, E., Chbeir, R., Dipanda, A. User Profile Matching in Social Networks. In: 13th International Conference on Network-Based Information Systems (NBiS), 2010, http://hal.archives-ouvertes.fr (accessed September, 2014).

CHAPTER 6

TECHNOLOGY ACCEPTANCE OF E-SERVICES IN INDIA: A PERCEIVED RISK PERSPECTIVE

ASHIS PANI[1] and RAKESH TIWARI[2]

[1]*Chairman, Information Systems Area, XLRI Jamshedpur, Jharkhand, 831001, India, Tel: +91-0657-3983144, Fax: +91-65722-27814, E-mail: akpani@xlri.ac.in*

[2]*Senior Manager, SSP India Private Limited, Gurgaon, Haryana, 122002, India; Research Fellow, Information Systems Area, XLRI Jamshedpur, Jharkhand, 831001, India, Tel: +91-99711-45616, E-mail: r10014@astra.xlri.ac.in*

CONTENTS

6.1   ABSTRACT

Aiming to improve public services delivery and information dissemination to public, governments all over the world have increased the use of Information and Communication Technologies, with the developed nations being at the advanced stage of e-governance implementation and developing nations having increased their focus only in the last decade.

India too has made significant progress in the e-governance technical implementation, utilizing the massive in-house world class technical resource pool, professional civil servants and committed leadership. Still, the adoption of e-governance is lagging, partly due to huge digital divide and largely due to issues related to the adoption of technology. This study examines the factors influencing the citizen's adoption of e-government services, focusing on the e-filing, the flagship implementation of e-governance in India having a huge potential for cost savings.

The Unified Theory of Acceptance and Use of Technology (UTAUT) model is extended to include Perceived Risk and system expertise.

The result supports that the system expertise has impact on effort expectancy and the Perceived Risk has impact on performance expectancy. The tax payers are uncomfortable providing personal information without face to face contact. The government, thus, has to improve the performance expectancy, effort expectancy and provide confidence to tax payers about the safe and risk free system and explore on the mechanism to improve the access and expertise of internet to larger mass.

6.2   INTRODUCTION

6.2.1   ABOUT E-GOVERNMENT

As per the World Bank's definition, "E-Government" refers to the use by government agencies of information technologies (such as Wide Area Networks, the Internet, and mobile computing) that have the ability to transform relations with citizens, businesses, and other arms of government. These technologies can serve a variety of different ends: better delivery of government services to citizens, improved interactions with business and industry, citizen empowerment through access to information, or more efficient government management. The resulting benefits can be less corruption, increased transparency, greater convenience, revenue growth, and/or cost reductions.

In India, it started with computerization of Government departments and gradually evolved into full scale e-governance with the introduction of

National e-Governance Plan (NeGP) in mid-2006 with vision of making all Government services accessible to the common man in his locality, through common service delivery outlets, and ensure efficiency, transparency, and reliability of such services at affordable costs to realize the basic needs of the common man.

6.2.2   ABOUT E-FILING

It is a legal obligation of every person, whose total income for the previous year has exceeded the maximum amount that is not chargeable for income tax under the provisions of the I.T. Act, 1961, to file the Income Tax returns.

Income Tax Department has introduced a convenient way to file these returns online using the Internet. The process of electronically filing the Income tax returns through the Internet is known as e-filing of returns.

E-filing of income tax was introduced in the year, 2004 via Electronic Furnishing of Return of Income Scheme, 2004, notified Vide Notification No. 253 of, 2004, dated 30th Sep., 2004 S.O.1073(E). Digital signature, e-return, e-return intermediary were later introduced in this scheme.

As per the site of Income tax department of India (https://incometaxindiaefiling.gov.in) the electronically filing of tax returns through the internet can be done in three ways.

Option 1: Use digital signature in which case no paper return is required to be submitted

Option 2: File without digital signature in which case ITR-V form is to be filed with the department. This is a single page receipt cum verification form.

Option 3: File through an e-return intermediary who would do e-Filing and also assist the Assesse file the ITR V Form.

6.3   PROBLEM STATEMENT

As per the OECD statistics, most of the developed nations have a very high rate of e-filing with countries like Singapore (96%), Australia (92%), Netherlands (95%) and Korea (87%) leading the race (2013). The complete list of the percentage of income tax returns files online is presented in Appendix Section 11. In, 2011, the number of tax payers who filed income tax returns through e-filing in India is 13.1 million (Income Tax, 2014). The detailed break-up of the income tax statement is presented in Appendix Section 12. With close to 40 million tax payers, 68% of taxes are filed manually and is thus error prone. It also requires extraordinary effort for the Income tax department to enter the data from hard copies into the system. As e-filing brings enormous potential to benefit both the government and the taxpayers, it is important to have majority of the tax-payers e-file.

While the implementation of e-filing has been simplified over years, the growth in the number of e-filing didn't increase proportionally. This requires understanding of the tax payer's perception towards the service.

To understand the user's perception towards e-filing, many studies, (e.g., Fu, J.R., Farn, C.K., Chao, W.P. 2006, Ludwig Christian Schaupp, Lemuria Carter, Jeff Hobbs. 2009), have taken place in developed nation, but there have been very limited studies in India.

Due to its unique social structure, it becomes important that a study specific to India is carried out considering the social aspect of technological acceptance.

6.4 LITERATURE REVIEW

Understanding why people accept or reject computers has proven to be one of the most challenging issues in the IS research (Swanson, 1988). To better predict, explain and increase user acceptance we need to understand why users accept or reject information systems. Investigators have studied impacts of user's internal belief, and attitudes on their usage behavior (DeSanctis, 1983; Ginzberg, 1981; Robey, 1979; Schultz and Slevin, 1975; Swanson, 1974, 1987) and how these internal beliefs and attitudes are, in turn, influenced by various external factors including: system's technical design characteristics (Benbasat and Dexter, 1986; Benbasat, Dexter and Todd, 1986); user involvement in system development (Baroudi, Olson and Ives, 1986; Franz and Robey, 1986); the type of system development process used (Alavi, 1984; King and Rodriguez, 1981); the nature of implementation process (Ginzberg, 1978; Vertinsky, Barth and Mitchell, 1975; Zand and Sorensen, 1975); and cognitive style (Huber and George, 1983).

Information Systems (IS) investigators have suggested intention models from social psychology as a potential theoretical foundation for research in the determinants of user behavior (Swanson, 1982, Fishbein and Ajzen's, 1975, Ajzen and Fishbein, 1980) theory of reasoned action (TRA) is an especially well researched intention model that has proven successful in predicting and explaining behavior across wide variety of domains. TRA is very general, "designed to explain virtually any human behavior" (Ajzen and Fishbein, 1980, p4) and should therefore be appropriate for studying the determinants of computer usage behavior as a special case.

Although numerous individual, organizational and technological variables have been investigated (Benbasat and Dexter, 1986; Franz and Robey, 1986), research has been constrained by the shortage of high-quality measures for key determinants of user acceptance (Davis et al., 1989).

Technology acceptance has been one of the most explored topics in IS Research for decades. The rationale behind the research is that in order to get the intended benefit from the IT investment, the users should accept and use the system.

Most prominent Information System theories, which were considered for this study, are described in the following subsection.

6.4.1 THEORY OF REASONED ACTIONS

The theory of reasoned action (Figure 6.1) (Ajzen & Fishbein, 1980) was first introduced in, 1967 by Martin Fishbein in an effort to understand the relationship between attitude and behavior. It is based on the assumption that human beings are rational and make systematic use of available information. People consider the implications of their actions before they decide whether or not to perform a given behavior. This theory is inspired by Social psychology and postulates that the behavior of an individual is driven by behavioral intentions which are function of two core constructs attitude toward behavior and subjective norm. Positive or negative feeling about performing behavior is termed as attitude while individual's perception of whether people important to the individual think the behavior should be performed is the subjective norms.

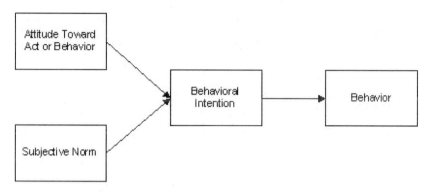

FIGURE 6.1   Theory of reasoned actions (Source: Fishbein, Ajzen, 1975).

6.4.1.1   Attitude Towards Behavior

People's likelihood of performing a given behavior will be strong if they hold a favorable attitude towards the performance of that behavior.

The relationship between behavior and intentions, is illustrated

$$B \approx I = f\{w_1 A_b + w_2 SN\}$$

where, B = behavior; I = intention to perform the behavior; A_b = attitude towards performing the behavior; SN = subjective norm with regard to the behavior; W_1 = weight (relative importance) of the attitude; W_2= weight (relative importance) of the normative component.

The first determinant of behavioral intention, attitude towards the behavior, is determined by a person's beliefs regarding the consequences of performing the behavior weighed against evaluation of these consequences. These beliefs, which underlie a person's attitude towards a given behavior, are termed behavioral beliefs (Ajzen & Fishbein, 1980; Montano & Kasprzyk, 2002). Beliefs are defined as the individual's subjective probability that performing the target behavior will result in specific consequence. Thus, a person who holds a belief that positively valued outcomes will result from performing a behavior (for example, smoking leading to socializing) will have a more positive attitude towards the behavior than one who has a strong belief that negatively valued outcomes will result.

The expectancy-value relationship between attitude and behavioral beliefs can be summarized below:

$$A_b = f\left\{\sum_i b_i e_i\right\}$$

where, A_b = attitude towards performance of a behavior; b = belief that performance of the behavior will lead to outcome i; e = evaluation of outcome i.

6.4.1.2 Subjective Norm

The second determinant of behavioral intention, subjective norm, refers to a person's perception of the social pressures to perform or not to perform a particular behavior. The subjective norm is determined by whether important referents approve or disapprove of the performance of a behavior, weighted by his/her motivation to comply with these expectations. These beliefs, which underlie a person's subjective norm, are termed normative beliefs.

The relationship between subjective norm and normative beliefs in

$$SN = f\left\{\sum_j b_j m_j\right\}$$

where, SN = subjective norm; b = normative belief that referent j thinks "I should or should not perform" a behavior; m = motivation to comply with referent j.

6.4.2 TECHNOLOGY ACCEPTANCE MODEL

Based on the theory of reasoned action (Figure 6.2) (Fishbein and Ajzen, 1975), Davis (Davis, 1989, Davis et al., 1989) adapted the causal chain to specifically predict the user acceptance of IT. Technology acceptance model (TAM) postulates that perceived usefulness and Perceived Ease of Use decide the individual's Intention to Use a system. Perceived usefulness is defined as "the degree to which a person believes that using a particular system would enhance his or her job performance" (Davis, 1989, p 320).

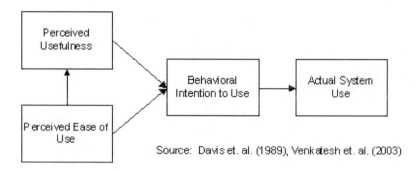

FIGURE 6.2   Technology acceptance model.

Perceived Ease of Use is defined as "the degree to which a person believes that using a particular system would be free of effort" (Davis, 1989, p 320).

Adoption of innovations research also has mention of Perceived Ease of Use in the form of complexity, defined by Rogers and Shoemaker (1971) as "the degree to which an innovation is perceived as relatively difficult to understand and use" (p 154). The meta-analysis of the relationship between the characteristics of an innovation and its adoption by Tornatzky and Klein (1982) found compatibility, relative advantage, and complexity having significant relationships across a broad range of innovation types.

6.4.3   THEORY OF PLANNED BEHAVIOR

Theory of Planned Behavior (TPB) (Figure 6.3) (Ajzen, 1985) postulates that the behavioral intentions are dependent upon attitude toward behavior, subjective norm and perceived behavioral control. The definitions of attitude toward behavior and Subjective norm were adapted from TRA. Behavioral control is defined as one's perception of the ease or difficulty of performing a behavior (Ajzen, 1991, p.188).

The extension was based on the idea that behavioral performance is determined by motivation (intention) and ability (behavioral control). According to Ajzen, actual behavioral control should be distinguished from perceived behavioral control. The latter, which is what distinguishes the theory of planned behavior from the theory of reasoned action, refers to people's perception of the ease or difficulty of performing a given behavior.

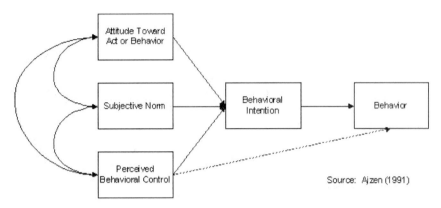

FIGURE 6.3   Theory of planned behavior.

Montano and Kaspyzyk (2002) regard perceived behavioral control as being similar to Bandura's concept of self-efficacy, which refers to an individual's judgment of how well he can perform a behavior under various conditions. Bandura (1977, 1982) proposed that there are two types of expectancies that influence people's decisions to engage in a particular behavior. These are efficacy expectancies and outcome expectancies. Efficacy expectancies refer to people's confidence in their ability to perform a behavior. Thus, people have positive efficacy expectancies if they are confident in their ability to perform a behavior which, in turn, motivates them to carry out the behavior. On the other hand, people may be reluctant to engage in a behavior if they doubt their ability to perform it (a negative expectancy scenario). Outcome expectancies refer to people's perception that the performance of a behavior will lead to a desired outcome. Thus, people will be reluctant to perform a behavior if they believe that the performance will not result in a desired outcome (negative outcome expectancy).

In trying to predict behavior, one would have to not only assess intentions, but also obtain an estimate of the extent to which individuals are apt to exercise control over the behavior in question. As Equation below illustrates, the strength of a person's attempt to perform a behavior (B_I) interacts with the degree of his/her control (C) to determine the likelihood of the actual performance of the behavior (B).

$$B \propto B_I * C$$

This implies that the harder the person tries, and the greater his/her control over personal and external factors that may interfere, the greater the likelihood that he/she will achieve his/her behavioral goal.

6.4.4 UNIFIED THEORY OF ACCEPTANCE AND USE OF TECHNOLOGY

Unified Theory of Acceptance and Use of Technology (UTAUT) (Figure 6.4) (Venkatesh, 2003) is the integration of eight prominent theoretical models – the theory of reasoned action (Davis et al., 1989), the technology acceptance model (Davis, 1989), the motivational model (Davis et al., 1992), the theory of planned behavior (Ajzen, 1991), a model combining the technology acceptance model and the theory of planned behavior (Taylor and Todd, 1995), the model of PC utilization (Thompson et al., 1991), the innovation diffusion theory(Rogers, 1995), and social cognitive theory (Compeau and Higgins, 1995).

The model comprises of four determinants: performance expectancy, effort expectancy, social influence, and facilitating conditions.

Performance expectancy is defined as the degree to which an individual believes that using the system will help him or her to attain gains in job performance and is the strongest predictor of intention. Effort

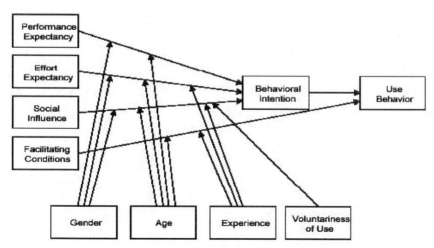

FIGURE 6.4   Unified Theory of User Acceptance (Source: Venkatesh et al.,, 2003).

expectancy is defined as the degree of ease associated with the use of the system. Social influence is defined as the degree to which an Individual perceives that important others believe he or she should use the new system. Facilitating conditions are defined as the degree to which an individual believes that an organizational and technical infrastructure exists to support use of the system.

6.5   RESEARCH MODEL

This study integrates the Unified Theory of Acceptance and Use of Technology (UTAUT) with Perceived Risk and system expertise to explore the adoption of e-filing services in India.

Perceived Risk (PR) is included in various studies of e-government (e.g., Carter et al., 2008; Fu et al., 2006) which resulted in contradictory results. In one such study (Fu et al., 2006), the Perceived Risk significantly influenced the behavioral intention of taxpayers while in other (Carter and Belanger, 2005) the Perceived Risk did not significantly impact the intentions. The effect of Perceived Risk and technology type on user's acceptance of technologies was studied using UTAUT as base and refining it to consider Perceived Risk on it (IlIma, Yongbeom Kim 2005b; Hyo-Joo Han, 2005c).

Some studies like Featherman and Pavlou (2003) have found that perceived overall risk plays an important role in adoption-related behaviors and integrated this construct into the technology acceptance model. Specifically, Perceived Risk was found to be a significant inhibitor of perceived usefulness (PU) and adoption intention. In the study of Featherman and Wells (2004), similar effects of Perceived Risk on PU and behavioral intention (BI) regarding e-services were observed. Therefore, in parallel to these studies, it is reasonable to assume that Perceived Risk has negative impact on performance expectancy.

In addition to the Perceived Risk, the system expertise will also be considered for impact on the intentions. The system expertise for this study is defined as perceived ability to apply the computer and tax knowledge. The support for social cognitive theory perspective on computing behavior is found in (Compeau and Higgins, 1995) where system expertise is found to play an important role in shaping individual's feeling and behavior.

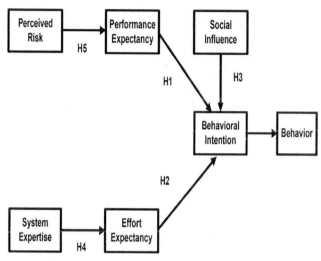

FIGURE 6.5   Research model.

Individuals with high self-efficacy used computers more, derived more enjoyment from their use and experienced less computer anxiety. Similar findings have been observed in many studies (Eastin and LaRose, 2000; Burkhardt, and Brass, 1990; Gist, Schwoerer, and Rosen, 1989). It is found that more technologically advanced a product, the more important a factor is personal efficacy in the decision to adopt the technology; and that people low in personal efficacy (as compared to people high in personal efficacy) with regard to computers tend to be more easily persuaded by expert communicators to try an advanced software product (Thomas Hill, Nancy D. Smith, and Millard F. Mann, 1986).

Thus, the following hypotheses are proposed:

H1: Performance expectancy of Indian tax payers will have significant influence on Intention to Use.

H2: Effort Expectancy of Indian tax payers will have a significant influence on Intention to Use.

H3: Social Influence of Indian tax payers will have a significant influence on Intention to Use.

H4: System Expertise of India tax payers will have a significant influence on Effort Expectancy.

H5: Perceived Risk of Indian tax payers will have a significant negative effect on Performance expectancy.

6.6   RESEARCH METHODOLOGY

6.6.1   SURVEY INSTRUMENT DESIGN AND MEASUREMENT

The survey instrument is a 5-point Likert scale questionnaire survey divided into two sections. Section A contains questions on tax payer's demography. Section B contains questions to measure tax payer's perception on e-filing system.

With respect to the measurement of the variables, the scale was adapted from the scale developed by Venkatesh Vishwanath, Michael G. Morris and Gordon B. Davis in their famous article on the Unified Theory of Acceptance and Use of Technology.

Each item was rated on a scale of 1 to 5 (Ranged from Strongly Disagree to Strongly Agree). High level of validity was ensured through extensive revision by experts and consultation of prior tested instruments. Based on the feedback the researchers received from the reviewers, any questions that caused confusion or where deemed potentially difficult to understand were dropped or replaced by new understandable items. The questionnaire used for this study is below.

Section A. Background of Respondent
1. Name
2. Age (Years): ( ) <30 ( ) 30–45   ( ) 45+
3. E-filing method: ( ) Manual ( ) Internet
4. Tax return filed personally? ( ) Yes ( ) No
5. Internet Experience (years) ( ) < 5 ( ) > 5
6. Gender: ( ) Male ( ) Female
7. Education: ( ) Undergraduate ( ) Graduate ( ) Post Graduate

Section B (To be answered in 5 Likert Scale)
Effort Expectancy
1. It is easy for me to learn the usage of e-filing system.
2. It is easy to find information required for e-filing.
3. It is overall easy for me to e-file.

Performance Expectancy
1. E-filing system will improve my performance.
2. E-filing system will improve my productivity in income tax filing.

3. E-filing system will enhance my effectiveness.
4. E-filing will help me better manage my returns by providing access to previous year's returns.
5. E-filing system will increase chances of early refund.

Behavior Intent
1. If I have access to the electronic tax-filing system, I intend to use it.
2. My first preference for next year would be e-filing.
3. I would recommend tax preparation software to my relatives and friends.

Perceived Risk
1. Using e-Filing system, I will lose control over the privacy of my personal information.
2. Using e-Filing system my personal information may be used without my knowledge.
3. Using e-Filing system will expose me to hackers.
4. E-Filing system doesn't have security strong enough to protect my account.
5. E-Filing system may not perform well and process data transmission incorrectly.

System Expertise (Computers and Taxation)
1. I use computer for variety of purposes.
2. I often struggle to complete a task using computer for some reasons.
3. I can complete a task using computer if there is no one around to tell me how to do.
4. I can e-file if I can call someone when I get stuck with computer or tax calculation.
5. I can e-file if instructions are provided on the usage of e-filing system and tax calculations.
6. Lot of domain knowledge is required to fill out the tax return.

Social Influence
1. People around me think that I should use tax preparation software.
2. People around me think I am trying to save money by e-filing.
3. People will make fun if I am not able to use e-filing.

6.6.2 POPULATION AND STUDY SAMPLE

Only the salaried person, and not the companies, will be the population while sample will be the salaried person employed in various companies.

The survey will be conducted based on Convenience sampling method with target sample size of 150. The survey will be floated to multiple companies of varying industry, varying size and at different location.

6.6.3 COLLECTION OF DATA

The most appropriate form for data collection is a web based question-naire. The survey was posted on Survey monkey and the link was made available to various friends and forums. In addition, the hard copy of the survey was made available for people without having access to computers. All the data will be gathered into Excel sheet for ease of analysis and then uploaded into SPSS for further analysis.

6.7 DATA ANALYSIS

The aim of the quantitative study is the empirical investigation of the research model and the hypothesis formulated. This section comprises of two parts. First part describes the sample characteristics and the second part verifies the reliability and validity of scale and studies the relationship between constructs and ends with regression analysis.

6.7.1 SAMPLE CHARACTERISTICS

The total response received was 203. Of these, 23 were invalid owing to incompleteness, resulting in final usable response from 180 respondents. The demographic profile of the sample is below.

6.7.1.1 Method of Filing

TABLE 6.1 E-filing Method Distribution

	Frequency	Percent	
Internet	75	41.7	■ Internet
Manual	105	58.3	■ Manual
Total	180	100.0	

In terms of method of filing, 105 (58.3%) respondents have filed e-returns manually while 75 (41.7%) of them have filed through internet.

6.7.1.2 Filed Tax Personally

TABLE 6.2 Tax Filed Personally Distribution

	Frequency	Percent	
No	106	58.9	■ No
Yes	74	41.1	■ Yes
Total	180	100.0	

Around 74 (41.1%) respondents have filed their taxes themselves while 106 (58.9%) did not file their own taxes. These must be using income tax experts or Tax Return Preparers. Tax Return Preparer scheme is provided by Income tax department for providing assistance to tax payers in filing their returns.

6.7.1.3 Years of Experience

TABLE 6.3 Experience Distribution

	Frequency	Percent	
< 5 years	41	22.8	■ < 5 years
> 5 years	139	77.2	■ > 5 years
Total	180	100.0	

With regards to computer experience, 139 (77.2%) respondents have more than 5 years of experience while 41 (22.8%) respondents have less than 5 years of experience. The majority of sample population is thus well versed with computer skills.

6.7.1.4   Age

TABLE 6.4   Age Distribution

	Frequency	Percent	
< 30	29	16.1	
30–45	111	61.7	
45 +	40	22.2	
Total	180	100.0	

As for the age, 29 (16.1%) of the respondents fall in the category of <30 years, 111 (61.7%) fall in the range 30–45 years and the remaining 40 (22.2%) were aged more than 45 years. Thus most of the sample population is relatively young with 76% of population being less than 45 years of age.

6.7.1.5   Gender

TABLE 6.5   Gender Distribution

	Frequency	Percent	
Female	31	17.2	
Male	149	82.8	
Total	180	100.0	

Most of the sample population was males (82.8%).

6.7.1.6   Academic Background

TABLE 6.6   Academic Distribution

	Frequency	Percent	
Undergraduate	5	2.8	
Graduate	43	23.9	
Post Graduate	132	73.3	
Total	180	100.0	

In terms of education, 5 (2.9%) were Undergraduate, 43 (23.9%) were graduates and 132 (73.3%) were post graduates.

6.7.2 SCALE RELIABILITY

Reliability is a measure of consistency of the scale for the construct it is measuring. In order to measure the reliability of the instrument, the Cronbach's coefficient alpha was examined.

Cronbach (1951) defined a way of measuring the consistency by splitting data in two in every possible way and calculating the correlation coefficient for each split. The average of these values is Cronbach's α. Hinton, Brownlow, McMurvay and Cozens have suggested four different points of reliability, excellent ranges (0.90 and above), high (0.70–0.90), high moderate (0.50–0.70) and low (0.50 and below) (Hinton, Brownlow, McMurvay, Cozens, 2004). The high Cronbach alpha coefficients obtained for all the study constructs indicates that they are internally consistent and are measuring the same construct.

This calculation was performed using SPSS 10.0.1.

The scales was refined by deleting items that did not load meaningfully on the underlying constructs and those that did not highly correlate with other items measuring the same construct.

The Table 6.7 below describes the Cronbach's alpha for the various constructs.

TABLE 6.7 Measurements of Reliability

Construct	Items	Cronbach's α
Effort Expectancy	3	0.82
Performance Expectancy	4	0.70
Behavior Intent	3	0.81
Perceived Risk	5	0.92
System Expertise*	2	0.58
Social Influence	2	0.64

*se2 was reverse coded.

6.7.3 FIT STATISTICS

Structural Equation Modeling using LISREL 8.80 (Jöreskog and Sörbom, 2000) was employed to test the hypothesis. As SEM estimates multiple and interrelated dependence relationships (Hair, Anderson, Tatham, & Black, 1998) it is an ideal technique to test the hypotheses.

The model fit is judged based on number of fit indices including chi-square tests, goodness-of-fit index (GFI), the adjusted goodness-of-fit index (AGFI), the normed fit index (NFI), the comparative fit index (CFI), the root mean square residual (RMR), and the root mean square error of approximation (RMSEA). The complete list of fit indices is available in the appendix. We will discuss the important ones in this section.

Chi-square is the traditional measure of overall fit of a model (Byrne, 1998). The norm is that the chi-square adjusted to the degree of freedom (chi-square/df) should be less than 5, whereas the most rigorous criteria have been set at 3, even 2 (Pedhazur and Schmelkin, 1991).

The Goodness-of-Fit statistic (GFI) was created by Jöreskog and Sorbom (2000) as an alternative to the Chi-Square test and calculates the proportion of variance that is accounted for by the estimated population covariance (Tabachnick, and Fidell, 2007). It represents the overall degree of fit; the squared residual from prediction compared with actual data, but is not adjusted for degrees of freedom. A GFI value higher than 0.90 indicates better fit of the model. AGFI is an extension of GFI, adjusted by the ratio of degrees of freedom for the proposed model to the degrees of freedom for the null model (Hair, Anderson, Tatham, & Black, 1998). The recommended acceptance level is a value greater than 0.90.

NFI assesses the model by comparing the $\chi 2$ value of the model to the $\chi 2$ of the null model (Hooper, Coughlan, and Mullen, 2008). Values for this statistic range between 0 and 1 with Bentler and Bonnet (1998) recommending values greater than 0.90 indicating a good fit.

The Comparative Fit Index (Bentler, 1990) is a revised form of the NFI, which takes into account sample size (Byrne, 1998) that performs well even when sample size is small (Tabachnick, and Fidell, 2007). The CFI value lies between 0 and 1.0, and the larger value of CFI indicates higher levels of goodness-of-fit.

The RMR and the SRMR are the square root of the difference between the residuals of the sample covariance matrix and the hypothesized covariance model (Hooper, Coughlan, and Mullen, 1998). Values for the SRMR range from zero to 1.0 with well fitting models obtaining values less than 0.05 (Byrne, 1998; Diamantopoulos, and Siguaw, 2000), however values as high as 0.08 are deemed acceptable (Hu, and Bentler, 1999).

The RMSEA tells us how well the model, with unknown but optimally chosen parameter estimates would fit the population covariance matrix (Byrne, 1998). RMSEA in the range of 0.05 to 0.10 was considered an indication of fair fit and values above 0.10 indicated poor fit. RMSEA between 0.08 and 0.10 provides a mediocre fit and below 0.08 shows a good fit (MacCallum, Browne, and Sugawara, 1996).

Through a series of scale purification processes, final acceptable CFA results were achieved.

All indicators loaded to a respective construct, providing unidimensionality and validity of the measurement.

6.7.4 CONFIRMATORY FACTOR ANALYSIS

Before the analysis of the hypotheses, the confirmatory factor analysis was performed to confirm the validity of each construct.

Confirmatory factor analysis is suggested as a more precise method to test the unidimensionality and validity of the measurements than an exploratory factor analysis and item-total correlations (Gerbing, & Anderson, 1988). CFA measures whether each factor exhibits convergent and discriminant validity. Convergent validity is defined as the agreement among measures of the same factor. Convergent validity is established when a CFA model fits satisfactorily and all factor loadings are significantly and preferably high (Bagozzi, Yi, & Phillips, 1991). Discriminant validity is the distinctiveness of the two conceptually similar constructs (Hair, Anderson, Tatham, & Black, 1998).

The Figure 6.6 is the validity coefficient of the indicator. Number below the coefficient is the standard error of the estimates. Below the standard errors are t-values. A t-value larger than 1.96 is significant.

TABLE 6.8   Fit Statistics

$\aleph^2$	df	$\aleph^2$/df	RMSEA	GFI	AGFI	NNFI	CFI	RFI
274.16	138	1.98	0.074	0.86	0.81	0.93	0.95	0.88

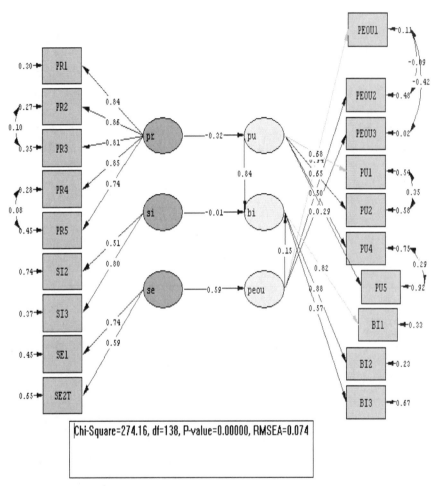

FIGURE 6.6 Path coefficients for research model.

6.7.5 HYPOTHESIS TESTING

The structural equations of the model are below:

pu = –0.32*pr, Errorvar. = 0.90, R^2 = 0.10
   (0.097)   (0.23)
   –3.33   3.92

$$bi = 0.84*pu + 0.15*peou - 0.0072*si, \text{ Errorvar.}= 0.25, R^2 = 0.75$$
$$(0.16) \quad (0.066) \quad (0.081) \quad (0.12)$$
$$5.37 \quad\quad 2.21 \quad\quad -0.088 \quad\quad 2.02$$

$$peou = 0.59*se, \text{ Errorvar.}= 0.65, R^2 = 0.35$$
$$(0.089) \quad (0.23)$$
$$6.63 \quad\quad 2.83$$

The estimated path coefficients appear in front of the * symbol before each variable. The positive values of these coefficients indicate that the contribution of these factors is positive. The figure below the * is the t value and is significant if the t-value is greater than 1.96.

TABLE 6.9 Confirmatory Factor Analysis

Construct	Item Description	Coeff.	t-Value
Effort Expectancy	PEOU1	0.89	
	PEOU2	0.60	5.91
	PEOU3	0.88	6.57
Performance Expectancy	PU1	0.59	5.81
	PU2	0.54	11.17
	PU4	0.37	5.30
	PU5	0.26	3.23
Behaviour Intent	BI1	0.63	5.96
	BI2	0.77	11.41
	BI3	0.47	7.62
Perceived Risk	PR1	0.94	13.37
	PR2	0.98	13.70
	PR3	0.88	12.39
	PR4	0.79	13.46
	PR5	0.70	11.07
System Expertise	SE1	0.62	8.97
	SE2T	0.77	7.23
Social Influence	SI2	0.50	4.74
	SI3	0.78	5.58

TABLE 6.10   Summary of Hypotheses Results

Hyp.	Path	Coeff.	t-Val	Result
H1	Performance expectancy → Intention	0.84	5.37	Supported
H2	Effort Expectancy → Intention	0.15	2.21	Supported
H3	Social Influence → Intention	0.0072	−0.088	Not Supported
H4	System expertise → Effort Expectancy	0.59	6.63	Supported
H5	Perceived Risk → Performance expectancy	0.32	−3.33	Supported

6.8   CONCLUSION

Performance expectancy, effort expectancy and Perceived Risk had significant impact on the Intention to Use. This means that to encourage the use of technology, it is important to consider the effort and performance of the system and eliminate the perception of risk associated with the use of technology.

It is also established that higher the risk perceived of the system, lower will be the performance expectancy. Similarly, more the ease associated with the system, lesser the system will be perceived as risky.

Social influence is not found to be having impact on the Intention to Use. This could be due to the bias in sampling. Studies have found that social influence on females were significantly higher than males (Güzin Mazman, Yasemin Koçak Usluel, and Vildan Çevik, 2009) but most of the respondents are in this study are males (82.8%). The UTAUT model suggests that older workers place more emphasize on Social Influence (Venkatesh, Morris, Davis, & Davis, (2003)) whereas majority of the respondents (78%) are less than 45 years. Since majority of the respondents are educated, 97% being graduate or post graduate and have good experience with computers, 77.2% having greater than 5 years of experience, the effect from social influence may be less (Waransanang Boontarig, Wichian Chutimaskul, Vithida Chongsuphajaisiddhi and Borworn Papasratorn, 2012).

Considering the benefit of e-filing, citizens should be encouraged to use e-filing. Government should address the risk perception of citizens. While there has been considerable advertisement in national news papers, this can be extended to local news papers and using radio stations.

From a long-term perspective, the awareness can be best spread by introducing the e-governance in the high school or college curriculum. This will ensure that the new generation will know of the existence, advantages and will encourage complete participation in not just e-filing, but also on other e-governance initiatives.

For the older generation who are averse to technology, the e-filing can still be promoted by organizing camps at the societies and colonies to ensure that the effort expectancy and system expertise factor can be eliminated.

Also, the Perceived Risk is due to the socio political life of contemporary India, where corruption is widespread as evident from the media around us. Even IT and computer technology have not been spared by unethical methodology of tapping and tampering. Hence people prefer to file the returns manually. Cultivation of trust in the government machinery can go long way in inculcating trust and hence reducing the Perceived Risk.

This first limitation of the study is the sample, which was concentrated on IT professionals and government employees. Hence the future research can be conducted to include, non-IT professionals in private sector and individuals having income from business. The second limitation of the study is that it did not incorporate actual behavior. Thus the relationship between behavioral intention and actual behavior needs further validation.

Although the study establishes the impact of Perceived Risk on the Intention to Use, this is done at the abstract level only. The detailed examination of different facets of risk, as given by Featherman and Pavlou (2002) can give insight as to which facets are salient for potential users of e-filing.

ACKNOWLEDGEMENTS

The authors acknowledge with gratitude the support provided by SSP (www.ssp-uk.com) for carrying out this research. The authors would also like to thank Professor Viswanath Venkatesh, who formulated UTAUT, for the suggestions received at the initial stage of research and the anonymous reviewers for their insightful comments that helped to refine this chapter.

KEYWORDS

- **adoption**
- **diffusion**
- **e-government**
- **technology adoption**

REFERENCES

1. Ajzen, I., "The Theory of Planned Behavior," Organizational Behavior and Human Decision Processes, 1991, 50(2), pp. 179–211.
2. Alavi, M., An assessment of the prototyping approach to information systems development, Communications of the ACM, 1984, 27(6), pp. 556–563.
3. Bagozzi, R. P., Yi, Y., Phillips, L. W. Assessing construct validity in organizational research. Administrative Science Quarterly, 1991, 36(3), 421–458.
4. Bandura, A., Social learning theory. Englewood Cliffs, NJ: Prentice-Hall, 1977.
5. Bandura, A., Self-efficacy mechanism in human agency. American Psychologist, 1982, 37, 122–147.
6. Baroudi, J. J., Olson, M. H., Ives, B., An Empirical Study of the Impact of User Involvement on System Usage and User Satisfaction, Communications of the ACM, 1986, 29(3), 232–238.
7. Benbesat, I., Dexter, A. S. An investigation of the effectiveness of color and graphical presentation under varying time constraints, MIS Quarterly, 1986, 59–84.
8. Benbesat, I., Dexter, A. S. Todd, P. An experimental Program investigating color-enhanced and graphical information presentation: an integration of the findings" Comm ACM, 1986, 29, 1094–1105.
9. Bentler, P. Comparative Fit Indexes in Structural Models. Psychological Bulletin, 1990, 107(2), 238–246.
10. Bentler, P. M., Bonnet, D. C., "Significance Tests and Goodness of Fit in the Analysis of Covariance Structures," Psychological Bulletin, 1980, 88(3), 588–606.
11. Burkhardt, M. E., Brass, D. J., Changing Patterns or Patterns of Change: The Effects of a Change in Technology on Social Network Structure and Power, Administrative Science Quarterly, 1990, (35:1), pp.104–127.
12. Byrne, B. Structural Equation Modeling with Lisrel, Prelis and Simplis: Basic Concepts, Application and Programming. Mahwah (N. J.) Lawrence Erlbaum Associates, (1998).
13. Carter, L., Belanger, F., The utilization of e-government services: citizen trust, innovation and acceptance factors. Information Systems Journal, 2005,15, 5–26.

14. Compeau, D. R., and Higgins, C. A., "Application of Social Cognitive Theory to Training for Computer Skills," Information Systems Research, 1995, 6(2), pp. 118–143.
15. Compeau, D., Higgins, C., Computer self-efficacy: Development of a measure and initial test. MIS Quarterly. 1995, 19, 189–211.
16. Cronbach, L. J., Coefficient alpha and the internal structure of tests. Psychometrika. 1951, 6, pp 297–334.
17. Davis, F. D. "Perceived Usefulness, Perceived Ease of Use, and User Acceptance of Information Technology," MIS Quarterly 1989, 13(3), pp. 319–339.
18. Davis, F. D., Bagozzi, R. P., and Warshaw, P. R. "Extrinsic and Intrinsic Motivation to Use Computers in the Workplace," Journal of Applied Social Psychology, 1992, 22(14), pp. 1111–1132.
19. DeSanctis, G., Courtney, J. F., "Toward Friendly User MIS Implementation," Communications of the ACM, 1983, 26(10), pp. 732–738.
20. Diamantopoulos, A., Siguaw, J. A., Introducing LISREL. London: Sage Publications, 2000.
21. Eastin, M. S., LaRose, R., Internet Self-Efficacy and the Psychology of the Digital Divide. Journal of Computer-Mediated Communication, 2000, 6: 0. doi: 10.1111/j.1083–6101.2000.tb00110.x.
22. Featherman, M. S., Wells, J. D., "The Intangibility of E-Services: Effects of Artificiality, Perceived Risk, and Adoption," Proceedings of the 37th Hawaii International Conference on System Sciences 2004. http://csd12.computer.org/comp/proceedings/hicss/2004/2056/07/2056701776.pdf/
23. Featherman, M. S., Pavlou, P. A., 'Predicting e-Services Adoption: A Perceived Risk Facets Perspective,' International Journal of Human Computer Studies, 2003, vol. 59, no.1, pp. 451–474.
24. Fishbein, M., Ajzen, I., Belief, Attitude, Intention and Behavior: An Introduction to Theory and Research, Addison-Wesley, Reading, MA, 1975.
25. Franz, C. R., Robey, D., Organizational Context, User Involvement, and the Usefulness of Information Systems, Decision Sciences, 1986, 17, 329–356.
26. Fu, J. R., Farn, C. K., and Chao, W. P., 'Acceptance of Electronic Tax Filing: A Study of Taxpayer Intentions,' Information & Management, 2006, vol. 43, pp. 109–126.
27. Gerbing, D. W., Anderson, J. C., An updated paradigm for scale development incorporating unidimensionality and its assessment. Journal of Marketing Research, 1988, 25(2), 186–192.
28. Ginzberg, M. J., A study of the implementation process. North-Holland/TIMS Studies in the Management Sciences. 1978.
29. Gist, M. E., Schwoerer, C. E., and Rosen, B, Effects of Alternative Training Methods on Selfefficacy and Performance in Computer Software Training, Journal of Applied Psychology, 1989, (74:6), pp. 884–891.
30. Hair, J. F., Anderson, R. L., Tatham, R. L., & Black, W. C., Multivariate Data Analysis (5th ed.). Upper Saddle River, NJ; Prentice Hall. 1998.
31. Hinton, P. R., Brownlow, C., McMurvay, I., Cozens, B. SPSS explained. East Sussex: Routledge Inc., 2004.

32. Hooper, D., Coughlan, J., Mullen, M. R. "Structural Equation Modeling: Guidelines for Determining Model Fit." The Electronic Journal of Business Research Methods, 2008, Volume 6 Issue 1 pp. 53–60.

33. http://law.incometaxindia.gov.in/DIT/Income-tax-acts.aspx Accessed on 11 September, 2014.

34. Hu, L. T., Bentler, P. M., "Cutoff Criteria for Fit Indexes in Covariance Structure Analysis: Conventional Criteria Versus New Alternatives," Structural Equation Modeling, 1999, 6 (1), 1–55.

35. Huber and, P. George, Cognitive Style as a Basis for MIS and DSS Design: Much Ado About Nothing?, Management Science, 1983, 29(5), 567–582.

36. Il Im a, Yongbeom Kim, Hyo-Joo Han, The effects of Perceived Risk and technology type on users' acceptance of technologies Science Direct Information & Management, 2008, 45, 1–9.

37. Income tax department, Government of India. (https://incometaxindiaefiling.gov.in) Accessed: Oct, 2014.

38. Jöreskog, K., & Sörbom, D. Lisrel: User's Reference Guide, Lincolnwood, IL: Scientific Software International, 2000.

39. King, W. R., J. I. Rodriguez, Participative design of strategic decision support systems: An empirical assessment, Management Science, 1981, 27(6), pp. 717–726.

40. Ludwig Christian Schaupp, Lemuria Carter, Jeff Hobbs. E-File Adoption: A Study of, U. S. Taxpayers' Intentions. Proceedings of the 42nd Hawaii International Conference on System Sciences, 2009, pp. 1–10.

41. MacCallum, R. C., Browne, M. W., and Sugawara, H., M., "Power Analysis and Determination of Sample Size for Covariance Structure Modeling," Psychological Methods, 1996, 1(2), 130–49.

42. Montano, D. E., Kasprzyk, D. The theory of reasoned action and the theory of planned behavior. In, K. Glanz, B. K. Rimer, & F. M. Lewis (Eds.), Health behavior and health education: Theory, research and practice, 2002, (pp. 67–98). San Francisco: Jossey Bass.

43. National E-governance Plan http://www.mit.gov.in/content/national-e-governance-plan Accessed on 15 September, 2014.

44. Organization for Economic Co-operation and Development, Tax Administration 2013, Comparative Information on OECD and Other Advanced and Emerging Economies, (2013).

45. Pedhazur and Schmelkin, Measurement, Design, and Analysis: An integrated approach. Hillsdale (N.J.) Erlbaum, 1991.

46. Robey, D., "User Attitudes and Management Information System Use, " Academy of Management Journal, 1979, (22:3), pp. 527–538.

47. Rogers, E. Diffusion of Innovations, Free Press, New York, 1995.

48. Rogers, E. M., Shoemaker, F. F. Communication of Innovations: A Cross-Cultural Approach, 1971, Free Press, New York, NY.

49. Güzin Mazman, S., Yasemin Koçak Usluel, and Vildan Çevik, Social Influence in the Adoption Process and Usage of Innovation: Gender Differences. International Journal of Behavioral, Cognitive, Educational and Psychological Sciences, 2009, 1, 4.

50. Schultz, R. L., Slevin, D. P. "Implementation and Organizational Validity: An Empirical Investigation, "in Implementing Operations Research/ Management Science, R. L. Schultz and, D. P. Slevin (eds.), American Elsevier, New York, 1975, NY, pp. 153–182.
51. Swanson, E. B., "Management Information System" Appreciation and involvement." Management Science, 1974, 21, 178–188.
52. Swanson, E. B., Measuring User Attitudes in MIS research: A review." OMEGA, 1982,10, 157–165.
53. Swanson, E. B. Information channel disposition and use" Decision science, 1987, 18, 131–145.
54. Swanson, E. B., Information system implementation: Bridging the gap between design and utilization, 1988, Irwin, Homewood, IL.
55. Tabachnick, B. G., Fidell, L. S., Using Multivariate Statistics (5th ed.). New York: Allyn and Bacon. 2007.
56. Taylor, S., & Todd, P. A. Understanding information technology usage: A test of competing models. Information Systems Research, 1995, 6(2), 144–176.
57. Thomas Hill, Nancy, D. Smith, and Millard, F. Mann, Communicating Innovations: Convincing Computer Phobics to Adopt Innovative Technologies, in NA – Advances in Consumer Research Volume 13, eds. Richard, J. Lutz, Provo, UT: Association for Consumer Research, 1986, Pages: 419–422.
58. Thompson, R. L., Higgins, C. A., and Howell, J. M., Personal Computing: Toward a Conceptual Model of Utilization," MIS Quarterly 1991, (15:1), pp. 124–143.
59. Tornatzky, G., Klein, K. J. "Innovation Characteristics and Innovation Adoption-Implementation: A Meta-Analysis of Findings, " IEEE Transactions on Engineering Management, 1982, (EM-29:1), pp. 28–45.
60. Venkatesh, V., Morris, M., Davis, G., Davis, F. D., User acceptance of information technology: toward a unified view. MIS Quarterly, 2003, 27(3), 425–478.
61. Vertinsky, I., Barth, R. T., Mitchell, V. F. A. study of OR/MS implementation as a social change process. In, R. L. Schultz & D. P. Slevin (Eds.), Implementing operations research/management science, New York: American Elsevier, 1975, pp. 253–270.
62. Waransanang Boontarig, Wichian Chutimaskul, Vithida Chongsuphajaisiddhi, Borworn Papasratorn, Factors Influencing The Thai Elderly Intention to Use Smartphone for e-Health Services. 2012 IEEE Symposium on Humanities, Science and Engineering Research 978-1-4673-1310-0/12.
63. World Bank definition of e-government. http://www1.worldbank.org/publicsector/egov/definition.htm. Accessed on 14 September, 2014.
64. Zand, D. E., Sorensen, R. E. Theory of change and the effective use of management science, Administrative Science Quarterly, 1975, 30(4), pp. 532–545.

APPENDIX I: PERSONAL INCOME TAX STATISTICS

Source: Comparative Information on OECD and Other Advanced and Emerging Economies, Tax Administration 2013.

TABLE 6.11   Personal Income Tax Country-Wide Statistics

Country	Use of electronic filing (e-filing)				
	Year begun	% of all returns e-filed			Mandatory for some/all in 2011
		2004	2009	2011	
OECD countries					
Australia	1990	80	92	92	X
Austria	2003	10	79	79	+/1
Belgium	2002	3	40	54	X
Canada	1993	49	58	62	X
Chile	1999	83	98	99	+
Czech Rep.	2004	< 1	1	1	X
Denmark	1994	68	96	98	X
Estonia	2000	59	92	94	X
Finland	2006	0	23	33	X
France	2001	4	27	33	X
Germany	1999	7	30	32	+
Greece	2001	4	13	49	+
Hungary	2003	3	30	17	X
Iceland	1999	86	92	92	X
Ireland	2001	62	67	81	+
Israel	2009	0	0	92	+
Italy	1998	100	100	100	+
Japan	2004	0	31	44	X
Korea	2004	43	80	87	X
Luxembourg	2009	0	< 1	1	X
Mexico	1998	48	96	99	+
Netherlands	1996	69	95	95	+
New Zealand	1991	56	63	71	X
Norway	1999	37	82	86	X

APPENDIX II: E-FILING TRENDS IN INDIA

Source: Income Tax department (https://incometaxindiaefiling.gov.in/por-tal/index.do)

TABLE 6.11 Continued

Country	Use of electronic filing (e-filing)			
	Year begun	**% of all returns e-filed**		**Mandatory for some/all in 2011**
		2004	**2009**	**2011**	
Poland	2008	0	1.4	11	X
Portugal	2000	24	80	83	+
Slovak Rep.	2005	0	n.a.	< 1	X
Slovenia	2004	0	77	n.a.	X
Spain/1	1999	23	36	74	X
Sweden	2002	15	55	63	X
Switzerland	Administered at sub-national level by cantons, some with their own e-filing systems				
Turkey	2005	30	99	99	+
United Kingdom	2000	17	73	77	X
United States	1986	47	65	76	X
OECD ave. (unw.)		31	59	65	
Non-OECD countries					
Argentina	1999	18	100	100	+
Brazil	n.a.	n.a.	n.a.	100	+
Bulgaria	2005	< 1	3	5	X
China	2005	0	n.a.	n.a.	n.a.
Colombia	n.a.	n.a.	n.a.	6	+
Cyprus	2004	< 1	6	22	+
Hong Kong, China	n.a.	n.a.	n.a.	14	x
India	n.a.	n.a.	17	(13.1 m)	x
Indonesia	n.a.				
Latvia	2008	0	10	15	x
Lithuania	2004	14	71	87	x
Malaysia	2004	33	56	69	x
Malta	2006	1	2	1	+
Romania	2007	0	< 1	n.a.	n.a.
Russia	2006	0	9	3	
Saudi Arabia	n.appl.				
Singapore/1	1998	67	91	96	x
South Africa	2001	4	46	99	x

TABLE 6.12   Trends in e-Filing

Form Type	FY 2011–12 (From 01/04/2011 to 31/03/2012)	FY 2012–13 (From 01/04/2012 to 31/03/2013)	FY 2013–14 (From 01/04/2013 to 31/03/2014)
ITR-1	4439001	6409881	10676604
ITR-2	1773659	2240995	3213262
ITR-3	522579	625890	721831
ITR-4S	1628312	2947568	4250709
ITR-4	6712032	7772962	9035055
ITR-5	765054	851327	960120
ITR-6	593047	638184	713736
TOTAL	16433684	21486807	29681794

TABLE 6.13   Year on Year Summary of e-Filed ITR Forms

Form Type	FY 2011–12 (From 01/04/2011 to 31/03/2012)	FY 2012–13 (From 01/04/2012 to 31/03/2013)	FY 2013–14 (From 01/04/2013 to 31/03/2014)
ITR-1	6409881	10676604	66.56
ITR-2	2240995	3213262	43.39
ITR-3	625890	721831	15.33
ITR-4S	2947568	4250709	44.21
ITR-4	7772962	9035055	16.24
ITR-5	851327	960120	12.78
ITR-6	638184	713736	11.84
TOTAL	21486807	29681794	38.14

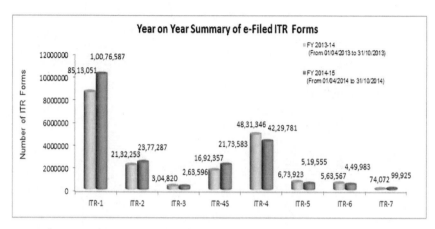

FIGURE 6.7   Year on Year Summary of e-filed ITR Forms.

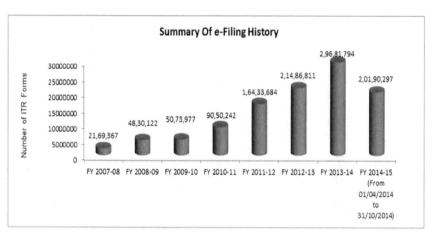

FIGURE 6.8   Summary of e-filing History.

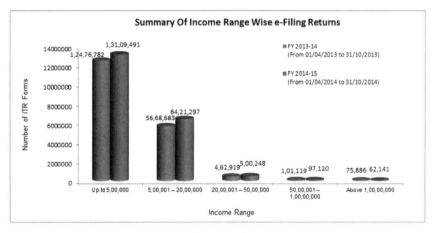

FIGURE 6.9   Summary of income range wise e-filing returns.

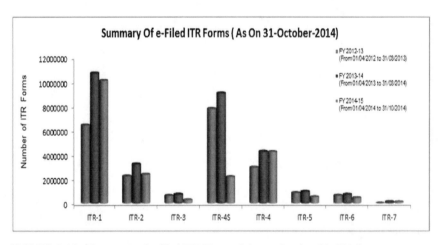

FIGURE 6.10   Summary of e-filed ITR Forms 9 (as on October 31, 2014).

PART 4:

E-HEALTH

CHAPTER 7

IMPLEMENTATION OF E-HEALTH PROJECT IN SLOVENIA: AN INTERNATIONAL PERSPECTIVE, KEY CHALLENGES, AND WAYS FORWARD

DALIBOR STANIMIROVIC

Assistant Professor, University of Ljubljana, Faculty of Administration, Gosarjeva ulica 5, 1000 Ljubljana, Slovenia, Tel.: 00386-1580-5551, E-mail: dalibor.stanimirovic@fu.uni-lj.si

CONTENTS

ABSTRACT

Fragmentation of information systems (ISs) and their limited interoperability significantly compromise further development of the health care system and adversely affect the quality of the health care services in Slovenia. Surmounting these challenges requires the intensified implementation of the e-health project including an integrated health information system (HIS), which is one of the most compelling objectives of the Slovenian public sector. Employing a case study approach, this chapter presents the review of e-health projects in Slovenia, Austria and Denmark, and provides a comparative analysis of the e-health development in designated countries. The case study is based on the structured interviews with fifteen prominent experts from the Slovenian health care system. Stemming from the Slovenian experience, this chapter summarizes the main deficiencies in the current e-health settings, and finally outlines a set of applicable guidelines for effective development and implementation of highly intricate e-health projects.

7.1   INTRODUCTION

Slovenian health care system has been facing serious structural problems in recent years. Due to objective circumstances, these problems cannot be avoided and will require fundamental changes in the current health care arrangements [1]. The health care system reform is becoming a social imperative, which calls for innovative approach in the next years. One of the fundamental tools that would allow for successful and effective tackling of challenges facing the Slovenian health care system is the comprehensive informatization, representing one of the key long-term goals of the public sector [2, 3]. Experience of the most developed countries shows that successful implementation of health care informatization projects is of immense strategic importance for further development of the health care system [4–6], but also displays important implications for the increase in social welfare [7, 8], economic growth [9, 10], and development of information society [11, 12]. Majority of existing ISs in the Slovenian health care have been developed within individual health care organizations and are designed specifically to meet their own needs, while they are not adequately interoperable and do not provide complete, relevant and timely information [3, 13], The already initiated national project of health care system informatization from 2005, known as e-health, should be able to integrate all fragmented ISs and offer a complete solution benefitting all interested parties [2]. e-health entails the inclusion of stakeholders into the functional network, reconstruction of health care system business model as well as integration and harmonization of many information subsystems at different levels [14, 15]. Informatization of the Slovenian health care system should provide opportunities for high quality and professional work with patients and long-term development, whereas relevant and reliable economic, administrative and medical data provided by e-health should facilitate better quality planning, control and management of individual health care organizations and health care system in general [1, 13].

The aim of this chapter is to provide a comparative assessment of the e-health progress in an international context, enable identification and analysis of the key e-health components and success factors, and propose a set of applicable guidelines for effective development and implementation of e-health projects. Accordingly, this chapter primarily focuses on the following interrelated research objectives:

1. Review of the e-health projects and related strategies in Slovenia, Austria and Denmark.
2. Analysis of the development of e-health projects in Slovenia, Austria and Denmark.
3. Identification of the main deficiencies in the current setting of e-health in Slovenia and provision of general guidelines for development and implementation of e-health projects.

Methodologically speaking, the chapter represents a case study, employing the comparative analysis of the e-health development in Slovenia, Austria and Denmark. The establishment of the comparative framework was based on the investigation of various electronic and written sources concerning and relating to the e-health projects in designated countries, and especially on the structured interviews with fifteen prominent experts from the Slovenian health care system. Selection of the research methods was adapted to the research field [16, 17], given the complexity and scope of the e-health initiatives.

Following the introduction, the Section 7.2 sketches the state of the art in the field and conceptual implications of e-health. The methodological framework comprising research design, sampling approach, data collection techniques and data analysis procedures is elucidated in the Section 7.3. The overall results of the case study are presented in the Section 7.4. This section outlines the e-health projects in Slovenia, Austria and Denmark, provides a comparative analysis of the e-health development in selected countries, and finally identifies the main deficiencies related to the e-health development in Slovenia. Lessons learned and general guidelines for more effective development and implementation of e-health projects are presented in the Section 7.5. The Section 7.6 contains the review of the case study results, discussion on their practical applicability, limitations and future work, and concludes by submitting final arguments and observations regarding the development prospects of the e-health in Slovenia.

7.2   STATE OF THE ART

Considering the multifaceted nature of health care systems and related IS, the body of knowledge in this field is relatively extensive, as well as the number of various definitions depicting the concept of e-health. Gaining international recognition in the last decade [18, 19], definitions of e-health are normally derived from the classifications of IS with the addition of certain features which are associated with the specific nature of health care services and health care system status, being essentially one of the most important segments of the public sector. While some definitions of e-health are rather general, others are more narrowly focused converging on individual aspects of ICT and health care interaction.

Eysenbach is referring to the term of e-health as a general "buzzword," which is used to characterize virtually everything related to computers and medicine, and interprets e-health as an emerging field in the intersection of internet-related medical informatics, public health and business [20]. In a broader sense, the term characterizes not only a technical development, but also a state-of-mind, a way of thinking, an attitude, and a commitment for networked, global thinking, to improve health care locally, regionally, and worldwide by using information and communication technology (ICT). Marconi defines e-health as the application of internet and other related technologies in the health care industry to improve the access, efficiency, effectiveness, and quality of clinical and business processes utilized by health care organizations, practitioners, patients, and consumers in an effort to improve the health status of patients [21]. According to Healthcare Information and Management Systems Society, being one of the leading authorities in the field, e-health represents a patient-focused framework including various dimensions such as: delivery of key information to health care partners, provision of health information delivery services, facilitation of interaction between providers and patients, acceleration of the integration of health care industry-related business processes, both local and remote access to health care information, support for employers and employees, payers and providers [22]. And conversely, there are a number of arbitrary and ad hoc definitions which are more narrowly targeted and focused on individual aspects of the e-health research. They are outlining the concept of e-health as the process of providing health care via electronic means, in particular over the internet, including web-based applications, cross-sectoral transfer of patient-related data, monitoring of health parameters (telemedicine solutions), and interaction with health care providers [23–25]. Extrapolating from different definitions, e-health concept can be generally regarded as a comprehensive mechanism based on the internet and other related ICTs, expected to facilitate integration of all stakeholders and evidence-based decision making at all levels, in order to improve quality of health care, administrative and managerial processes as well as related outcomes in the health care system.

As noticed by several authors [26–28], the term e-health has a highly variable and interchangeable usage, which significantly complicates its substantive characterization and distinction from other related concepts.

In addition, there are hardly any extensive empirical studies systematically identifying and analyzing the general implications of e-health projects on the health care system transformation, its impacts on public health and public finance aspects, and issues related to the long-term development of the health care systems [8, 29]. Majority of the studies in the field are usually focused on the selected aspect of e-health, its implications on certain health care service/product or particular institution within the health care system. The latter reasons considerably hinder the research of the very concept of e-health on the one hand [19, 30], and on the other hand, they prevent the evaluation of the actual effects of e-health on business and health outcomes of the health care system.

7.2.1   E-HEALTH IMPLICATIONS

Regardless of their definition and research perspective, virtually all authors emphasize that the main goal of e-health should be the contribution to a high-quality, efficient patient care and effective performance of the health care system [31–33]. e-health could empower patients and help in exceeding information asymmetry between main stakeholders while ensuring that reliable and timely health care information is available for operational and strategic decision making providing better health care services and enhancing public health [34]. e-health systems and services combined with organizational changes, process reengineering and development of new skills can act as key enabling tools facilitating considerable enhancements in access to care, quality of care, as well as efficiency and productivity of the health care system [35, 36]. Implementation of e-health is expected to reduce costs and improve productivity in such areas as: (i) billing and record-keeping, (ii) reduction in medical error, (iii) alleviation of unnecessary care, and (iv) savings achieved by business-to-business e-commerce [37, 38].

Given the innate complex nature of the health care activities and specificity of the health care-related IS, there are a reasonable number of requirements, constraints and risks associated with the implementation of e-health project. The quality introduction and performance of e-health depend not only on technical determinants such as ICT infrastructure, data quality, system design, or adequate use of ICT [14, 39]. Other factors are also involved, such as: (i) organizational policies and environmental determinants that relate to the information culture within the country context [40, 41], the structure of e-health [42, 43], the roles and responsibilities of the different actors and the available resources for e-health [29, 44], and (ii) the behavioral determinants such as the knowledge and skills, attitudes, values, and motivation of those involved in the production, collection, collation, analysis, and dissemination of information [45, 46].

Attributable to these highly challenging preconditions, practice reveals that planning, development and implementation of e-health are riddled with major problems, even in countries with relatively well-developed health care systems [47, 48]. Furthermore, the information generated and retrieved from inadequately conceptualized e-health is often not helpful for health care management decision-making [49], because information is not applicably clustered; it is frequently disparate with predefined indicators, while modalities and jurisdiction on management and transaction of information can be ambiguous and unrelated to priority tasks and functions of health care professionals [50, 51]. In other words, poorly defined and unstructured e-health projects have a tendency to be data and information driven, instead of action driven [52]. In order to avoid these threats, the entire e-health project, including its long-term and wide-ranging implications, must be well thought out, while its contextual role and functions within the health care system must be clearly defined [53, 54], yet flexible and adaptable to requirements and continuous changes in health care ecosystem and broader social environment.

7.3   METHODS

7.3.1   RESEARCH DESIGN

The study employs a single explanatory/exploratory case study design. The case study on the development of e-health projects in Slovenia, Austria and Denmark, and subsequent provision of general guidelines for effective development and implementation of e-health projects was conducted in the first half of 2014. A selection of the research method was adapted to the particularities of the research problem [16, 17], given that quantitative empirical research could not yield a satisfactory and credible picture, since this complex field of research is still in an early development stage, and it would be difficult to ensure the representativeness of the research sample. An iterative, structured interview process complemented by the literature review and observations were used as the main data collection techniques during the formative research phase.

7.3.2   SAMPLE

Selection of the potential interviewees was based primarily on their experience and expertise in the field of e-health. Good knowledge of informational, structural and contextual characteristics of the health care system by the selected interviewees would ensure credibility and validity of their views and recommendations. The final sample size comprised fifteen (n = 15) prominent experts from a cross-section of areas strongly related to the concept of e-health.

All 15 interviewees came from different health care-related institutions. Five experts were chosen from each key area: health care professionals (3 specialists from public hospitals and 2 general practitioners from public health care centers), ICT experts from health care and government institutions (2 analysts from government institutions and 3 ICT consultants from public clinics and public hospital), and health care managers (3 managers of public health care center and public hospitals and 2 managers of public clinics). The interviewees occupied senior positions in different levels of the health care system (information, medical and financial directors of health care centers/clinics, heads of government sectors and departments, etc.). Age and gender of the interviewees reflect the situation in the Slovenian health care system management practice. The participants were aged between 43–62 years, and the ratio between men and women was 4:1.

7.3.3   DATA COLLECTION

Before commencement of the interviewing process, three pilot interviews were carried out, including one expert from each of the designated areas. The final set of questions was revised in line with their comments and suggestions, which helped to resolve some conceptual weaknesses and ambiguities. The response rate was 100%, as all invited experts responded to the invitation and participated in the interviews. The interviews were carried out in the period from February to May 2014. Since the new information and knowledge on the subject appeared after certain interviews, interviewees went through the iterative interview process consisting of several interview meetings. The interviews, which lasted approximately 90–120 minutes, were conducted by the author in person at the official premises of the interviewees. All participants were told the purpose and objectives of the study. Interviewees gave informed consent and were provided anonymity and assured confidentiality of the information obtained.

The interview comprised 8 open-ended questions, which were based on the review of existing literature (journal articles, papers, strategy documents, project reports, online resources, etc.) and related to the study objectives. The questions were broadly focused on exploring the different dimensions of the Slovenian health care system, the concept and potentials of e-health, the operative aspects of e-health and its applications, and specification of the necessary preconditions and measures for the effective development and implementation of e-health in the upcoming years. Unresolved issues, existing limitations and potentials, and future directions were further analytically discussed with the interviewees in the iterative manner. Namely, the role of the health care experts within the proposed case study was threefold. First, they had to summarize the recent developments concerning e-health in Slovenia and partly Austria and Denmark. Second, they had to take part in the structural decomposition of e-health and subsequent analysis and evaluation of e-health development in Slovenia, Austria and Denmark. And third, drawing from their own experience and knowledge of the health care system, they had to identify main operative deficiencies of the e-health in Slovenia, and ultimately provide their vision of further expansion of e-health, and propose a set of general guidelines for minimizing risk and securing enhanced development and implementation of e-health projects in the forthcoming years. Given the iterative nature of the overall interview scheme and the active role of the interviewees in the later stages of the research, special content authentication (authorization) of their responses was not required. Responses of the interviewees to the questions were recorded in writing by the interviewer.

7.3.4   DATA ANALYSIS

This research phase included the analysis of two datasets, namely: (i) literature (journal articles, papers, strategy documents, project reports, online resources, etc.), and (ii) interview data (contained in the interview transcripts). The data obtained through the theoretical and empirical qualitative research, were analyzed by the author in accordance with the guidelines proposed by the case study methodological framework.

After an extensive review of the literature and investigation of primary and secondary online resources and other forms containing e-health-related content, author systematically analyzed the e-health concept, different aspects of the e-health research, and the necessary steps for overcoming the existing technical, organizational, policy and other constraints and obstacles. The analysis of the case study evidence served as a platform for the identification of the current deficiencies of e-health in Slovenia, and final provision of general guidelines for the effective development and implementation of the e-health projects in the future.

Multiple analysis of the interview data and their interpretations (re-analysis after each round of interviews), structural decomposition of e-health and evaluation of its development in Slovenia, identification of its main operative deficiencies, and the following provision of guidelines for the effective development and implementation of e-health were carried out in collaboration with the experts from the health care system, who took a constructive part throughout all phases of the study. The final identification of deficiencies and provision of substantially harmonized general guidelines were achieved through the joint efforts, whereas the resolution of inconsistencies and the reaching of consensus demanded a great deal of patience, and extensive communication and coordination between the author and the participating experts.

The application of the specific methodological framework and the respective data collection techniques have been instrumental for the overall data analysis. The latter provided a platform for the interpretation of data obtained by the review of e-health-related sources, and synthesis of data obtained by the conducted interviews, which ultimately facilitated an establishment of coherent guiding principles for effective development and implementation of e-health projects.

7.4 RESULTS

7.4.1 E-HEALTH IN SLOVENIA

Ministry of Health has been dealing with the informatization of the Slovenian health care system for almost two decades. e-health project from 2005 in its latest form consists of 17 sub-projects aiming at extensive renewal and integration of information and communication systems in health care domain. Strategic objectives within the e-health strategy should be implemented by the year 2023 facilitating fully integrated national IS enabling monitoring of the on-going treatments and related costs, faster access to medical data, medical services as well as cost evaluation, online ordering and coordination of waiting lists, increase of efficiency and transparency of the health care system and optimization of the business processes taking place in health care institutions [2, 3]. Based on the Strategy for informatization of the Slovenian health care system 2005–2010 [2] and the Resolution on the National Health Care Plan for the period 2008–2013 [3] all activities in the field of Slovenian health care system informatization are aiming at realization of e-health, whereas summary of its development goals is presented below:

1. The establishment of basic ICT infrastructure including: network used for communication and data exchange, Diagnosis Related Groups (DRG) and standardized definitions of health and social data required for development and management of Electronic Health Records (EHR) and e-prescription as well as improvement of the health care Smart card functionalities (Smart card allows access to medical data containing information on: the cardholder, the person liable for health insurance contribution, compulsory health insurance, voluntary health insurance, selected personal physician and General Practitioner (GP), issued medication, issued prosthetic equipment, potential organ and tissue donation for transplantation, etc. After all functionalities of e-health are implemented, smart card will allow all users to remotely access to their own health data via Personal Health Record – PHR). Currently, EHR content is still not defined explicitly, while its structure comprises free text, preventing its full exploitation. Existing diagnosis as well as medical

procedures are standardized and structured according to ICD 10 AM[1] classification, whereas EDIFACT[2], HL7[3] and XML[4] are the current data standards for transfer of messages.

2. Integration and merging health and social IS into a national HIS and establishing a central, unified health information portal that will allow all stakeholders within the health care system secure and reliable exchange of data, execution of electronic services as well as standardized and transparent information and interoperability with similar systems in the European Union (EU).

3. Introduction of e-business as standard way of conducting operations and processes in the Slovenian health care system and promoting and encouraging the use of e-health applications by all health care system stakeholders.

E-health project is thus divided into three substantially separate, yet related areas. The first area is the establishment of a national IS, comprised of Health Network (hNET), a health portal (hAOP) and EHR. The second area represents the establishment and operation of Center for health care informatics, undertaking the central role in governing of IS. This area also includes upgrading and maintenance activities of the entire project after its completion. The third area will enable the improvement of health care processes, access to health care services as well as education and training of target groups. Although the e-health project is still

---

[1] International Statistical Classification of Diseases and Related Health Problems (ICD) is a medical classification list developed by the WHO. It codes for diseases, signs and symptoms, abnormal findings, complaints, social circumstances, and external causes of injury or diseases [55].

[2] Electronic Data Interchange for Administration, Commerce and Transport (EDIFACT) is the international standard developed under the United Nations. It comprises a set of internationally agreed standards, directories, and guidelines for the electronic interchange of structured data between independent computerized information systems [56].

[3] Health Level Seven (HL7) is a set of international healthcare informatics interoperability standards developed by the Health Level Seven International. HL7 network provides a framework and related standards for the exchange, integration, sharing, and retrieval of electronic health information [57].

[4] Extensible Markup Language (XML) is a markup language that defines a set of rules and standards for encoding documents in a format that is both human-readable and machine-readable. It is developed by the World Wide Web Consortium (W3C) [58].

deep in the implementation phase, the Figure 7.1 presents the projected infrastructure of e-health, which should become fully operational sometime after 2020 [2].

The implications of e-health will presumably be twofold. First, significant changes can be expected in the field of informing, empowerment and inclusion of patients in the health care process, and second, well-designed e-health should facilitate timely access to relevant data and information and consequently initiate better supported decision-making at all health care, administrative and management levels. According to project objectives, the fully functional version of e-health should provide standardized

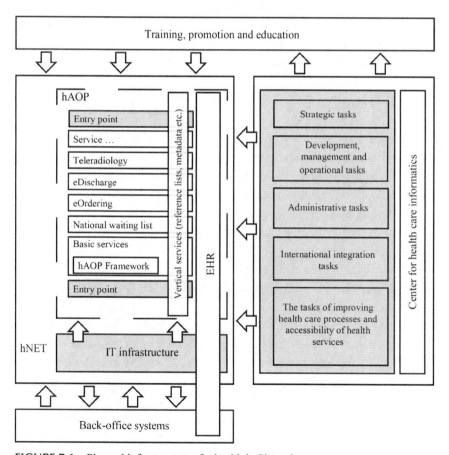

FIGURE 7.1 Planned infrastructure of e-health in Slovenia.

bi-directional connections between the designated entities of the health care system, network synergies and substantial improvements in information and resource flows. Nevertheless, despite ambitious e-health strategy and objectives, most of the project goals presented above, have remained unfulfilled. Namely, the current infrastructure of e-health includes components facilitating only a few peripheral functionalities (Smart card, Professional card), which do not yield tangible benefits neither for patients nor for health care workers and health care system managers. Due to leadership issues and lack of coordination as well as financial restrictions and technical problems, the e-health development has stagnated in the recent period on almost all key areas, while the main project deliverables in the form of infrastructure building blocks have not reached the desired level of development according to the schedule. Consequently, the current infrastructure of the Slovenian e-health is not fully functional and causes considerable time and resource losses.

7.4.2   E-HEALTH IN AUSTRIA

Development of e-health in Austria has been founded on the eGovernment Act from 2004 [59] and the Health Care Reform Act from 2005 [60] including the Health Telematics Law [61], which focuses on the secure exchange of medical data. The Health Care Reform Act emphasizes the role of ICT in the future development of Austrian health care system and outlines the informatization of the health care system as one of the public sector priorities [60]. The main coordinating body responsible for promoting the use of ICT and mechanisms for planning, financing and management of informatization projects is the Ministry of Health (Bundesministerium für Gesundheit). Strategic framework for the health care system reform has defined e-health as a set of business models and information tools, which should provide improved health care and health care system performance in general while facilitating effective implementation of the priorities listed in the strategic documents from the field. In accordance with the objectives of the i2010 initiative and some other documents, issued by the European Commission [62–64], Austria has established the Information Society Development Program, comprising activities for the harmonization of projects and mechanisms within e-health and e-government areas. Significant results in this program have been achieved especially in the management of e-identities and electronic signatures. Austrian Citizen Card issued by the federal government in 2008 is considered one of the leading e-identity projects implemented in the EU. In accordance with the informatization strategy, the e-card (health insurance card) was delivered to more than 8 million policyholders and more than 12.000 GPs since 2005 [5]. As in Slovenia, the e-card initially contained only information about the health care insurance of citizens, in the second phase, however, which lasts from 2006 onwards, e-card contains an integrated suite of medical information which is complemented and updated sequentially.

Despite significant achievements in the field, the most important sub-project of e-health remains development and implementation of a national EHR called Elektronische Gesundheitsakte [65, 66]. Development of EHR began in 2006 when a thorough analysis of the Austrian health care system and a feasibility study were conducted. In 2009 the institutional framework for the project was established, and a national health care portal (www.gesundheit.gv.at) was created in 2010. In parallel with development activities, the technical standards, interoperability framework and guidelines for further development of health care enterprise architecture were established and adopted. Actual implementation of national EHR started in 2011 through the realization of three pilot projects which were carried out at the regional level. In its first implementation phase, EHR will be mainly focused on e-prescribing and dispensing of e-prescriptions, along with gradual integration of the increasing number of medical data on e-card, in the years ahead. This should lead to the greater exploitation of medical data and higher quality of medical treatments as well as considerable elimination of the contraindications, reduction of allergic reactions and side effects. On the other hand, the implementation of e-prescribing should facilitate control over costs of medical treatments, prevent duplication of prescriptions, establish transparent functioning of the pharmacy market and provide an overview of the types and quantities of prescribed pharmaceuticals, as well as simplify their supply and distribution.

Notwithstanding the legitimate caveats highlighting primarily the protection of personal data and privacy as the most problematic areas of Austrian health care system informatization, development of EHR is undoubtedly an important asset for all policyholders and the entire health care system, while its long-term benefits will only be seen in the following years, when all planned applications and functionalities of e-health become operative. Planned infrastructure of the e-health project in Austria and the main relations between its components are depicted in Figure 7.2 [64].

Effective implementation of national e-health strategy in the coming years will require a parallel restructuration of the entire health care system and coordinated action at the medical, information and legislative level, as well as the execution of activities to raise awareness of the health care professionals and citizens, and to promote greater use of ICT solutions in health care. Under the latter assumptions, the Austrian health care policy is focusing particularly on the following activities:

- Achieving the overall interoperability and definition of standards for technical, semantic and organizational interoperability within the strategy for informatization of the health care system as well as the development of information society in general;
- Building of trust, protection of patient rights and ensuring responsibility of physicians and personal data protection in the process of EHR management;
- Development of national strategy and specific guidelines in order to facilitate safe and long-term archiving of EHR;
- Gradual elimination of semantic problems and implementation of common terminology for communication and data exchange, as well

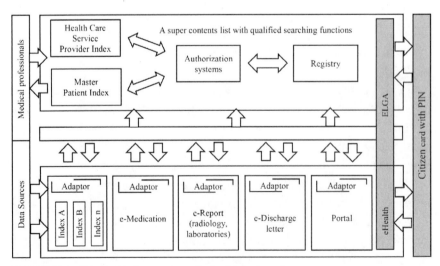

FIGURE 7.2   Planned infrastructure of e-health in Austria.

as monitoring and control of treatment procedures, patient requirements and satisfaction, and comparison of health care services and
their quality nation-wide;
• Promotion and upgrading of a user-friendly health care portal accen-
tuating prevention and citizens' participation in caring for their own
health and well-being;
• Establishment of the health care provider networks and ICT infrastructure for the provision and execution of integrated social and
health care services;
• Implementation of telemedicine projects for home therapy and assistance for disadvantaged population groups.

7.4.3 E-HEALTH IN DENMARK

Denmark has a long history in the development of e-health, which dates
back to 1996, when implementation of the strategy for informatization
of the health care and development of EHR began [67]. Other initiatives
for health care reform and introduction of advanced ICT solutions in
health care are embodied in National Strategy for Information Technology
in Hospitals from 1999 [68] and National Strategy for Information
Technology in Health Service 2003–2007 [69]. The last set of strategic
guidelines in the field of e-health development and implementation is contained in Danish Policy Strategies with e-health relevance, which refers
to the Action Plan from 2003 and includes 29 projects connecting many
different public institutions. MedCom is national institution responsible
for realization of the e-health project and acts as coordinator of project
activities between health care policy, health care professionals, citizens
and ICT service providers. MedCom manages the process of informatization in the Danish health care system, issues certificates of safety and
quality in health data exchange and promotes integration of HIS in hospitals and pharmacies [70]. Within the Danish health care system 4 million
messages are exchanged every month, including 80 percent of all prescriptions. MedCom also controls electronic data interchange and manages
patient identities and safety of personal data through integrated three-tier
system, which includes the public key infrastructure and allows the
traceability of each entry to the system. A key part of the strategy and
the ultimate goals of the Danish health care system informatization are the

development of integrated HIS and implementation of HER [71], whereas the future activities within e-health are focused primarily on:

- Extension of existing applications in the e-health scheme, more effective integration of the local HIS and e-prescriptions with the aim of developing a personal health profile of the patient, which would be stored on a national medical data server, further improvement of e-prescribing;
- Promotion, upgrading and enhancement of national health portal Sundhed.dk, awareness-raising between citizens and health care professionals, facilitating the full functionality of the national health portal and general accessibility by using the digital signature;
- Upgrading of health data networks, information infrastructure and personal data protection system, effective intersectoral communication that includes the exchange of more than 40 different types of standard documents (e-prescriptions, e-referrals and e-lab tests, specialist e-referrals, etc.);
- Effective further implementation of the electronic health card project (Common Medication Card – FMK) throughout the country, inclusion of wider range of medical and administrative data on the electronic health card and promotion of its functionality for both patients and health care professionals;
- Effective transfer of medical and administrative patient data across regional boundaries in order to ensure the quality of health care throughout the country, further development and implementation of telemedicine projects for chronic patients, and deployment of crossborder health care networks in the region.

Danish health policy makers have managed to attract a wide range of stakeholders collaborating in the development and implementation of e-health. Political will and stakeholders' commitment as well as their coordinated action have provided necessary resources, professional and technical support and adaptation of legislation, being some of the reasons for their success. For example, in 2005 the tax legislation was adjusted, which allowed a separate agreement between the government and owners of the regional hospitals, who required equal access to EHR and e-prescriptions throughout the country. The regulatory framework of health care has been adapted as well, since the Act on health care from

2008 [72] had to take into account the specific requirements in the area of confidentiality and protection of personal data, referring to the implementation and use of e-health. Among other factors, which influenced the development and intensive use of e-health applications in Denmark, some other aspects could be exposed, such as: the construction of high-quality ICT infrastructure and health information network, which was built on existing infrastructure building blocks of e-government and the establishment of the National Health Portal (Sundhed.dk), which provides uniform access point to health care services for both citizens and health care professionals. Planned infrastructure of the e-health project in Denmark and the main relations between its components are presented in Figure 7.3 [73].

The National Patient Index and The National Health Record will provide health care professionals and patients with access to more complete

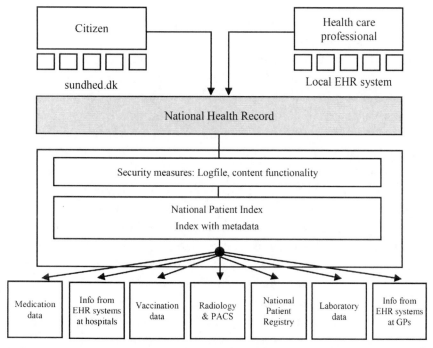

FIGURE 7.3   Planned infrastructure of e-health in Denmark.

overview of existing patient data. This will benefit health care profession-als and patients in several ways by facilitating [73]:

- A clinical tool that enables digital sharing of data across borders and sectors in the health care system;
- A tool for gaining digital access to patient data not already stored in local EHR systems;
- Support in decision-making in relation to referral, elucidation and treatment of a patient;
- Giving citizens access to a broader range of own health data thereby establishing a foundation for improved dialog, better insight in their own health condition and improved possibility for active involve-ment in their own treatment.

7.4.4 COMPARATIVE ANALYSIS OF THE E-HEALTH DEVELOPMENT IN SLOVENIA, AUSTRIA AND DENMARK

Given the substantial scope and complexity of the e-health projects, as well as the asymmetrical development of the individual thematic and orga-nizational areas within them, comparing the development of the entire e-health projects was unfeasible. Research team therefore applied struc-tural decomposition techniques, through which we identified and extracted 12 relatively autonomous and comparable infrastructure components from designated e-health projects. Subsequently, by evaluating the development level of the selected components, we transformed these components into 12 equally weighted indicators. Given the two fundamental contextual dimensions, which reflect the general development degree of the e-health projects and their alignment with other relevant factors within the health care ecosystem, the indicators were categorized in two groups, namely: operative and technological indicators, and policy and performance indi-cators (see Table 7.1). The aggregate of the reasonably evaluated indica-tors should reflect the actual overall development level of the e-health projects in the selected countries.

Development level of the individual components, and ultimately the overall development of e-health, was evaluated applying the following grades (see their explanations in parentheses):

TABLE 7.1   Indicators for the Comparative Analysis of e-health Development in Slovenia, Austria and Denmark

Indicators	
Operative and technological indicators	**Policy and performance indicators**
EHR/PHR	Integration of stakeholders (policy guidelines, reports, data exchange, education and training, etc.)
Interoperability framework	Legal regulation
Data standards	Health care system performance indicators
E-prescription	Performance evaluation of e-health
Smart card	
Professional card	
Telemedicine	
National Health Portal	

Overall rating of the e-health development in the selected countries.

1. Conceptual phase (Component and its operations are based only on the conceptual design; its development, sourcing and implementation procedures have not yet been defined or started).
2. Development phase (There is a concrete blueprint for the construction of the component encompassing all planned operations. Development, sourcing and implementation procedures have been defined, initiated and monitored).
3. Partly functional (Some of the planned component operations are implemented, functional and applied in practice within the health care environment).
4. Functional (All of the planned component operations are implemented, functional and applied in practice within the health care environment).

Finally, based on the assigned grades, the calculation of the average score of the components' development level was carried out, facilitating the determination of overall development level and associated comparative ranking of e-health projects in the selected countries. The nominated components within e-health projects were defined and selected partly on the basis of EU research and guidelines [5, 6, 74] striving to identify the most important factors for development of comprehensive e-health projects. Comparative analysis was conducted combining different techniques of qualitative research methods [17]. The initial part of the comparative analysis has focused on the document analysis through in-depth investigation of existing e-health-related sources, whereas deriving from obtained investigation results, the conclusive part of the comparative analysis is striving to integrate theoretical and practical aspects regarding the research subject, and provide generally applicable guidelines for more effective development and implementation of e-health projects.

Despite the fact that, unlike the Danish project, both the Slovenian and Austrian e-health projects are still deep in the implementation phase and will not become fully functional for some time [1, 64, 71], the research revealed some interesting findings. Comparative analysis confirms the undisputed supremacy of Denmark (overall average score 3.67, std. deviation 0.65) in the field of overall e-health development in comparison with Austria (overall average score 2.75, std. deviation 0.75) and Slovenia (overall average score 2.33, std. deviation 0.98) (Figure 7.4). Considering the particular groups of indicators (operative and technological indicators, policy and

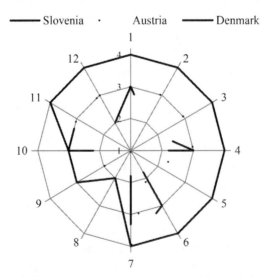

FIGURE 7.4 Development of individual e-health components in Slovenia, Austria and Denmark.

performance indicators) ranking has remained unchanged. Based on the operative and technological indicators, Denmark has won the highest ratings (average score 3.50, std. deviation 0.58), Austria is in the second place (average score 3.00, std. deviation 0), and Slovenia was third (average score 2.25, std. deviation 0.96). Taking into account the policy and performance indicators, the evaluation yielded rather similar results (Denmark – average score 3.75, std. deviation 0.71, Austria – average score 2.63, std. deviation 0.92, and Slovenia – average score 2.38, std. deviation 1.06). Although relatively successful in the field of e-health development, according to our comparative analysis Austria achieved comparatively lower results than Denmark in most of the categories compared. Slovenia showed the least progress in the field of e-health development and achieved the lowest score in the comparative analysis, considerably lagging behind the Denmark, and Austria as well. Danish e-health project achieved superior results in all comparative categories, except two (performance evaluation of e-health, health care system performance indicators).

The most visible gap involving the comparison of strategic documents related to the development of e-health is reflected in the very start of the e-health project in Denmark, which was initiated nearly 10 years before e-health projects in Slovenia and Austria. In addition, the number of strategies and documents concerning the national project of e-health and general promotion of ICT application in health care exceeds the number of similar documents from Slovenia and Austria. Relating to the number of stakeholders involved, which is comparable to the numbers in Slovenia and Austria, Denmark has obviously managed to dispel the conflicting views and other barriers between them, and establish their quality cooperation, coordination and commitment to the e-health initiative, proving that development and implementation of such complex and important projects require broad social consensus and close interdepartmental collaboration. From the comparative perspective, e-health in Denmark achieved notable results in almost all categories, except in the areas of Telemedicine, Performance evaluation of e-health and Health care system performance indicators, which comparatively accomplished relatively lower results. Denmark is producing excellent comparative results at the level of the EU28 as well, often dominating the top rankings in various classifications of e-health development [5, 6, 74].

Slovenian e-health project has encountered a series of obstacles and setbacks in the course of development and implementation, consequently the date of its completion, earlier planned for 2023, is rather difficult to determine. However, the comparative analysis revealed significant deficiencies in the overall up-to-date development of e-health in Slovenia and a large gap between Slovenia on one hand, and Austria and Denmark on the other hand, especially regarding the development of individual components of e-health. Namely, 5 out of 12 selected and compared components of e-health in Slovenia reached a lower development level than in Austria; however there is even bigger difference compared to Denmark, where Slovenia has achieved a lower development level in 10 of the 12 components compared. Especially concerning is the fact that according to some estimates, e-health development in Slovenia considerably lags behind the EU28 average, as well [1, 5, 6]. Based on the comparison of e-health development in Slovenia, Austria and Denmark (Table 7.1), the explicit deficiencies related to particularly underdeveloped e-health components (listed components were graded with scores less than 3 – partly functional) are summarized and defined below:

- EHR/PHR – Two of the most important components of e-health are in the development phase and currently do not provide required functionality enabling database connectivity for patients migrating from primary to secondary and tertiary health care level);
- Interoperability framework – Component is in the development phase and currently does not facilitate operationalization of adopted standards and integration of existing IS within health care, laboratory and radiology departments (lab results, Picture Archiving and Communication System – PACS, etc.);
- E-prescription – Component is in the conceptual phase and the time frame for its construction and subsequent inclusion in the e-health infrastructure is still indeterminate;
- Telemedicine – Component is in the conceptual phase and although contained in the Slovenian e-health strategy from 2005, development activities in the telemedicine field have not been specified, let alone launched;
- Performance evaluation of e-health – Component is in the conceptual phase, since health policy in Slovenia has not established a methodology including appropriate indicators for evaluating the performance of already implemented operational components of

e-health and monitoring of the components in the development process;

- Legal regulation – Component is in the development phase lacking several important regulations for the e-health application, especially regarding the transfer of medical data, personal data protection, privacy, interoperability standards, liability and risk issues within the usage of EHR/PHR, Telemedicine and E-prescription. Given the existing political debate focused predominantly on economic issues and stringent austerity measures, lack of support and incentives for legislative amendment in the field of e-health is likely to remain unchanged for some time.

Listed components are in the early development stages. Taking into account the complexity of developing such components, time required for their transfer into operational use and the current budgetary restrictions, it is clear that operations depending on these components, and consequently the entire e-health project, will not become fully functional for a long time. This is certainly a broader systemic problem and given the scope of health care system, its relations and interdependencies with other segments of the society [6], it should be noted that e-health is only a part of the complex social system [63], while its perception and subsequent application are deeply rooted in the social mode of behavior and working practices of organizations and people [49, 75, 76].

The deficiencies within the development of e-health in Slovenia, which obviously extend to several areas, such as policy-regulatory, financial, institutional and technological area, could have been mitigated by taking appropriate measures in the course of its conceptualization, planning and implementation. Nevertheless, exposed deficiencies have significant impact on overall performance of e-health, and consequently do not allow its effective utilization for improvement of health care services and evidence-based management of the health care system. The most significant deficiencies revealed by our research are summarized below:

- Absence of top-down support for implementation of e-health;
- Poorly defined health care policies and e-health project objectives;
- Unadjusted and hyper-regulated normative framework;
- Insufficient funding, lack of management skills and human resources;
- Fragmentation and large number of diverse legacy IS on all three levels of health care system;

- Partially defined communication network standards and data exchange standards;
- Lack of standardized definitions of health and social data required for development and management of EHR, PHR and DRG;
- Disregarding interoperability perspective while procuring an increasing number of narrowly specialized IS;
- Inadequate and vague evaluation practice in the field of major ICT projects;
- Lack of experience in the execution of complex and long-term national (ICT) projects;
- Unawareness of the potential benefits of e-health and lack of skills within the scope of ICT by the health care professionals;
- Lack of consensus on development priorities as well as cooperation and coordination between key stakeholders.

Deriving from the comparative analysis, issues listed above have not been properly and fully addressed, while they seem to be very important elements of the effective strategy for development and implementation of e-health.

7.5   LESSONS LEARNED AND GUIDELINES FOR FUTURE ACTIONS

Assessing the development and future trajectory of the e-health has proven to be a very difficult task, given the complexity of the e-health projects and lack of appropriate evaluation metrics. Therefore, it is not surprising that in Slovenia, as well as in the international arena, there are only a very small number of research attempts concerning evaluation of the e-health development, especially through the international comparison. Notwithstanding the state of affairs in the research field, certain preliminary conclusions can be drawn. It is evident that deficiencies concerning the progress of the Slovenian e-health extend to the various areas. Identified problems are reflected in the unsatisfactory development level of individual infrastructure components and e-health project as a whole, whereas on-going financial and economic crisis just revealed the magnitude of pertaining deficiencies, additionally undermining public trust and stakeholders' engagement. Health care systems which strive for the successful development and implementation of e-health projects generally have to overcome difficulties with the political, legal/regulatory and technical constraints, provide appropriate funding for material and immaterial resources, and precisely specify the course and objectives of the e-health projects.

Analyzing the current situation in the field of e-health, we identified various deficiencies, which have in our opinion substantially affected the development of e-health in Slovenia. Some of the problems associated with e-health development and implementation have been expected, given its scope and complexity, while the other complications appeared unpredictably and were merely the results of poor planning and insufficient project analysis. Synthesis of the research results and derived deductions, based on the identified deficiencies, are presented in the form of general guidelines, which could facilitate a more effective and structured approach to the realization of the e-health projects:

- Obtain political support, bring together stakeholders from the public sector, not-for-profit organizations and the private sector, and prepare viable strategy documents and action plans (assess the current ICT infrastructure, departmental ISs, legacy ISs, interoperability issues, specify the health information standards, education and training of the medical staff, analyze different informational needs of primary, secondary and tertiary health care level, check the financial construction and financial projections related to the budget of e-health in the medium and long-term, examine the potential obstacles to e-health realization and conduct a sensitivity analysis, etc.);
- Examine current and projected health care issues, incorporate country specificities, determine national health care priorities, and provide an action plan clearly specifying how e-health will contribute to the solution of national health care priorities, as well as enable desired reorganization and restructuration of the health care system itself;
- Select a top manager and a quality project team with experience in large ICT projects, clearly structure the project plan, project phases and deliverables for each phase, determine the timeline of the project by reaching mutual consensus with all stakeholders, distribute the assignments and strictly monitor and inspect the work on the project;
- Ensure adequate resources before the start of each phase of the project and make realistic plans within both temporal as well as financial terms;

- Mobilize all stakeholders to ensure commitment, material and moral support, encourage their participation and constructive criticism, provide an inclusive plan for permanent education of the stakeholders and communication between the project team;
- Enhance the preparation and implementation of public tenders (materially and procedurally) related to procurement of ICT equipment and realization of smaller individual ICT projects within the overall e-health project;
- Perform a constant supervision and strict control of the already executed project tasks with respect to the substantive and temporal objectives, and ensure close monitoring of the tasks which are in the execution phase;
- Inform and sensitize the public, promote project achievements so far, organize marketing campaign to popularize the e-health project and increase user acceptance of e-health services, gain support from the media, experts and citizens; e-health is a socio-technical project.

Presented research results cannot be easily transferred into action, while the poor progress in development of e-health in Slovenia is related to several factors. Delays in e-health development require a detailed analysis of the current situation, accommodation of new resources and well-coordinated implementation of operational tasks, which will gradually bring the development of e-health to its final phase. These measures usually necessitate a radical change in the project management and government financial stimulus. Alarming socio-economic situation could jeopardize the latest efforts and compel the government to focus on predominantly short-term economic issues and lower the investments for development of e-health and health care system in general, which could result in far-reaching and irreversible implications for public health in the future. Better exploitation of ICT in health care and eventual provision of medical and economic benefits as well, will therefore require the mobilization of all stakeholders and experts in the field, definition of clear and measurable objectives and a broad consensus about the necessary public expenditures.

7.6 CONCLUSIONS

Considering the almost unparalleled role of ICT in the modern health care systems, e-health currently represents a very hot topic, which could determine the main trajectory of the health care system behavior in the future. Conducting a case study on the e-health development has emerged as a very challenging mission. So far there is no universally-acknowledged methodology for evaluating the development of overall e-health projects or their individual components, while the efforts trying to provide at least some kind of comparative framework or conduct international comparative analysis of e-health development are extremely limited. Although reasonably susceptible to subjectivity and arbitrary interpretations, case study in hand provides a valuable insight into the underlying forces and specific characteristics of e-health projects in Slovenia, Austria and Denmark, and can hopefully contribute to theory building in the field. In addition, stemming from the deficits identified in Slovenia, the findings and guidelines could represent useful starting points for more informed and consistent decision-making throughout the whole process of development and implementation of e-health projects in other countries, which have been facing similar problems, despite different contextual features and system dynamics.

Main limitations of the study probably concern the adequacy of performed weighting process and the fact that development level of individual e-health component was actually defined on the basis of primary and secondary sources investigation without empirical testing and practical validation of each component in the health care environment. Accordingly, the issues of equal weights assigned to designated indicators and objective definition of development level raise some questions of principle, while the results of the comparative analysis may therefore be arguable. These issues should be properly resolved in further research and succeeding experiments trying to establish a theory-based and balanced framework for evaluation and comparative analysis of the e-health development in national and international context.

Despite certain methodological dilemmas and limited resources, conducted case study reveals the intricate dynamics of the e-health development and potential deficiencies and barriers. Moreover, the comparative analysis, including designated guidelines, may eventually provide the groundwork for more effective development and implementation of the intractable and costly e-health projects, and useful assistance for enhanced allocation of project management resources.

KEYWORDS

- **case study**
- **comparative analysis**
- **deficiencies and barriers**
- **development guidelines**
- **e-health in Slovenia, Austria and Denmark**

REFERENCES

1. Ministry of Health. Upgrade of the healthcare system by 2020. Government of the Republic of Slovenia: Ljubljana, 2011.
2. Ministry of Health. Strategy for informatization of the Slovenian healthcare system 2005–2010 (e-health 2010). Government of the Republic of Slovenia: Ljubljana, 2005.
3. Ministry of Health. Resolution on the National Healthcare Plan for the period 2008–2013. Government of the Republic of Slovenia: Ljubljana, 2008.
4. Chaudhry, B., Wang, J., Wu, S., Maglione, M., Mojica, W., Roth, E., Morton, S. C., Shekelle, P. G. Systematic Review: Impact of Health Information Technology on Quality, Efficiency, and Costs of Medical Care. *Annals of Internal Care.* 2006, *144(10)*, 742–752.
5. European Commission. Benchmarking ICT use among General Practitioners in Europe. Final Report. European Commission, Information Society and Media Directorate: Brussels, 2008.
6. European Commission. e-health Benchmarking III. Final Report SMART 2009/0022. Deloitte & Ipsos Belgium. European Commission, Information Society and Media Directorate: Brussels, 2011.
7. Nyamtema, A. S. Bridging the Gaps in the Health Management Information System in the Context of a Changing Health Sector. *BMC Medical Informatics and Decision Making.* 2010, *10(36)*, 1–6.
8. Bardhan, I. R., Thouin, M. F. Health information technology and its impact on the quality and cost of healthcare delivery. *Decision Support Systems.* 2013, *55(2)*, 438–449.
9. Goldzweig, C. L., Towfigh, A., Maglione, M., Shekelle, P. G. Costs and benefits of health information technology. *Health Affairs.* 2009, *28(2)*, 282–293.
10. Valeri, L., Giesen, D., Jansen, P., Klokgieters, K. Business models for e-health. European Commission, Information Society and Media Directorate: Brussels, 2010.
11. Walsham, G. Integrated health information architecture: power to the users. *Information Technology for Development.* 2013, *19(3)*, 264–266.
12. Stanimirovic, D. Empirical analysis of IT outsourcing in municipal governments – drivers, inhibitors and predictive factors. *Information.* 2014, *17(6A)*, 2041–2054.

13. Stanimirovic, D., Vintar, M. Development of e-health at a national level – comparative aspects and mapping of general success factors. *Informatics for Health and Social Care*. 2014, *39(2)*, 140–160.

14. Haux, R. Medical informatics: past, present, future. *International Journal of Medical Informatics*. 2010, *79(9)*, 599–610.

15. Iveroth, E., Fryk, P., Rapp, B. Information technology strategy and alignment issues in health care organizations. *Health Care Management Review*. 2013, *38(3)*, 188–200.

16. Patton, M. In: Qualitative Evaluation and Research Methods (2nd ed.). Sage Publications: Thousand Oaks, 1990.

17. Yin, R. In: Case study research: design and methods (4th ed.). Sage Publications: Thousand Oaks, 2009.

18. Black, A. D., Car, J., Pagliari, C., Anandan, C., Cresswell, K., Bokun, T., McKinstry, B., Procter, R., Majeed, A., Sheikh, A. The impact of e-health on the quality and safety of health care: a systematic overview. *PLoS Medicine*. 2011, *8(1)*, 1–19.

19. Gillies, A. C., Howard, J. Information as Change Agent or Barrier in Health Care Reform? *International Journal of Healthcare Information Systems and Informatics*. 2011, *6(1)*, 19–35.

20. Eysenbach, G. What is e-health? *J Med Internet Res*. 2001, *3(2)*. http:// www.jmir. org/2001/2/e20/ (accessed February 22, 2014).

21. Marconi, J. E-Health: Navigating the Internet for Health Information Healthcare. Advocacy White Paper. Healthcare Information and Management Systems Society: Chicago, 2002.

22. Healthcare Information and Management Systems Society – HIMSS. E-Health Defined. White Paper. E-Health Special Interest Group (SIG): Chicago, IL, 2003.

23. Alpay, L. L., Henkemans, O. B., Otten, W., Rövekamp, T. A., Dumay, A. C. E-health applications and services for patient empowerment: directions for best practices in The Netherlands. *Telemedicine and e-Health*. 2010, *16(7)*, 787–791.

24. Dedding, C., van Doorn, R., Winkler, L., Reis, R. How will e-health affect patient participation in the clinic? A review of e-health studies and the current evidence for changes in the relationship between medical professionals and patients. *Social science & medicine*. 2011, *72(1)*, 49–53.

25. Gibbons, M. C., Fleisher, L., Slamon, R. E., Bass, S., Kandadai, V., Beck, J. R. Exploring the potential of Web 2.0 to address health disparities. *Journal of health communication*. 2011, *16(1)*, 77–89.

26. Ahern, D. K., Kreslake, J. M., Phalen, J. M. What Is e-health: Perspectives on the Evolution of e-health Research. *J Med Internet Res*. 2006, *8(1)*. http:// www.jmir.org/ 2006/1/e4/ (accessed February 15, 2014).

27. International Telecommunication Union – ITU. Implementing e-Health in Developing Countries: Guidance and Principles. ITU Telecommunication Development Sector: Geneva, 2008.

28. Oh, H., Rizo, C., Enkin, M., Jadad, A. What is e-health? A systematic review of published definitions. *World Hosp Health Serv*. 2005, *41(1)*, 32–40.

29. Murray, E., Burns, J., May, C., Finch, T., O'Donnell, C., Wallace, P., Mair, F. Why is it difficult to implement e-health initiatives? A qualitative study. *Implementation Science*. 2011, *6(6)*, 1–11.

30. Nykänen, P., Brender, J., Talmon, J., de Keizer, N., Rigby, M., Beuscart-Zephir, M. C., Ammenwerth, E. Guideline for good evaluation practice in health informatics (GEP-HI). *International Journal of Medical Informatics*. 2011, *80(12)*, 815–827.

31. Haux, R. Health information systems – past, present, future. *International Journal of Medical Informatics*. 2006, *75(3–4)*, 268–281.

32. Li, L., Ge, R. L., Zhou, S. M., Valerdi, R. Integrated healthcare information systems. *IEEE Trans Inf Technol Biomed*. 2012, *16(4)*, 515–527.

33. Trudel, M. C., Paré, G., Laflamme, J. Health information technology success and the art of being mindful: Preliminary insights from a comparative case study analysis. *Health Care Management Review*. 2012, *37(1)*, 31–42.

34. Leung, R. C. Health information technology and dynamic capabilities. *Health Care Management Review*. 2012, *37(1)*, 43–53.

35. Arndt, M., Bigelow, B. Evidence-based management in health care organizations: A cautionary note. *Health Care Management Review*. 2009, *34(3)*, 206–213.

36. Hunter, R. L. Health information technology costs and patient safety concerns. *Osteopathic Family Physician*. 2011, *3(4)*, 154–160.

37. Stroetman, V. e-health for safety: impact of ICT on patient safety and risk management. Report prepared for ICT for Health Unit. European Commission, Information Society and Media Directorate, RAND Europe, Capgemini Consulting: Brussels, 2007.

38. Vest, J. R., Jasperson, J. S. How are health professionals using health information exchange systems? Measuring usage for evaluation and system improvement. *Journal of Medical Systems*. 2012, *36(5)*, 3195–3204.

39. Lucas, H. Information and communications technology for future health systems in developing countries. *Social Science and Medicine*. 2008, *66(10)*, 2122–2132.

40. Lluch, M. Healthcare professionals' organizational barriers to health information technologies – A literature review. *International Journal of Medical Informatics*. 2011, *80(12)*, 849–862.

41. Piette, J. D., Lun, K. C., Moura, L. A. Jr., Fraser, H. S., Mechael, P. N., Powell, J., Khoja, S. R. Impacts of e-health on the outcomes of care in low-and middle-income countries: where do we go from here? *Bulletin of the World Health Organization*. 2012, *90(5)*, 365–372.

42. Winter, A., Brigl, B., Funkat, G., Haeber, A., Heller, O., Wendt, T. 3LGM2-Modeling to support management of health information systems. *International Journal of Medical Informatics*. 2007, *76(2–3)*, 145–150.

43. Jensen, T. B. Design principles for achieving integrated healthcare information systems. *Health Informatics J*. 2013, *19(1)*, 29–45.

44. Bush, M., Lederer, A. L., Li, X., Palmisano, J., Rao, S. The alignment of information systems with organizational objectives and strategies in health care. *International Journal of Medical Informatics*. 2009, *78(7)*, 446–456.

45. Walker, J. The value of health care information exchange and interoperability. *Health Affairs*. 2005, *25(6)*, 5–10.

46. Jaana, M., Tamim, H., Paré, G., Teitelbaum, M. Key IT management issues in hospitals: Results of a Delphi study in Canada. *International Journal of Medical Informatics*. 2011, *80(12)*, 828–840.

47. Lapointe, L., Mignerat, M., Vedel, I. The IT productivity paradox in health: A stakeholder's perspective. *International Journal of Medical Informatics.* 2011, *80(2)*, 102–115.

48. Protti, D. J. A Comparison of Information Technology in General Practice in Ten Countries. *Electronic Healthcare in Healthcare Quarterly.* 2007, *10(2)*, 107–116.

49. Kaye, R., Kokia, E., Shalev, V., Idar, D., Chinitz, D. Barriers and success factors in health information technology: A practitioner's perspective. *Journal of Management and Marketing in Healthcare.* 2010, *3(2)*, 163–175.

50. Heeks, R. Health information systems: Failure, success and improvisation. *International Journal of Medical Informatics.* 2006, *75(2)*, 125–137.

51. Ibrahim, R., Ayazi E., Nasrmalek, S., Nakhat, S. An Investigation of Critical Failure Factors In Information Technology Projects. *Journal of Business and Management.* 2013, *10(3)*, 87–92.

52. Karsh, B. T., Weinger, M. B., Abbott, P. A., Wears, R. L. Health information technology: fallacies and sober realities. *J Am Med Inform Assoc.* 2010, *17(6)*, 617–623.

53. Haux, R., Howe, J., Marschollek, M., Plischke, M., Wolf, K. H. Health-enabling technologies for pervasive health care: on services and ICT architecture paradigms. *Informatics for Health and Social Care.* 2008, *33(2)*, 77–89.

54. Kanjo, C. Pragmatism or policy: Implications on health information systems success. *The Electronic Journal of Information Systems in Developing Countries.* 2011, *48(1)*, 1–20.

55. World Health Organization – WHO. International Classification of Diseases (ICD). 10th Revision, World Health Organization: Geneva, 2012.

56. United Nations – UN. Introducing UN/EDIFACT. http://www.unece.org/trade/untdid/welcome.html (accessed April 20, 2014).

57. Health Level Seven International – HL7. About HL7. http://www.hl7.org/about/index.cfm (accessed April 8, 2014).

58. World Wide Web Consortium – W3C. Extensible Markup Language (XML). http://www.w3.org/XML/ (accessed April 15, 2014).

59. Government of Austria. The Austrian E-Government Act. Federal Act on Provisions Facilitating Electronic Communications with Public Bodies. Republik Österreich: Vienna, 2004.

60. Government of Austria. The Austrian Health Reform. The improvement of the efficiency of the health care system and ensuring sustainable financing, Ministry of Health and Women: Vienna, 2005.

61. Government of Austria. Health Telematics Law. GTelG. BMGF, Hofmarcher, M.M. Abteilung IV/6 – Gesundheitstelematik. Republik Österreich: Vienna, 2005.

62. European Commission. i2010 A European Information Society for growth and employment. European Commission, Information Society and Media Directorate: Brussels, 2005.

63. European Commission. e-health priorities and strategies in European countries. Towards the Establishment of a European e-health Research Area. European Commission, Information Society and Media Directorate: Brussels, 2007.

64. Pfeiffer, K. P., Giest, S., Dumortier, J., Artmann, J. e-health Strategies. Country Brief: Austria. European Commission, Information Society and Media Directorate: Brussels, 2010.

65. Government of Austria. ELGA. Elektronische Gesundheitsakte. Bundesgesundheit-sagentur. Republik Österreich: Vienna, 2009.

66. Government of Austria. The Austrian Internet Declaration. Rundfunk und Telekom Regulierungs-GmbH. Republik Österreich: Vienna, 2010.

67. Government of Denmark. Strategy for the development of Electronic Patient Records. Ministry of Health and Prevention: Copenhagen, 1996.

68. Government of Denmark. National Strategy for Information Technology in Hospitals. Ministry of Health and Prevention: Copenhagen, 1999.

69. Government of Denmark. National Strategy for Information Technology in the Healthcare System 2003–2007. Ministry of Health and Prevention: Copenhagen, 2003.

70. Jensen, H. B., Pedersen, C. D. MedCom: Danish Health Care Network in Current Situation and Examples of Implemented and Beneficial E-Health Applications. IOS Press: Copenhagen, 2004.

71. Doupi, P., Renko, E., Giest, S., Dumortier, J., e-health Strategies. Country Brief: Denmark. European Commission, DG Information Society and Media, ICT for Health Unit: Brussels, 2010.

72. Government of Denmark. National Strategy for Digitalisation of the Danish Healthcare Service. Ministry of Health and Prevention: Copenhagen, 2008.

73. Government of Denmark. e-health in Denmark. e-health as a part of a coherent Danish health care system. Ministry of Health and Prevention. National Board of e-health and Sundhed.dk: Copenhagen, 2012.

74. European Commission. ICT for better Healthcare in Europe. e-health – Better Healthcare for Europe. European Commission, Information Society and Media Directorate: Brussels, 2009.

75. Anwar, F., Shamim, A., Khan, S. Barriers in adoption of health information technology in developing societies. *Int. J Adv. Comput. Sci. Appl.* 2011, *2(8)*, 40–45.

76. World Health Organization – WHO. HMN Framework and Standards for Country Health Information System Strengthening. Health Metrics Network: Geneva, 2009.

CHAPTER 8

ACCEPTANCE OF E-HEALTH BY HEALTHCARE PROFESSIONALS IN DEVELOPING COUNTRIES: CASE OF MOROCCO

RACHID OUMLIL

Associate Professor, Department of Management, ENCG-Agadir, Ibnou Zohr, Agadir, Morocco, Africa, Tel: (212) 6-61-16-90-87, E-mail: r.oumlil@uiz.ac.ma

CONTENTS

8.1   INTRODUCTION

Healthcare organizations include medical and administrative units. Medical ones deploy skills and knowledge to promulgate patient care. As for the administrative units, they provide necessary tools to support medical activities. Both of these two units generate diverse information. Thus, usage of Information Technology (IT) becomes a crucial strategic challenge for these organizations (Ammenwerth et al., 2004). Some authors noted that IT usage offers opportunities ranging from cost reduction and clinical errors, support of care professionals and improving the quality of care efficiency (Scott et al., 2007).

Since 1997, Morocco, as a developing country, has undergone several issues related to financing arrangement, care access, productivity and quality of healthcare organizations. These issues led the Moroccan government to a process of structural reforms enabling health sectors to improve management processes, quality and performance of healthcare organizations.

In 2001, the government launched an integrated project to support healthcare reforms: the Project Management and Financing of Health Sector (PMFHS). This aimed to introduce managerial tools to support changes in healthcare system. It provided the Moroccan population with integrated and standardized care services. It gave better support to healthcare professionals and to improve the quality and efficiency of the promulgated care services. Implementing IT in the administrative and healthcare processes was considered as one of the main goals of this project. Despite these efforts to enhance the care system, the project still faces to two major shortcomings: (i) an unequal distribution of health care promulgation, and (ii) an inadequate hospital management.

In 2005, Moroccan government launched E-health project. This later aimed to provide timely reliable indicators to health decision makers and to improve availability and access to information and data related to citizens and health professionals. Moreover, a strategy 2008–2012 was proposed, it aimed ensuring fairness of care offer between regions and at facilitating access to the rural population. It also aimed to provide citizens confidence toward health system by reducing hospital cost and improving health quality.

Furthermore, another strategy is proposed for the period 2012–2016. It is considered as a consolidation and a continuum of the past strategies. Improvement of the health system governance was the pillar axe of this strategy. It is focused on IT usage and its contribution to improve health quality within healthcare organizations.

As part of this initiative, Provincial Hospital (PH) Hassan Ist of Tiznit city, located in south of Morocco expected to integrate IT within its departments. Many objectives are behind this integration: improve the care quality and communication, reduce waiting time for patients, and reduce hospital costs. However, this integration is impeded by to cultural barriers related to the reluctance of health professionals (nurses and physicians) toward IT acceptance.

In fact, acceptance and integration of e-health into the everyday life of healthcare workers became a reality in developing as well as developed countries (World Health Organisation, 2004). It is almost impossible to make a strategic decision without involving this technology in healthcare organizations (Arlotto and Oakes, 2003). Hence, healthcare organizations invested heavily in IT expecting to improve their service quality and performance (Davaraj and Kohli, 2003).

Yet, the expected benefits cannot be harvested unless this technology is accepted by health professionals. Thus, the main objective of this chapter is to identify factors predicting E-health acceptance by physicians and nurses working in developing countries. It suggests a model based on Technology Acceptance Model (TAM) (Davis et al., 1989) extended by two added constructs "Image" and "Trust." TAM is considered a salient model that was widely used to predict IT acceptance in many contexts; nevertheless, it is not sufficiently used to predict Healthcare information technologies in developing countries. Hence, determined factors will help stakeholders involved in E-health and healthcare information systems projects especially in developing countries.

The study was conducted in the Provincial Hospital (PH) Hassan I[st] of Tiznit city, located in south of Morocco. The choice of this hospital is due essentially to: (1) It belongs to a region considered as the pilot site for the E-health integration in Morocco (2) This hospital is considered as the second organization in the region that plans to integrate E-health.

The rest of the chapter is organized as follows: In the theoretical background section we review relevant theories and models of technology acceptance, we add studies based on TAM to predict E-health acceptance. Specific hypotheses are developed in the research model and Hypothesis Development section. The methodology section describes the quantitative study and the results are presented in the results section. We then go over our results in the discussion section. The chapter ends with a brief conclusion.

8.2 THEORIES AND MODELS PREDICTING INFORMATION TECHNOLOGY ACCEPTANCE

Research in information systems studied why and how individuals accept IT (Venkatesh et al., 2003). It refers to theories and models that deal with IT acceptance issue (Kukafka et al., 2007). Many theoretical models attempt to predict the level of IT acceptance and even explain why and how this technology is used. The following section reviews briefly these theories and describes TAM model considered as the basis to build up a framework allowing prediction of IT acceptance.

8.2.1 THEORIES PREDICTING INDIVIDUAL BEHAVIOR TOWARD INFORMATION TECHNOLOGY

The behavioral approach related to acceptance and usage of technologies provides a general framework to understand and predict individual behavior toward IT. It mobilizes theories from several fields: Sociology, Psychology and Management Information Systems. The main objective of this subsection is to review the relevant theories predicting IT acceptance by individuals.

8.2.1.1   Innovation Diffusion Theory

In 1983, Rogers developed Innovation Diffusion Theory (IDT) to understand diffusion and adoption of innovations. This theory provides a conceptual framework to identify factors influencing IT acceptance by individuals
and organizations. It describes the process through which new ideas, practices, or technologies are spread into a social system (Rogers, 2003).

In the beginning, IDT attempted to understand innovation adoption by a social group only. Furthermore, several studies showed that it
could be used for individual and collective decisions toward IT adoption
(Compeau et al., 1999). In 1995, Rogers identified five factors predicting
diffusion innovation adoption: (1) perceived attributes; (2) type of decision (3) communication channel; (4) social system; and (5) change agent.
He suggested that to accept an innovation, the individual should first be
informed of its existence. After then, he could develop an attitude towards
this innovation, and then decide whether to accept or reject it. Hence, an
individual with a positive attitude towards an innovation should, first,
adopt it and then use it. Rogers states that the perceived attributes include
five theoretical constructs to accept a new technology.

1. Relative Advantage: Degree to which an innovation is perceived as
being better than the idea it supersedes. It is not necessary if an innovation has more objective advantages. What actually matters is whether and
individual perceived this innovation as being beneficial (advantageous).

2. Compatibility: Degree to which an innovation is perceived as consistent with existing values, style of work, past experiences, social practices
and standards users. An innovation that is incompatible with the values
and norms of a social system will not be accepted as rapidly as the one
that is compatible.

3. Complexity: Degree to which an innovation is perceived as difficult to
use. New ideas that are simple to understand will be accepted much faster
than others that require development of new skills for their comprehension.

4. Trialability: Degree to which an innovation may be experimented with
a limited basis. The opportunity to test an innovation allows potential
users to have more confidence; and,

5. Observability: Degree to which results and benefits of innovation are visible and accessible. Individuals are more likely to accept an innovation whenever its results are clear.

Furthermore, accepted innovations are those that have a strong comparative advantage, an ability to try, a satisfactory visibility, a low complexity, and finally a high compatibility.

IDT aims to understand the phenomenon of innovations acceptance (Rogers, 1995), however it has many limits: It does not explain clearly IT acceptance, or how to avoid users reluctances toward this acceptance (Chau and Tam, 1997). Hence the need for other theories that would predict technology acceptance.

8.2.1.2   Theory of Reasoned Action

In 1975, Fishbein and Ajzen developed Theory of Reasoned Action (TRA). It is a psychosocial theory that explains behavior in an interaction perspective between individual and his social environment. It provides a solid basis for understanding and predicting the individual behavior (Benbasat and Barki, 1995). It proposes a model predicting individual intention. This model suggests that individual behavior is directly determined by Behavioral Intention that is influenced by attitude and subjective norms.

Unlike the IDT based on the innovation itself, the TRA examines perceptions of the actual use of an innovation (Moore and Benbassat, 1991). It is seen as a general model that can be used for predicting individual behavior. Many limits of this theory could be listed: The first one concerned the relationship between attitudes and subjective norms. Miniard and Cohen (1981) noted that the TRA did not distinguish effectively between attitude and subjective norm impacts on Intention. As for the second limit, it is due to the fact that the behavior is under individual control, something that is highly questionable.

Faced with these limits, Ajzen (1991) modified the TRA and proposed an alternative theory to predict individual behavior: Theory of Planned Behavior (TPB).

8.2.1.3   Theory of Planned Behavior

TRA asserts that individual behavior is totally controlled by himself (Fishbien and Ajzen, 1975). Eleven years later, in 1986, Ajzen and Madden underlined the importance of individual feelings to conduct and control an action criticized this statement. Hence the development of TPB proposed by Ajzen (1991). This theory stated that the individual behavior is directly determined by his intention.

TPB is characterized by the introduction "perceived behavioral control" that determines "usage behavior." Ajzen (1991) stated that behavioral intention is a combination of three variables: (1) attitude, (2) subjective norm, and (3) behavioral control. The later refers to perceptions constraints behavior (Ajzen, 1988). It explains individual's perception toward difficulty or easy of performing such behavior (Ajzen, 1991).

TPB proves that attitude, subjective norm and perceived behavioral control influence positively the individual Intention to accept a technology. Its main interest is to provide a model that could be used in all situations where the behavior is intentional. The main limit of this theory was absence of factor analysis explaining the link between beliefs and intentions.

8.2.1.4   Social Cognitive Theory

Bandura (1977) developed the Social Cognitive Theory (SCT). He considered contributions of behaviorism and socio-psychology. SCT placed individuals at the center of cognitive, behavioral and contextual factors interaction. It differed from IDT, TRA and TPB that emphasized the influence of environmental factors on the individual behavior. It explains how people acquire and maintain their behaviors by providing a basis for intervention strategies. In 1980, Bundura stated two main objectives of the (SCT):

- To understand and predict individual and groups behavior; and
- To identify methods and ways to change behavior.

For Bandura (1986), human behavior is seen as the product of dynamic interactions between personal and environmental influences. He underlined the importance of interrelation between three factors: personal, environmental and behavioral. Thus the influence of the environment on behavior remains prominent.

Social cognitive theory is related to a set of variables:

- **Social environments**: It includes family members, friends and colleagues. It provides a framework for understanding the behavior;

- **Situation**: It explains environment perception. It refers to mental or cognitive representations that can influence individual behavior;

- **Behavioral capacity**: It includes knowledge and skills to perform a given behavior;

- **Expectations**: They explain the results of proactive behavior;

- **Self-efficacy**: It is defined as a set of beliefs about the ability of a person to adopt such a behavior; and

- **Reciprocal Determinism**: It explains the dynamic interaction of the person behavior and environment.

SCT argues that technology acceptance is not based on convincing individuals of its profits only; it requires their skills and their trust toward this technology.

8.2.1.5   Interpersonal Behaviors Theory

Interpersonal Behavior Theory (IBT) referred to various disciplines (Sociology, Psychology and Anthropology) to understand and predict the individual behavior. It integrates most of variables presented in the behavior theories (TRA and TPB). It introduces two sets of factors: (i) factors related directly to behavior (genetic factors, social status, Habits, Attitudes, Intentions and Behavior); (ii) Factors related to the environment (culture, social factors and facilitating Conditions).

Triandis (1980) referred to principles of the TRA. To develop his model, he integrated "habit" and "External conditions" that facilitate or complicate individual behavior. Furthermore, he distinguishes between "beliefs" that bind emotions to behavior, and those that bind the act to future consequences.

IBT states that individual behavior is determined by three main factors: (i) intention to accept this behavior, (ii) habit and (iii) facilitating Conditions. IBT model showed that Habit influences both of behavior and attitudes (affect), while facilitating conditions include contexts and situations that could foster behavior accept. Intention as the most important variable of this model is determined by individual feeling toward a behavior

(Attitude of Affect), by social factors and by outcomes expected of his behavior (perceived consequences).

IDT, TRA, TPB, SCT and IBT provide a theoretical basis to predict the individual behavior toward Information technology.

8.2.2 TECHNOLOGY ACCEPTANCE MODEL

Researches conducted on IT acceptance (Karahanna, Straub and Chervany, 1999) showed that a large number of models were derived from TRA (Ajzen and Fishbien, 1980). One of these models is the Technology Acceptance Model (TAM) (Davis et al., 1989).

TAM is the most used by researchers in Information Systems field.

It is considered the most robust model to identify variables influencing individual acceptance to use particular technology or not (Venkatesh and Davis, 2000).

It was introduced to explain individual behavior toward computers (Davis et al., 1989). It mobilized TRA to identify key relations between its main constructs: (i) perceived usefulness, (ii) perceived ease of use, (iii) attitude, (iv) intention and (v) usage. It is a causal model that studies relationships between these five constructs. According to TAM, IT use is explained by the Behavioral Intention.

TRA gathered all beliefs in one construct: subjective norms (Ajzen, 1991). TAM enriched TRA by incorporating explicitly external variables to model user behavior. It shows how these variables impact two specific beliefs: PEOU, PU, and how they influence Attitude and Intention to Use Technology to predict actual use.

TAM model suggested that IT use could be determined by behavioral intention. This later is influenced by attitude toward IT usage. PU and PEOU are the prominent components of this model. They were inspired from IDT (Rogers, 1995) that considered complexity and relative advantage as two main determinants of innovation adoption. Complexity joined PEOU, while the Relative Advantage indicated PU.

PU explained the degree to which a person believes that using a particular technology would improve his work performance (Davis, 1989). In other words, it's a subjective probability indicating that an individual agrees technology usage would improve his performance in an organizational context. Direct relation between PU and Behavioral Intention is based on the assumption that IT use decision is made on the IT use performance perceived by the user. Furthermore, PU is influenced by the PEOU and many other external variables.

PEOU indicated the degree to which the user thinks that the use of a particular technology would be without effort. It implies that a technology perceived as ease is more likely to be used. Benbasat et al. (1986) noted that the PEOU is influenced by characteristics of the technology itself.

These two main constructs, PU and PEOU are affected positively or negatively by external factors (Agarwal and Prasad, 1998, Hong et al., 2002). The model provides a basis for tracing the impact of external factors on internal beliefs, "Attitude" and "Intention".

In summary, TAM was the most widely model, the most robust and the most parsimonious model used to predict technology acceptance across various contexts (Holden and Karsh, 2010; Egea and González, 2011). It is due to its attribute of being a generic model that can be applied to any context using IT. TAM posits that two fundamental beliefs, PU and PEOU, determined an attitude, behavioral intention (BI), and actual usage of Information Technology (Davis et al., 1989).

8.3   ACCEPTANCE OF E-HEALTH

The ubiquitous usage of Information Technology in healthcare context engendered e-health concept. Referring to Eysenbach (2001), e-health is an emerging field in the intersection of medical informatics, health and business. It is driven from the expansion of health care informatics and biomedical computer (Bennani et al., 2008). In this chapter, E-health is seen as the usage of IT healthcare activities. To understand E-health acceptance, many studies have been developed testing TAM. The following section revues studies predicting E-health by healthcare professionals for the period between 2007 and 2013.

In 2007, Wu et al., aimed to develop a conceptual framework for significant factors influencing the medical professional behavioral intention to accept Mobile Healthcare Systems (MHS). TAM and IDT served as the theoretical basis to develop their research model. Confirmation factor analysis was performed to test the reliability and validity of the measurement model, and the structural equation modeling technique was used to evaluate the causal model. Their results indicated that compatibility, PU and PEOU significantly affected healthcare professional intention. Furthermore, MHS self-efficacy had strong indirect impact on healthcare professional acceptance intention through the mediators of perceived usefulness and Perceived Ease of Use. However, the hypotheses for technical support and training effects on the PU and PEOU were not supported.

Kim and Chang (2007) observed the importance of health information websites, and noted the limit studies presenting significant solutions to problems related to effective operations of health information websites. To overcome this limitation, they explored factors affecting health information websites acceptance from the perspective of both the user and the provider. To do, they employed TAM and extend it by four exogenous variables: (i) information search, (ii) usage support, (iii) customization, and (iv) purchase and security. The developed model was tested on a target population of 250 online consumers. Only 228 respondents were retained. Results from the structural equation analysis suggested that PU had a significant effect on Customer Satisfaction. Moreover PEOU turned out to have an indirect effect on this satisfaction. Healthcare services are radically different from ordinary services. Self-Service Technology (SST) is considered as a means to reduce costs and improve quality in the health care sector. Lanseng and Andreassen (2007) examined the introduction of this technology in health diagnosis. For this issue, they mobilized TAM to identify factors that may influence the public's acceptance and adoption of a self-diagnosis system. The study was conducted on a population of 470 inhabitants in an affluent suburb of Oslo the capital of Norway. Results revealed that TAM has an excellent ability to predict SST acceptance in the health care services context. As for Park and Chen (2007), they investigated human motivations affecting medical doctors and nurses' smartphone acceptance. They referred to TAM, users' perceived adoption under the self-efficacy, and innovation attributes to develop their model. To test this later, 823 questionnaires were mailed to medical doctors and nurses, and healthcare providers stationed throughout clinics and the main hospital in the healthcare network. Only 135 were received and 20 were returned undeliverable. Results showed that medical doctors and nurses intention to accept smartphone was influenced by PU and attitude toward using smartphone.

In 2008, Wu et al., noted that several healthcare organizations invested in adverse event reporting systems to prevent adverse events and medical errors. However, a number of these systems have failed. Hence, the authors, tried to understand factors that might influence healthcare professional's acceptance of adverse event reporting system. To do, they extended TAM by integrating trust and management support to develop their model. After, they tested it on a population of 940 health professionals from 144 Taiwanese hospitals implementing reporting systems. The outcomes indicated that PU, PEOU, subjective norm, and trust had a significant effect on a professional's intention to accept an adverse event reporting system. Changes in health insurance policy engendered operating expenses within Taiwanese hospitals. Hence the latters attempted to and improve their efficiency throughout integration of the electronic logistics information system within department of central warehouse and the nursing stations. Tung et al. (2008) noted that the first step to achieve this integration is to examine the intention of nurses to accept e-logistics information systems. They combined IDT, the main TAM constructs (PU and PEOU), Trust and Perceived Financial Cost to propose hybrid technology acceptance model. After testing this model on 350 nurses, results revealed that compatibility, PU, PEOU, and trust had a great positive influence on the nurses' intention to adopt e-logistics information system.

Carter (2008) observed the importance of physicians' Perceived Risks associated to usage of novel medical technologies. Hence, he attempted to investigate physicians' attitudes toward the use of smart fabric technology. Otherwise, he examined how PEOU and PU affect physicians' attitudes and intention to accept this technology. Empirically, Carter, collected data using a web-based survey instrument delivered via email to a randomly selected group of physicians. The study concerned 2210 physicians, physician's assistants and medical students. Results showed significant effects of attitude on physicians 'intention to accept smart fabric technology. As the web-based learning technology has been widely recognized and become a valuable and legitimate learning tool for health care professionals, Chen et al. (2008) tried to understand its importance for nursing. They tried to understand public health nurses' intention to accept this technology in pre-implementation stage. They based on TAM constructs to predict this intention. From 369 health centers in Taiwan, 202 public health nurses randomly selected participated in this study. Results of the multiple regression analyzes indicated that PU is the most significant factor for public health nurses' intention to accept web-based learning.

In 2009, Aggelidis and Chatzoglou attempted to test the applicability and effectiveness of technology acceptance models in health care area in Greece. They developed and tested the modify Technology Acceptance Model to examine factors affecting intention of staff (Medical, Nurses and Administrative personnel) using Hospital Information System (HIS) and working in public health institutions. Their results showed that Attitude, PU, Ease of Use, Social Influence, Facilitating Conditions and Self-Efficacy influenced positively personal intention to accept HIS, whereas no significant influence is found between anxiety and this intention. Authors concluded that their modified model can explain 87% of the variance of behavioral intention. Yu and Li (2009) applied a modified Technology Acceptance Model (TAM2) to examine factors explaining Health IT applications by caregivers. Their results revealed that PU, PEOU and computer skills had a strong positive impact on intention of caregivers to accept Health IT. However, Image had the significant negative impact on this intention. Authors pointed out that the research model explained 34% of the variance caregiver's behavioral intention. Lishan et al. (2009) explored differences in perception and acceptance levels of Female-focused Healthcare Applications (FHA). They aimed at detecting factors that would encourage acceptance of these applications. They referred to TAM to develop Female-focused Acceptance Model (FAM). Afterwards, they tested the developed model on a sample comprising of 241 healthcare workers (HC) and 830 non-healthcare (NHC) workers. Results of the regression analyzes showed that PU and PEOU had a significant influence on the intention of female healthcare workers to accept FHA.

Whetstone and Goldsmith (2009) investigated three main questions: (1) what motivates consumers to consider Personal Health Records (PHR) acceptance? (2) What consumer characteristics promote intended PHR usage? (3) Do privacy and security (PS) concern act as impediments to intended PHR usage? Their purpose was to examine factors that bear on consumers' intention to create and to use (PHRs). They laid on TAM to develop their model. This later was tested on a total of 542 U.S. college students with an online questionnaire. Results found that PU influenced significantly PHR acceptance by consumers. Djamasbi, et al. (2009) examined the acceptance of a specialized telemedicine system for microbiology consultation and diagnostics by microbiology laboratory assistants. They referred to TAM constructs and affect variable to develop their model. This later was tested using a laboratory experiment on Thirty-nine microbiology laboratory assistants. Results of the regression analysis revealed that PEOU might not be so important in the healthcare field. As for PU and attitude they showed a significant positive influence on the intention of microbiology laboratory assistants to accept Healthcare Information System.

In 2010, Veer and Francke stated that a great number of health care organizations are integrating electronic patient records (EPR). This integration implied a change in nurses work routines and allowed quality improvement. They explored determinants of nursing staff attitude towards using EPR by extending the TAM. Empirically, the model was tested on 685 nursing staff members working in Dutch hospitals, psychiatric organizations, care organizations for mentally retarded people, home care organizations and nursing homes or homes for the elderly. One of the relevant results of this study was that attitudes towards EPR were primarily associated with job-related characteristics and PU with respect to care quality.

The low rate of mobile healthcare acceptance in Taiwan pushed Wu et al. (2011) to propose a model to understand this acceptance. For this issue, they referred to TAM and TPB. A survey was conducted to examine mobile acceptance healthcare by health professionals in hospitals. 140 valid questionnaires were retained, representing a response rate of 17.5%. Results indicated that Attitude, PU, Perceived behavioral control and Subjective norm had a significant impact on healthcare professionals to accept mobile healthcare. Egea and González (2011) analyzed physicians' individual acceptance decisions of electronic health care records (EHCR) systems. For this purpose, they referred to the original version of the TAM and extended it by trust and risk-related factors. The final sample consists of 254 physicians with private medical practices located in southern Spain representing response rate of about 18%. Results stressed the special importance of Attitude, Trust and PU in predicting EHCR systems by physicians. Melas et al. (2011) underlined the qualitative nature and usage of small convenience samples by researches attempting to understand doctors' and nurses' technology acceptance in the workplace. Hence, they mobilized TAM and introduce physicians' specialty as a moderator to develop their research model explaining clinical information systems acceptance. They tested their model in a random sample of 604 medical staff (534 physicians) working in 14 hospitals in Greece. Their results showed that TAM constructs, especially PU, PEOU and Attitude predicted a substantial proportion of the intention of doctors and nurses to accept clinical information systems. H. M. Mohamed et al. (2011) conducted a research to understand population behavior as well as the position of factors affecting e-Health acceptance. To do they employed TAM extended by two sociocultural factors 'trust' and 'tangibility' and proposed e-Health Technology Acceptance Model (e-HTAM). To empirically assess its validity, a pilot study was conducted in UAE and UK. The participants were randomly selected with no prior preferences. The data was collected through online questionnaire. Results revealed the positive correlation between the dependent variable intention and the independent variables: 'trust' and 'tangibility' and the three constructs TAM constructs (PU, PEOU and SN).

The rapid development of healthcare information system attracted the intention of more and more health professionals. By the way, Pai and Huang (2011) proposed a model that predicts acceptance of these systems. The proposed model is inspired from TAM and the Information system success model (DeLone and Mclean, 1992). Empirically, authors conducted questionnaire survey among district hospital nurses, head directors, and other related personnel. Results of the Structural Equation Modeling (SEM) analysis showed that the proposed factors, especially TAM constructs (PU and PEOU) positively influence users' intention to accept a healthcare system.

In 2012, Aldosari underlined the importance of studying archiving and communication systems (PACSs) acceptance by users. He noted that this acceptance is needed given the growing financial investment associated with these systems and its influence on the expected benefits. He used TAM to assess the level of acceptance of the host PACS by staff in the radiology department in Saudi Arabia. He administrated a questionnaire survey of 89 PACS users to collect data. His results found that the three constructs of PU, PEU, and change explained 41% of the variation in PACS user acceptance.

Asua et al. (2012) conducted a pilot research of a telemonitoring system for the management of chronic patients in primary care in the Bilbao Primary Care Health Region. They aimed to (1) to understand factors related to healthcare professionals' acceptance of the telemonitoring system and (2) to apply an adapted theory-based instrument to assess healthcare professionals acceptance of this system. For these issues they relied on TAM and extended it by Compatibility, and Facilitators. To test their model, they distributed a questionnaire to a total of 605 nurses, general practitioners and pediatricians. Results of the logistic regression analysis indicated that PU predicted significantly intention of health professionals to accept telemonitoring system. Moreover, the extended model was more powerful in predicting this intention.

Yan and Wang (2012) noted the studies limit exploring physician's willingness to accept an e-health care system. Hence, they tried to better understand and predict physicians' acceptance of asthma care mobile service (ACMS) in Taiwan. They based on the TAM and included "subjective norm," "innovativeness," and "managerial support" to predict this acceptance. Of the 700 questionnaires distributed to physicians using ACMS, 504 completed were received. After their analysis by SEM method, results found that attitude, PU, SN and PEOU were the most critical factors affecting physicians intention to accept ACMS. Chen and Hsiao (2012) attempted to better understand factors affecting physician hospital information systems (HIS). To do they referred to TAM and extended it by three quality antecedents (system quality, information quality, and service quality) as the exogenous variables. The model was tested using a survey that targeted physicians with over 1 year's experience using HIS in the selected case hospital. Results of the partial least square approach showed that PU and PEOU significantly affected HIS acceptance by Taiwanese physicians. Moores (2012) referred to TAM to develop an integrated model predicting clinical management system acceptance by hospital workers. They retained PU and PEOU as the fundamental predictors of this acceptance. They targeted clinical management system (CMS) used in a public, regional hospital with approximately 900 clinical staff, which includes physicians, nurses, and allied health workers. To test their research model, they used the partial least squares (PLS) and SmartPLS. Results showed that in a mandatory use, Attitude, PU and PEOU had no significant influence on system use. Ketikidis et al. (2012) applied a modified version of the TAM to assess the relevant beliefs and acceptance of Healthcare Information Technology (HIT) by medical doctors and nurses systems. They used the two-stage cluster sampling method. The first stage concerned three randomly selected clinics in the city of Skopje. As for the second concerned nurses and medical doctors employed in the selected clinics. Only 133 questionnaires of the 200 distributed were analyzed using multiple linear regression analysis. Results found that PEOU, but not PU, relevance and subjective norms directly predicted health professionals' intention to accept HIT. Jimoh et al. (2012) investigated the potential of Healthcare information technology (HIT) acceptance within maternal and child health workers in rural Nigeria. They aimed to propose a model that could predict IT acceptance in health in sub-Saharan Africa. They referred to TAM to develop their research model. To test the later, a prospective, quantitative survey design was mobilized to collect data from quasi randomly selected clusters of 25 rural health facilities in 5 of the 36 states in Nigeria. Results of the Regression analysis revealed that PU and PEOU were the significant factors influencing HIT acceptance by midwives and health extension workers in rural Nigeria.

In 2013, Cheng attempted to explore nurses' intention to accept e-learning system. Otherwise, he studied the role and the relevance of interaction factors, intrinsic motivators and extrinsic motivators, PU and PEOU in explaining this intention. Empirically, he gathered data from nurses working at two regional hospitals in Taiwan. Analysis of 218 responses by the SEM method showed that effects of PU, PEOU and Flow on nurses' Intention to Use e-learning system were very significant. Chow et al. (2012) pointed out the role of clinical imaging in disease diagnosis and patient care. Moreover, they underlined the need for nurses to acquire skills and knowledge in radiological examination requests and image interpretation. After describing the clinical imaging to facilitate independent learning in image interpretation, they explored factors affecting intentions to accept the portal. They referred to TAM extended by the computer self-efficacy construct as an external variable. The study was conducted on 188 nursing students. Results of the SEM method revealed that Attitude had a strong effect on this intention, followed by PEOU and self-efficacy. As for Bennani and Oumlil (2013), they attempted to identify factors influencing Health Information Technology (HIT) acceptance by nurses in Morocco. They referred to TAM and extended it by two added constructs "Trust" and "Image." The two authors conducted their study on a sample of 250 nurses working in health public and private organizations located in Agadir city, south of Morocco. Their results revealed that the proposed model explained 44.1% of total variance in the intention to accept HIT by nurses. Moreover, PU, Attitude, Trust and Image showed significant influence the nurses' intention to accept HIT.

Table 8.1 synthetises studies mobilized TAM model to predict E-health acceptance by healthcare professionals. Most of these studies were conducted in developed countries (Wu et al., 2007; Lanseng and Andreassen, 2007; Park and Chen, 2007; Carter, 2008; Yu et al., 2009; Asua et al., 2012). Nevertheless, few studies have been conducted in developing countries (Jimoh et al., 2012; Bennani and Oumlil, 2013). The majority of studies concerned nurses acceptance of E-health (Tung et al., 2008; Chen et al., 2008; Veer and Francke, 2010; Asua et al., 2013; Cheng, 2012; Bennani and Oumlil, 2013) followed by physicians (Carter, 2008; Egea and González, 2011; Yang and Wang, 2012; Chen and Hsiao, 2012). Concerning e-health applications, table, shows variation in these applications (Mobile healthcare systems (MHS), electronic patient records (EPR), PACS, personal health records (PHRs)…). Furthermore, the highest variation in predicting E-health acceptance was found in southern Spain (96.4%) where the respondents were only physicians.

8.4   RESEARCH MODEL AND HYPOTHESES

The research model is inspired from TAM (Davis et al., 1989). It includes three principal TAM constructs: Perceived Usefulness, Perceived Ease of Use and Attitude. Two added constructs "Image" and "trust" extended this model. These constructs are considered as external variables supposed to enhance Moroccan healthcare professionals' intention to accept E-health (Figure 8 1). We chose TAM, because it is considered a salient model predicting IT acceptance in various contexts. Moreover, researches referring to TAM to study E-health acceptance are very limited.

TABLE 8.1 Research Works Applying TAM and Its Extensions in E-health Context

Authors	Year	Significant TAM factors	E-health Applications	Population studied	Country	R^2
Wu et al.	2007	PU and PEOU	Mobile healthcare systems (MHS)	Physicians Nurses and Medical technician	Taiwan	0.70
Kim and Chang	2007	PU, PEOU	Health information websites	Online consumers	Korea	NA
Lanseng and Andreassen	2007	PU, PEOU	Self-service technology internet-based medical self-diagnosis	Inhabitant of Oslo	Norway	NA
Park and Chen	2007	PU and Attitude	Smartphone	Doctors and nurses	USA	0.69
Wu et al.	2008	PU, PEOU, SN	Event reporting systems	Healthcare professional	Taiwan	NA
Tung et al.	2008	PU, PEOU	E-logistics information system	Nurses	Taiwan	0.66
Carter	2008	Attitude	Smart fabric medical innovation	Physicians, physician's assistants and medical students	USA	NA
Chen et al.	2008	PU	Web-based learning	Public health Nurses	Taiwan	0.078
Aggelidis and Chatzoglou	2009	Attitude, PEOU, Social	Hospital Information System	Healthcare professional	Greece	0.87
Yu et al.	2009	PU, PEOU, Image	Health IT applications	Healthcare staff	Australia	0.46
Lishan et al.	2009	PU and PEOU	Telemedicine and e-health applications	Female healthcare and non-healthcare workers	Singapore	54–68%
Whetstone and Goldsmith	2009	PU	Personal health records (PHRs)	Students	USA	0.38

TABLE 8.1 Continued

Authors	Year	Significant TAM factors	E-health Applications	Population studied	Country	R²
Djamasbi et al.	2009	PU and Attitude	Healthcare Information System: laboratory telemedicine system	Microbiology laboratory assistants	USA	0.61
Veer and Francke	2010	PU	Electronic patient records (EPR)	Nurses	Netherlands	0.32
Wu et al.	2011	PU	Mobile healthcare	Physicians and nurses	Taiwan	0.63
Egea and González	2011	PU and PEOU	Electronic health care records (EHCR systems)	Physicians	Southern Spain	0.964
Melas et al.	2011	Attitudes PEOU and PU	Clinical information systems (CIS)	Doctors' and nurses	Greece	NA
H. M. Mohamed et al.	2011	PU, PEOU, SN	e-Health	Participants randomly selected	UK and UAE	NA
Pai and Huang	2011	PU and PEOU	Healthcare Information System	Nurses head Directors, and other related personnel	Taiwan	NA
Aldosari	2012	PU, PEOU	PACS	Staff in the radiology department	Saudi Arabia	0.38
Asua et al.	2012	PU	Telemonitoring system for chronic care patients	Nurses	Spain	R1 = 0.63
Yang and Wang	2012	Attitude, PU, PEOU, SN	Asthma care mobile service (ACMS)	Physicians	Taiwan	0.86

TABLE 8.1 Continued

Authors	Year	Significant TAM factors	E-health Applications	Population studied	Country	R^2
Chen and Hsiao	2012	PU and PEOU	Hospital information systems (HIS)	Physician	Taiwan	0.545
Moores	2012	PU	IT	Physicians, Nurses and allied health	France	More than 0.3
Ketikidis et al.	2012	PEOU and SN	Health information technology (HIT)	Nurses and medical doctors	Macedonia	0.679
Cheng	2012	PU, and PEOU	E-learning system	Nurses	Taiwan	NA
Chow et al	2012	PEOU	Clinical imaging portal	Students (nurses)	Hong Kong	0.77
Jimoh et al.	2012	PU and PEOU	ICT	Midwives and health extension workers	Nigeria	NA
Bennani and Oumlil	2013	PU and Attitude	IT	Nurses	Morocco	0.44

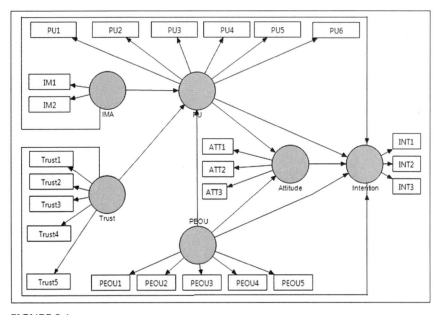

FIGURE 8.1

Perceived usefulness is defined as the degree to which a person believes that using a particular system would enhance his or her job performance. In the Information System area, it means, whether if IT usage can improve user' performance, productivity and efficiency, it will affect his intention to accept it. Therefore, Perceived usefulness is supposed to influence positively directly and indirectly (via attitude) the intention of healthcare professionals to accept E-health. Hence, these relationships are hypothesized as follows:

H1: Perceived Usefulness is positively and directly related to the Moroccan healthcare professionals' intention to accept E-health.

H2: Perceived Usefulness is positively and indirectly (through attitude) related to the Moroccan healthcare professionals' intention to accept E-health.

As for Perceived Ease of Use, it refers to the extent to which a person believes that using a particular system would be free of effort. In Information system arena, user intention is affected by his feelings about whether this system is easier to use. Referring to Davis et al. (1989), Perceived Ease of Use is supposed to influence positively directly and indirectly through-out attitude, the intention of Moroccan healthcare professionals to accept E-health. Therefore, this relationship is hypothesized as follows:

H3: Perceived Ease of Use is positively and directly related to the Moroccan healthcare professionals' intention to accept E-health.

H4: Perceived Ease of Use is positively and indirectly (via Attitude) related to the Moroccan healthcare professionals' intention to accept E-health.

TAM considered Attitude as an important factor predicting behavioral intention. In healthcare context, it is supposed that professionals with a more positive attitude toward IT are likely to be more satisfied with E-health and view it as more beneficial. Therefore, we hypothesized:

H5: Attitude toward E-health is positively related to the Moroccan healthcare professionals' intention to accept E-health.

As mentioned above, an external variable (image) is expected to influence acceptance of E-health by Moroccan healthcare professionals. This construct is drawn from Venkatesh and Davis (2000), and it is due to the need of Moroccan health professionals to improve their image toward other health actors. They thought that using IT could improve their image towards other professionals or even all actors belonging to health arena.

Image is defined as the degree to which usage of an innovation is per-ceived to enhance one's status in one's social system (Moore and Benbasat, 1991). In the Moroccan healthcare context, professionals with a higher need for social recognition and reputation are likely to accept E-health. Hence we hypothesized:

H6: Image is positively related to the Moroccan healthcare professionals' intention to accept E-health.

H7: Image is positively related to PU.

Trust is defined as the measure of the belief and goodwill that poli-cymakers feel in and for trusted people (Doney et al., 1998). Garbarino and Johnson (1999) added that Trust is the customers' confidence in qual-ity and reliability of the services offered by an organization. Prior studies showed the positive direct relation between Trust and Intention (Gefen et al., 2003). In healthcare context Shibl et al. (2013), Bennani and Oumlil

(2013) pointed out the importance of the trust to predict IT adoption by healthcare professionals. For this chapter it supposed that:

H8: Trust is positively related to the Moroccan healthcare professionals' intention to accept E-health.

H9: Trust is positively related to the Perceived usefulness.

8.5   RESEARCH METHODOLOGY

Consistent with previous researches studying IT acceptance by healthcare professionals (Aggelidis and Chatzoglou, 2009), a quantitative method was performed to test the hypotheses above. This study concerned healthcare professionals (physicians and nurses) working in the Provincial Hospital (PH) Hassan Ist of Tiznit city, located in south of Morocco. The choice of this city is due to the fact of belonging to the region considered as the Moroccan leading site for integration E-health.

The public health infrastructure of Tiznit province consist of provincial hospital Hassan Ist and 84 health centers. These latter include 78 rural centers and 6 urban centers. This chapter is limited to the provincial hospital Hassan Ist. This hospital was created in 1981. It has a capacity of 402 beds; It provides basic medicine services. Moreover, it ensures the following specialties: Ophthalmology, dermatology, infectious diseases, Ear Nose and Throat, Psychiatry, Pneumo-phtisiology, Cardiology, Gastroenterology, Rehabi-litation, Neurology, Urology, Endocrinology, Trauma and medical-surgical resuscitation.

Since 2009, the Provincial hospital Hassan Ist developed internal applications by the IT department. These applications covered the following services:

- Admissions Office: Since 2009, this office used software called BAF which deals with entry tickets, exit, billing, death and birth notices.
- Radiology department: It is equipped with "Visual PACS" system that manage results of different units: echography, mammography, Radio and OS-Lung and Scanner.
- Laboratory service: It integrates IS Laboratory system. It displays results of analysis performed by different units, prints them with barcodes.

- Contracts service: It contains an information system called "Gide." This system allows linking the hospital to the provincial treasury.

Data were collected via a survey questionnaire that was divided into two parts. The first one captured perception of model constructs by participants. All the measurements items were anchored on a 7-point Likert scale from "strongly disagree" (1) to "strongly agree" (7) in which respondents indicated an appropriate response. As for the second part, it captured demographic information about each participant. Questionnaires were randomly distributed. Respondents were volunteers asked to complete the questionnaire according to their perception towards IT acceptance to perform their tasks.

The concerned population here was 219 individuals and involved 36 physicians, 138 nurses and 25 administrative staff. Only 160 questionnaires were distributed. This limitation is mainly due to the difficult access to all hospital wards, and to the teams shift. The data collection process took nine weeks, from the middle of June 2013 till the middle of August 2013. This delay is due to three reasons: (1) Low education level of some nurses leads difficulties in understanding the French version of the questionnaire (2) Translation of the questionnaire from French to Arabic or Berber (3) Physicians and nurses were always annoyed with tasks that prevent them to complete the questionnaires.

Of the 160 questionnaires distributed, 125 were completed and retained for the analysis representing a response rate of 78.125%. It consisted of 26 physicians, 71 nurses and 28 administrative staff. Most of the respondents were men (68%) and aged less than 40 years old (55.2%) (Table 8.2). Note that this chapter concerned physicians and nurses only.

8.6   DATA ANALYSIS AND RESULTS

The research model represented in Figure 8.2 was evaluated using Partial Least Squares (PLS) approach. It is classified as a powerful method to study structural models involving multiple constructs with multiple items (Rigdon, 1998).

For this chapter, we chose PLS approach as the main data analysis technique due to the following three reasons:

TABLE 8.2   Demographic Profile of the Sample (N = 125)

Variable	Content	N	(%)
Gender	Men	85	68.0
	Women	40	32.0
Activity	Physicians	26	20.8
	Nurses	71	56.8
	Administrative Staff	28	22.4
Age	20–25 years	9	7.2
	25–40 years	60	48.0
	40–65 years	56	44.8
Education Level	Baccalaureate (Bac) level	21	16.8
	Bac + 1-Bac + 3	63	50.4
	Bac + 4-Bac + 5	15	12
	Bac + 7 and more	26	20.8

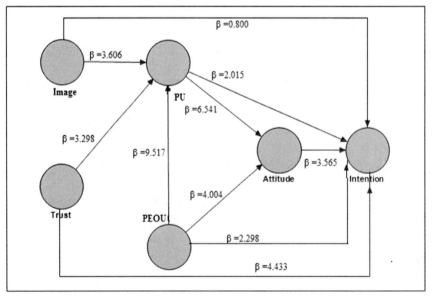

FIGURE 8.2

1. It places minimal demands on sample sizes and data distribution assumptions (Hair et al., 2012);
2. It is a variance-based technique that is oriented towards the predictive aspects of the model (Barclay et al., 1995); and
3. It takes into account the measurement error when testing the structural model (Fornell and Cha, 1994).

However, we should note that PLS method does not provide any global goodness-of fit criterion, which implies a lack of measures for overall model fit (Henseler et al., 2012).

PLS approach consists of two steps to validate predictive models using reflective latent constructs (Henseler et al., 2009). The first step allows the assessment of the measurement model representing the relationships between items and the constructs that they measure. The second step includes the assessment of the structural model depicting the relationships between the constructs as specified by the research model. Authors used SmartPLS software version 2.0M3 to assess both of the measurement and the structural models. They run PLS-Graph using a nonparametric test of significance known as bootstrapping method with 200 resamples to determine the significance levels for loadings, weights, and path coefficients (Gil-Garcia, 2008).

8.6.1   MEASUREMENT MODEL ASSESSMENT

Measurement model called also the outer model. It describes the relationships between constructs and their corresponding manifest indicators. Assessment of the measurement model is performed by: Internal Consistency Reliability (ICR), Convergent validity and discriminant validity (Henseler et al., 2009).

To assess the ICR, we used the Cronbach's Alpha coefficient (α). It is verified when the alpha is above the minimum acceptable threshold of 0.7. In Table 8.3, all constructs indicated composite reliability (CR) above 0.7. Moreover, internal consistency (ICR) of the scales is verified for all constructs because their Cronbach's Alpha doesn't exceed the threshold value and confirmed a satisfactory reliability.

Convergent validity explains the extent to which the indicators of a construct are extremely correlated to each other, and also to indicators of other

TABLE 8.3 Items Loading, Construct Composite Reliability (CR), Internal Consistency Reliability (ICR)

Construct	Items	Item loading	Construct CR	ICR (Cronbach's alpha)
Perceived Usefulness (PU)	PU1	0.834694	0.918462	0.893978
	PU2	0.767659		
	PU3	0.823145		
	PU4	0.839033		
	PU5	0.820333		
	PU6	0.759187		
Perceived Ease of Use (PEOU)	PEOU1	0.688947	0.870839	0.819217
	PEOU2	0.77931		
	PEOU3	0.774069		
	PEOU4	0.775837		
	PEOU5	0.768362		
Attitude (ATT)	ATT1	0.906851	0.866762	0.77026
	ATT2	0.801303		
	ATT3	0.769395		
Image	IMA1	0.919302	0.94197	0.882284
	IMA 2	0.967275		
Trust	Trust 1	0.782884	0.910356	0.877237
	Trust 2	0.862741		
	Trust 3	0.833123		
	Trust 4	0.813936		
	Trust 5	0.798663		
Intention (INT)	INT1	0.909933	0.936178	0.897912
	INT2	0.89988		
	INT3	0.923528		

constructs. Referring to Fornell and Larcker (1981), convergent validity assessment implies to calculate item reliability for each measure, composite reliability for each construct and the average variance extracted (AVE).

Item reliability involves the factor loading as that measure should be 0.70 or greater, which indicates a well-defined structure. As shown in Table 8.4, all of the factor loadings were greater than 0.70, and most of

TABLE 8.4 Diagonal Elements Are the Square Root of the Shared Variance Between the Constructs and Their Measures (AVE); Off-Diagonal Elements Are Correlations Between Constructs (N = 125)

	AVE	Attitude	PEOU	Image	Intention	Trust	PU
Attitude	0.685	**0.8279**					
PEOU	0.574	0.5877	**0.7580**				
Image	0.890	0.2297	0.1518	**0.9435**			
Intention	0.830	0.3938	0.1871	0.0696	**0.9111**		
Trust	0.670	0.5376	0.4116	0.1436	0.4666	**0.8187**	
PU	0.652	0.6481	0.6144	0.2556	0.3113	0.4330	**0.8079**

them were well above that level. Moreover, composite reliability, 0.70 value or higher was recommended by Bacon et al. (1995).

As for AVE, it measures the percentage of variance captured by a construct showing the ratio of the sum of the variance captured by the construct and measurement variance. It is acceptable if it is greater than 0.5.

Concerning discriminant validity, it is assessed following Fornell and Larcker (1981) guidelines. We first use the average variance extracted (AVE), then capture the average variance shared between a construct and its measures. These measures should be greater than the variance shared between the construct and other constructs in the model (Barclay et al., 1995). Moreover, the diagonal elements in the matrix should be significantly greater than the off-diagonal elements in the corresponding rows and columns (Gefen and Straub, 2005). Note, that AVE is generated automatically using the bootstrap technique by the PLS-Graph. Table lists the correlation matrix for the construct. The diagonal elements in the "correlation construct" (in bold) are the square roots of the AVE. Off-diagonal elements are the correlation among construct. Results revealed that AVE for each construct is larger than the correlation of that construct with all the constructs in the model.

Moreover, Discriminant validity is assessed by extracting the factor and cross loadings of all indicators to their respective constructs. Each indicator should be significantly related to the construct, and should not have a stronger connection with another construct (Chin, 2010). Results (Table 8.4) indicated that all indicators loaded on their respective construct are more highly than on any other, confirming that the constructs are distinct. Hence the discriminate validity was confirmed.

8.6.2 STRUCTURAL MODEL ASSESSMENT

The structural model is also called the inner model. It indicates the causal relationships among constructs in the conceptual model (Hair et al., 2012). It includes estimates of the path coefficients, which indicate the significance of the hypothesized relationship (relationship between the dependent and independent variables). It also provides R^2 value, which determine power of the model (the variance explained by the independent variables). The R^2 value and the path coefficients indicate how well the data support the model (Wixom and Todd, 2005). We run PLS-Graph using a nonparametric test of significance known as bootstrapping method to determine the significance levels for loading weights and path coefficients (Gil-Garcia, 2008). Figure 8.2 shows that the research model could explain 26.3% of total variable (R2) in the intention of the healthcare professionals to accept E-health. As expected, Perceived Usefulness showed a significant impact on the healthcare professionals intention ($\beta = 2.015$) (H1). Moreover, Perceived Ease of Use had a significant influence on the healthcare professionals intention ($\beta = 2.298$) (H3). Likewise, Attitude depicted a significant positive influence on this intention ($\beta = 3.565$) (H5). Concerning the added constructs, only Trust showed a significant impact on the healthcare professionals intention ($\beta = 4.433$) (H8). However, Image showed no influence on this intention to accept E-health ($\beta = 0.8$) (H6). Results also revealed the significant impact of PU ($\beta = 6.541$) and PEOU ($\beta = 4.004$) on the Attitude confirming H2 and H4. Furthermore, Trust ($\beta = 3.298$) and PU ($\beta = 3.606$) showed a significant influence on PU, confirming H9 and H7.

8.7 DISCUSSION

The main objective of this chapter is to identify factors predicting E-health acceptance by Moroccan professionals working at the provincial hospital Hassan I[st] located at Tiznit city, south of Morocco. Technology acceptance model extended by Image and trust constructs was adopted for this issue.

As indicated in results above, Perceived Usefulness influences positively the intention of Moroccan health professionals to accept E-health. This result confirms the pragmatic behavior of the interviewed healthcare professionals, and tends to focus on the usefulness of the information technology itself. Moreover, positive impact of PU on the Moroccan healthcare professionals intention to accept E-health confirms beliefs in using a particular system that would enhance their job performance. Hence, Moroccan health professionals accept this technology whenever it demonstrates its ability to meet their needs. This finding is similar to those of (Wu et al., 2007; Kim and Chang, 2007; Lanseng and Andreassen, 2007; Park and Chen, 2007; Tung et al., 2008; Yu and Li, 2009; Whetstone and Goldsmith, 2009; Veer and Francke, 2010; Egea and González, 2011; Pai and Huang, 2011; Melas et al., 2011; Aldosari, 2012; Jimoh et al., 2012; Yang and Wang 2012; Bennani and Oumlil, 2013).

In agreement with Wu et al. (2007), Kim and Chang (2007), Lanseng and Andreassen (2007), Park and Chen (2007), Tung et al. (2008), Yu and Li (2009), Egea and González (2011), Melas et al. (2011), Pai and Huang (2011), Yang and Wang (2012), Chow et al. (2012) the Perceived Easy of Use influences positively the intention of health professionals to accept E-health. This may suggest that they are ready to use E-health whenever it would be free of effort. Otherwise, these professionals will use technologies if they are relatively effortless and less complicated.

Furthermore, our research showed that the added construct "image" does not influence positively and significantly the intention of the healthcare professionals to accept E-health. Hence, using E-health would not be an opportunity to improve their Image toward other health professionals.

Prior studies stressed the positive direct relation between Attitude and Behavioral intention to accept E-health by healthcare professionals (Park and Chen, 2007; Aggelidis and Chatzoglou, 2009; Gaylin et al., 2010; Melas et al., 2011; Yang and Wang, 2012). Similarly, our research showed that attitude was significant in the Moroccan context. This emphasizes the importance of health professionals' attitude and their beliefs in the benefits of using E-health.

Concerning Trust, results showed its significant impact on the intention of health professionals to accept E-health. This explained that these professionals' confidence in quality and reliability of the services provided by E-health would enhance their intention to accept this technology.

In order to more explain E-health acceptance by Moroccan professionals, a comparison considering type of activity was performed. Results revealed that: contrary to nurses ($R^2 = 28.3\%$) physicians showed a high variance value in Behavioral intention ($R^2 = 71\%$). Moreover, Attitude, PU, PEOU, Trust and Image impacted significantly E-health acceptance for physicians. However, Attitude, PEOU and Image were the only main construct determining this acceptance for nurses.

8.8   CONCLUSION

This chapter aimed to identify factors predicting E-health acceptance by healthcare professional in developing countries. The study was conducted in the provincial hospital Hassan I[st], located in Tiznit south of Morocco. It studied intention of physicians and nurses toward acceptance of E-health. Results showed that Perceived Usefulness, Perceived Ease of Use, Attitude and trust explained this acceptation, while Image showed no significant effect.

This study proposed a valuable tool to managers and decision makers working for healthcare organizations in developing countries. It helps them to identify factors determining E-health adoption and ensuring its success within these organizations.

The sample size represents the main limitation to this research work and could bias results. Future work should consider a more accurate representation of the sample, and give more intention to specific constructs such as 'culture' that could help predicting E-health in developing countries.

KEYWORDS

- **developing countries**
- **e-health**
- **health professionals**
- **healthcare organizations**
- **IT acceptance**
- **morocco**
- **Technology Acceptance Model (TAM)**

REFERENCES

1. Agarwal, R., Prasad, J. The antecedents and consequents of user perceptions in information technology adoption. *Decision Support Systems.* 1998, 22, 15–29.
2. Aggelidis, V. P., Chatzoglou, P. D. Using a modified technology acceptance model in hospitals. *International journal of medical informatics.* 2009, *78*,115–126.
3. Ajzen, I. (1985). *From intentions to actions: A theory of planned behavior.* In J. Kuhi J. Beckmann (Eds.), *Action Control: From cognition to behavior.* Heidelberg: Springer; 1985, 11.39.
4. Ajzen, I. (1988). *Attitudes, personality, and behavior.* Chicago: Dorsey Press, 1988.
5. Ajzen, I. (1991). The theory of planned behavior. *Organizational Behavior and Human Decision Processes.* 1991, 50, 179–211.
6. Ajzen, I., Fishbein, M. *Understanding Attitudes and Predicting Behaviors, 1ˢᵗ ed* Prentice-Hall, Englewood Cliffs, 1980.
7. Ajzen, I., Madden, T. J. Prediction of Goal-Directed Behavior: Attitudes, Intentions, and Perceived Behavioral Control. *Journal of Experimental Social Psychology.* 1986, 20, 453–474.
8. Aldosari, B. User acceptance of a picture archiving and communication system (PACS) in a Saudi Arabian hospital radiology department. *BMC medical informatics and decision making.* 2012, 12, 1–44.
9. Ammenwerth, E., Brender, J., Nykänen, P., Prokosch, H. U., Rigby, M., Talmon, J. Visions and strategies to improve evaluation of health information systems: Reflections and lessons based on the HIS-EVAL workshop in Innsbruck. *International journal of medical informatics.* 2004, *73*, 479–491.
10. Arlotto, P., Oakes, J. *Return on investment: maximizing the value of healthcare information technology Healthcare Information and Management Systems Society.* Chicago, Illinois, 2003.
11. Asua, J., Orruño, E., Reviriego, E., Gagnon, M. P. Healthcare professional acceptance of telemonitoring for chronic care patients in primary care. *BMC medical informatics and decision making.* 2012, *12*, 139.
12. Bacon, D. R., Sauer, P. L., Young, M. Composite reliability in structural equations modeling. *Educational and Psychological Measurement.* 1995, 55, 394–406.
13. Bandura, A. *Social Foundations of Thought and Action*, Englewood Cliffs, N. J. Prenctice-Hal. 1986.
14. Bandura, A. *Social Learning Theory,* Prentice-Hall, Englewood Cliffs. 1977.
15. Bandura, A. *L'apprentissage social.* Bruxelles: Mardaga. 1980.
16. Barclay, D., Higgins, C., Thompson, R. (1995). The partial least squares (PLS) approach to causal modeling: Personal computer adoption and use as an illustration. *Technology studies.* 1995, 2, 285–309.
17. Barclay, D., Higgins, C., Thompson, R. The partial least squares (PLS) approach to causal modeling: Personal computer adoption and use as an illustration. *Technology studies.* 1995, 2, 285–309.
18. Benbasat, I., Barki, H. (2007). Quo vadis TAM? *Journal of the association for information systems.* 2007, *8*, 1–7.
19. Benbasat, I., Dexter, A. S., Todd, P. An Experimental Program Investigation Color-enhanced and Graphical Information Presentation: An Integration of the Findings. *Communications of the ACM.* 1986, 29, 1094–1105.

20. Bennani, A. E., Oumlil, R. (2013, June). IT acceptance by nurses in Morocco: Application of Technology Acceptance Model. *In Information Society (i-Society), IEEE*. 2013. 79–84.

21. Bennani, A., Belalia, M. Oumlil, R. As a human factor, the attitude of healthcare practitioners is the primary step for the e-health: first outcome of an ongoing study in Morocco. *Commun IBIMA*. 2008, *3*, 1–28.

22. Carter, E. Marketing "smart" medical innovation: physicians' attitudes and intentions. *International Journal of Pharmaceutical and Healthcare Marketing*. 2008, *2*, 307–320.

23. Chau. P. Y. K., Tam. K. Y. Factors Affecting the Adoption of Open Systems: An Exploratory Study. *MIS Quarterly*. 1997, 21, 1–24.

24. Chen, P. S. D., Lambert, A. D., Guidry, K. R. Engaging online learners: The impact of Web-based learning technology on college student engagement. *Computers & Education*. 2010, *54*, 1222–1232.

25. Chen, R. F., Hsiao, J. L. (2012). An investigation on physicians' acceptance of hospital information systems: a case study. *International journal of medical informatics*. 2012, *81*, 810–820.

26. Cheng, Y. M. Exploring the roles of interaction and flow in explaining nurses' e-learning acceptance. *Nurse education today*. 2013, *33*, 73–80.

27. Chin, W. W. *How to write up and report PLS analyzes*. In Handbook of partial least squares. Springer Berlin Heidelberg; 655–690, 2010.

28. Chow, M., Chan, L., Lo, B., Chu, W. P., Chan, T., Lai, Y. M. Exploring the Intention to Use a clinical imaging portal for enhancing healthcare education. *Nurse Education Today*. 2013, *33*, 655–662.

29. Compeau, D., Higgins C. A., Huff, S. Social Cognitive Theory and Individual Reactions to Computing Technology: A Longitudinal Study. *MIS Quarterly*. 1999, 23, 145–158.

30. Davis, F. D., Bagozzi, R. P., Warshaw, P. R. User Acceptance of Computer Technology: A Comparison of two Theoretical Models. *Management Science*, 1989, 35, 982–1003.

31. Davis, F. D. Perceived Usefulness, Perceived Ease of Use, and User Acceptance of Information Technology. *MIS Quarterly*, 1989, 13, 329–340.

32. Veer, A. J., Francke, A. L. Attitudes of nursing staff towards electronic patient records: a questionnaire survey. *International Journal of Nursing Studies*. 2010, *47*, 846–854.

33. DeLone, W. H., McLean, E. R. Information systems success: the quest for the dependent variable. *Information Systems Research*, 1992, *3*, 60–95.

34. Devaraj, S., Kohli, R. Performance impacts of information technology: is actual usage the missing link? *Management Science*. 2003, *49*, 273–289.

35. Djamasbi, S., Fruhling, A. L., Loiacono, E. The influence of affect, attitude and usefulness in the acceptance of telemedicine systems. *Journal of Information Technology Theory and Application (JITTA)*. 2009, *10*, 1–16.

36. Doney, P. M., Barry, J. M., Abratt, R. Trust determinants and outcomes in global B2B services. *European Journal of Marketing*. 2007, *41*, 1096–1116.

37. Eysenbach, G. What is e-health? *Journal of Medical Internet Research*. 2001, *3*, 1–12.

38. Fishbein, M., Ajzen, I. Attitudes towards objects as predictors of single and multiple behavioral criteria," *Psychological Review*.1975, 82, 59–74.

39. Fornell, C., Cha, J. Partial least squares. *Advanced Methods of Marketing Research*. 1994, 407, 52–78.
40. Fornell, C., Larcker, D. F. Structural equation models with unobservable variables and measurement error: Algebra and statistics. *Journal of Marketing Research*. 1981, 382–388.
41. Garbarino, E., Johnson, M. S. The different roles of satisfaction, trust, and commitment in customer relationships. *The Journal of Marketing*. 1999, 2, 70–87.
42. Gefen, D., Straub, D. A practical guide to factorial validity using PLS-Graph: Tutorial and annotated example. *Communications of the Association for Information systems*. 2005, 16, 5.
43. Gefen, D., Straub, D. A practical guide to factorial validity using PLS-Graph: Tutorial and annotated example. *Communications of the Association for Information systems*. 2005, 16, 5.
44. Gil-Garcia, J. R. *Using partial least squares in digital government research*. Handbook of research on public information technology. 2008, 239–253.
45. Hair, J. F., Sarstedt, M., Ringle, C. M., Mena, J. A. An assessment of the use of partial least squares structural equation modeling in marketing research. *Journal of the Academy of Marketing Science*. 2012, 40, 414–433.
46. Henseler, J., Ringle, C. M., Sarstedt, M. 12 *Using Partial Least Squares Path Modeling in Advertising Research: Basic Concepts and Recent Issues*. Handbook of research on international advertising, 2012, p252.
47. Henseler, J., Ringle, C. M., Sinkovics, R. R. The use of partial least squares path modeling in international marketing. *Advances in International Marketing*. 2009, 20, 277–319.
48. Holden, R. J., Karsh, B. T. The technology acceptance model: its past and its future in health care. *Journal of Biomedical Informatics*. 2010, *43*, 159–172.
49. Hong W., Thong J., Wong W. M. and Tam K. Y. Determinants of user acceptance of digital libraries, *Journal of Management Information System*. 2002, 18, 97–124.
50. Jimoh, L., Pate, M. A., Lin, L., Schulman, K. A. A model for the adoption of ICT by health workers in Africa. *International Journal of Medical Informatics*. 2012, *81*, 773–781.
51. Karahanna, E., Straub, D. W., Chervany, N. L. Information Technology Adoption Accross Time: A Cross-Sectional Comparison of Pre-Adoption and Post-Adoption Beliefs, *MIS Quarterly*. 1999, 23, 183–213.
52. Ketikidis, P., Dimitrovski, T., Lazuras, L., Bath, P. A. Acceptance of health information technology in health professionals: An application of the revised technology acceptance model. *Health Informatics Journal*. 2012, *18*, 124–134.
53. Kim, D., Chang, H. Key functional characteristics in designing and operating health information websites for user satisfaction: An application of the extended technology acceptance model. *International Journal of Medical Informatics*. 2007, *76*, 790–800.
54. Kukafka, R., Ancker, J. S., Chan, C., Chelico, J., Khan, S., Mortoti, S., Stephens, K. Redesigning electronic health record systems to support public health. *Journal of Biomedical Informatics*. 2007, *40*, 398–409.
55. Lanseng, E. J., Andreassen, T. W. Electronic healthcare: a study of people's readiness and attitude toward performing self-diagnosis. *International Journal of Service Industry Management*. 2007, *18*, 394–417.
56. Lishan, X., Chiuan, Y. C., Choolani, M., Chuan, C. H. The perception and intention to adopt female-focused healthcare applications (FHA): A comparison between

healthcare workers and non-healthcare workers. *International journal of medical informatics*. 2009, *78*, 248–258.

57. Melas, C. D., Zampetakis, L. A., Dimopoulou, A., Moustakis, V. Modeling the acceptance of clinical information systems among hospital medical staff: an extended TAM model. *Journal of biomedical informatics*. 2011, *44*, 553–564.

58. Miniard, P. W., Cohen. J. B. An examination of the Fishbein-Ajzen behavioral intentions model's concepts and measures. *Journal of Experimental Social Psychology*. 1981, 7, 309–339.

59. Mohamed, A. H. H., Tawfik, H., Al-Jumeily, D., Norton, L. MoHTAM: A technology acceptance model for mobile health applications. In: *Developments in E-systems Engineering (DeSE)*, December (pp. 13–18). IEEE. 2011.

60. Moore. G. C. and Benbasat. I. Development of an instrument to measure the perception of adopting and information technology innovation, *Information systems Research*. 1991, 2, 192–223.

61. Ortega Egea, J. M., Román González, M. V. Explaining physicians' acceptance of EHCR systems: an extension of TAM with trust and risk factors. *Computers in Human Behavior*. 2011, *27*, 319–332.

62. Pai, F. Y., Huang, K. I. Applying the Technology Acceptance Model to the introduction of healthcare information systems. *Technological Forecasting and Social Change*. 2001, *78*, 650–660.

63. Park, Y., Chen, J. V. Acceptance and adoption of the innovative use of smartphone. *Industrial Management and Data Systems*. 2007, *107*, 1349–1365.

64. Rigdon, E. E. The equal correlation baseline model for comparative fit assessment in structural equation modeling. *Structural Equation Modeling: A Multidisciplinary Journal*, 1998, *5*, 63–77.

65. Rogers, E. M. *The Diffusion of innovations. 3rd Edition*. New York: The Free Press. 1983.

66. Rogers, E. M. *Diffusion of innovation.* 4th Edition. Free Press. 1995.

67. Rogers, E. M. *Diffusion* of *innovations, 5th edition*, New York: Free Press. 2003.

68. Scott, R. E. e-Records in health—preserving our future. *international journal of medical informatics*. 2007, *76*, 427–431.

69. Shibl, R., Lawley, M., Debuse, J. Factors influencing decision support system acceptance. *Decision Support Systems*. 2013, *54*, 953–961.

70. Triandis, H. C. *Values, Attitudes and Interpersonal Behavior*, In M. M. Page (Ed.): *Nebraska Symposium on Motivation, 1979: Beliefs, Attitudes and Values*. Lincoln: University of Nebraska Press. 1980.

71. Tung, F. C., Chang, S. C., Chou, C. M. An extension of trust and TAM model with IDT in the adoption of the electronic logistics information system in HIS in the medical industry. *International Journal of Medical Informatics*. 2008, *77*, 324–335.

72. Venkatesh, V., Davis, F. D. A theoretical extension of the technology acceptance model: four longitudinal field studies. *Management science*. 2000, *46*, 186–204.

73. Venkatesh, V., Morris, M. G., Davis, G. B., Davis, F. D. User acceptance of information technology: Toward a unified view. *MIS quarterly*. 2003, 425–478.

74. Whetstone, M., Goldsmith, R. Factors influencing Intention to Use personal health records. *International Journal of Pharmaceutical and Healthcare Marketing*. 2009, *3*, 8–25.

75. Wixom, B. H., Todd, P. A. A theoretical integration of user satisfaction and technology acceptance. *Information Systems Research.* 2005, *16*, 85–102.

76. Wu, J.-H., Wang, S.-C., Lin, L.-M. Mobile computing acceptance factors in the healthcare industry: A structural equation model. *International Journal of Medical Informatics,* 2007, 76, 66–77.

77. Wu, J. H., Shen, W. S., Lin, L. M., Greenes, R. A., Bates, D. W. Testing the technology acceptance model for evaluating healthcare professionals' Intention to Use an adverse event reporting system. *International Journal for Quality in Health Care.* 2008, *20*, 123–129.

78. Wu, W. H., Jim Wu, Y. C., Chen, C. Y., Kao, H. Y., Lin, C. H., Huang, S. H. Review of trends from mobile learning studies: A meta-analysis. *Computers and Education,* 2012, *59*, 817–827.

79. Yan, H. Y., Wang, M. J. What factors affect physicians' decisions to use an e-health care system? *Health,* 2012, *4*, 1023–1033.

80. Yu, P., Li, H., Gagnon, M. P. Health IT acceptance factors in long-term care facilities: a cross-sectional survey. *International Journal of Medical Informatics.* 2009, *78*, 219–229.

CHAPTER 9

E-HEALTH FOR DEVELOPING COUNTRIES: A THEORETICAL MODEL GROUNDED ON LITERATURE

HARRY FULGENCIO

Leiden Institute of Advanced Computer Science, Niels Bohrweg 1 Leiden, 2333CA, The Netherlands, Tel: +31-71-5275778, Fax: +31-71-5276985, E-mail: h.fulgencio@liacs.leidenuniv.nl

CONTENTS

ABSTRACT

This chapter is related to the societal challenge of managing Health through e-health. Providing health care using information and communication technology in both developed and developing countries, has already shown some promising implementations. However, there is a lack of research on the major challenges of implementing e-health solutions for developing countries. We identified e-health implementation challenges for developing countries, together with a corresponding set of solution. The systematic literature review of 46 articles resulted in six themes of challenges: (i) lack of skilled stakeholders, (ii) inadequate infrastructures, (iii) lack of acceptance, (iv) limited resources, (v) inadequate information communication, and (vi) inadequate process guidance. Based on the findings, we grounded our proposed theoretical model to focus on three aspects: (i) guidance, (ii) acceptance, and (iii) building base. We theorized that this model will improve or will help a developing country's e-health implementations.

9.1 INTRODUCTION

At the time of this research, the European Union is composed of 28 member states. These are predominantly developed countries, although there are four developing countries: Bulgaria, Lithuania, Latvia, and Romania. Outside of the European Union, majority of countries are categorized as developing countries [1]. Developing countries are confronted with health problems such as occupational diseases and preventable chronic diseases [2]. People in rural areas of developing countries often live in poverty; and they have limited access to appropriate health care. In aid of better health care provision, e-health can serve as a supplementary tool to reach wider

Keeping this in mind, let us focus on the European Union's Horizon 2020 program, a sub section of this program elaborates on societal challenges [3]. One out of the seven major societal challenge is health; a challenge to enhance the health management. By management we mean monitoring, prevention, detection, treatment and support. A leading business trend and a recognized research topic that is directly contributing to the improvement of the health societal challenges through information communication technology is e-health. E-health is defined as the "use of information and communication technology in health" [4]. E-health implementations in developing countries can improve communication between institutions, assist in providing medication, and help monitor and detect patients [5]. Details about the definition of e-health can be found in other articles [6, 7]. Related studies include: (a) an anecdotal, based on personal experience research and is non-referential article about contemporary issues, challenges and opportunities for hospitals [4]; (b) non rigorous theoretical article and was focused on opportunities as a challenge in global e-health implementation [8]; (c) relating to e-health is a well-argued study exploring Health informatics but again is non-rigorous and does not detail the method used [9]; (d) an empirical and rigorous research on feasibility of e-health by exploring the determinants, opportunities and challenges in the African Region's Infrastructure capabilities [10].

Existing models were formed on the basis of partiality and heuristics in addition it is not an up-to-date compilation of situation [11]. We argue that Unified Theory of Acceptance and use of Technology [12] is too generic and does not identify the practical challenges that implementing e-health. It is apparent that there hasn't been a state-of-the-art review and problem solution recommendations based on available research articles. Therefore, we think that it is important to contribute on the societal challenge of Health by focusing on e-health. This led to the research question: What should Developing Countries focus on when implementing e-health? The research objectives were to: (a) provide a comprehensive analysis of the challenges and solutions of e-health implementation in Developing Countries; and (b) develop a grounded framework for implementing e-health initiatives in Developing Countries. The next section details the research design.

9.3   METHODOLOGY

The research question was addressed by first evaluating the challenges and proposed solutions for e-health in developing countries through a systematic literature review, and then used the results for the proposed theoretical model. After having decided on the research question and approach, several steps had to be taken for our systematic literature review: search for literature, apply exclusion and inclusion criteria, apply quality assessment, and synthesis [13].

9.2.1   SEARCH STRATEGY

Literature was obtained using Leiden University Library Catalogue and Google Scholar. In the Leiden University Library catalog, a structured search was done using the keyword combination "e-health developing countries." The advanced search option allows the inclusion criteria to be enforced. The Google Scholar keywords used were "e-health," "healthcare," "developing countries," and "global health" in different combinations. Google Scholar is less structured, because entering keywords in Google Scholar can literally result in thousands of possible titles, which is too much to process. So we opted to enforce a subjective sampling to collect papers or articles. Subjective means that the search was stopped when it was no longer expected to find useful results. Note the Leiden

University Catalogue and Google Scholar are not databases themselves, but provide links to databases, search portals, or alternative sources.

9.2.2 INCLUSION AND EXCLUSION CRITERIA

Included literatures were focused on e-health in developing countries, written from 2008 until 2013, and published in English. The time scope was chosen to cover the most recent challenges and proposed solutions for e-health in developing countries. To determine which countries are considered developing countries, a list of the American Mathematical Society [14] was used. These criteria were enforced through advanced search options or were done manually. The Leiden University Library Catalogue, were a source of peer-reviewed articles and Google Scholar, the year, language, and quality of the articles are checked manually.

9.2.3 QUALITY ASSESSMENT

Quality assessment was not enforced with the articles obtained using the Leiden University Library Catalogue because articles were filtered using the advanced search options of peer-reviewed articles. The articles obtained using Google Scholar, had to undergo a more rigorous manual assessment for the existence of detailed reference list, appropriate structure, academic writing style, and appropriate title. If the articles were considered relevant, the researcher proceeded to reading the sections abstract, discussion and conclusion. Then finally, the articles were fully read and it decided whether challenges to or suggested measures for e-health in implementations developing countries can be identified.

9.2.4 SELECTION OF CHALLENGES AND SUGGESTED MEASURES

Challenges and suggested measures are identified through means of coding using a tool called "QDA Miner Lite." Challenges were included in the results only if, in literature, it was specifically mentioned that these were challenges to e-health implementations in Developing Countries. Any measure, however, that was considered useful and applicable to address any of the challenges is included. Note that a measure such as "Improve infrastructure" as a measure against infrastructural problems was not considered to be useful as it was too broad and straightforward. An attempt was done to find solutions on a more detailed level, though not too project-specific. Similar concepts were combined and were grouped in more general themes. These themes helped in identifying the most critical areas of attention for e-health implementations in developing countries.

9.3   RESULTS

The numbers enclosed in parenthesis – (..) were all part of the of the analyzed data set and the details were provided in Appendix 1: Data Set Analyzed.

9.3.1   INCLUDED STUDIES

Utilizing the keywords and enforcing the inclusion criteria using the advanced search option of the Leiden University's Library catalog in the 3rd week of July 2013 resulted in 403 articles, of which 32 titles were considered to be relevant and in closer inspection 13 articles were excluded for any of the following reasons: no full-text available, focus of article was not on e-health in Developing Countries, no mentioned challenges, or solutions couldn't be could not be identified. Therefore, the structured search in the Leiden University's Library catalog resulted in 19 articles with a focus on e-health in developing countries from which challenges and solutions could be identified. The less structured search in Google Scholar resulted to an additional 27 articles. There were 46 articles in total, 19 articles from the Leiden University Catalogue and 27 articles from the Google Scholar that satisfied the inclusion criteria.

9.3.2   CHALLENGES: LACK OF SKILLED STAKEHOLDERS

First of the six themes for challenges to e-health implementation in developing countries is the lack of skilled users. The users that lack skills in using the system are health practitioners or patients (39), developers and maintainers, related ICT professionals. These maybe due to poor literacy and poor technological skills – internet and computer literacy. The stakeholders with low level of education are technical staff but are the primary users of the Health Information Systems in developing countries (22). A major challenge is training users in patient confidentiality (12). Some challenges facing ICT and e-health policy makers in developing countries is a lack of human resources at all levels (23). In effect, this may lead to shortage of ICT-professionals (37). There is also an issue on the qualification standards for ICT-knowledgeable health care professionals, even though professionals are expected to have a certain degree of extra knowledge (28), and lack of security awareness, that poses risks security breaches with regard to computer security perception (32). Below are the details:

- Limited technology skills or knowledge of technology (1, 2, 4, 14, 15, 18, 22, 23, 26, 30, 32, 33, 34, 37, 39, 40, 42, 46);
- Lack of general skills (1, 3, 4, 9, 16, 25, 26, 32, 37);
- Lack of awareness, familiarity or knowledge on e-health (1, 9, 10, 18, 21, 32, 37, 41);
- Poor or lack of education or training (2, 3, 12, 16, 18, 22, 32, 37, 41, 42);
- Lack of workforce (7, 23, 37, 43);
- Uneven distribution of workforce (46);
- Lack of research (15);
- Problems with setting qualification standards (28).

9.3.3   SOLUTIONS: BUILDING A SKILLS BASE

The straightforward approach to deal with most of the challenges is training or education, either in general or specifically in technology skills about e-Learning and tele-training. Partnering with universities and institutions is essential for emerging educational programs (15). During needs assessment and before e-health implementation: customs, culture, and health care needs should have been taken into account. In the aspect of support and maintenance of the systems contracting to other a more specialized company may be opted (33). Suggested measures found with regard to a lack of skilled stakeholders include:

- Teletraining (13, 14, 16, 19, 27, 29, 33, 45, 46);

- e-Learning (2, 14, 27, 29, 33, 46);
- Partnering with universities and institutions (15, 27, 29, 43);
- In-service training (32), Contracting ICT professionals (33);
- Needs assessment and workforce research (15);
- Available software application such as RightChoice (4);
- Emphasize importance (37) of security risks or any priority issues;
- Seminars and workshops (1), Standardization of learning (15).

9.3.4   CHALLENGES: INADEQUATE INFRASTRUCTURES

Second of the six themes of challenges is inadequate infrastructures. An often mentioned example of an inadequate ICT infrastructure is inadequate network infrastructure and connectivity. Lack of network infrastructure (24), can result in a set of security flaws, and others are lack of backup systems. What is meant by the term supporting infrastructure in relation to e-health is well explained by Ouma and Herselman (33), for example, Health Information Systems infrastructure (30). Supporting infrastructures are directly related to the utilities within the community or location of the e-health system and these are transport system in rural areas can be inadequate, lack of power generators makes it unable to put the right ICT infrastructure in place, and power and electricity problems. These can even extend to a lack of basic amenities such as safe drinking water (9). Below are the details:

- Inadequate ICT infrastructure (1, 2, 3, 7, 10, 11, 19, 21, 22, 24, 25, 28, 30, 32–35, 39, 40, 42, 43);
- Inadequate supporting infrastructure (1, 9, 10, 12, 21, 22, 25, 33, 35, 42, 43);
- Limited infrastructure in general (3, 10, 16, 20, 25, 26, 30);
- Inadequate maintenance and support (3, 20, 22, 25, 39, 42);
- Security flaws (7, 13, 24, 42), Trouble with designing ICT infrastructure (13).

9.3.5   SOLUTIONS: BUILDING AN INFRASTRUCTURE BASE

The solutions recommended can be categorized into an institutional level or a community level. The difference between the two is that the former requires a great deal of investment while the latter requires less. The institutional level solutions are establishing mobile telephone technology and promotion of Telemedicine to overcome distance barriers, lack of human capital in the local areas, and building solar and wind power capacities. The community level or on a smaller scale: software infrastructure should include both computerized and manual solutions. To counteract power failures solutions maybe be any of the following: network-connected and disconnected solutions (30), offsite data system and offline data entry allowing storage of data locally (12), or off-site data backup (35). The network architecture can be a mesh networks (24). An alternative to the expensive investment with wind power is solar power. However e-health stakeholders need to ensure compatibility of physical infrastructure to run on battery and solar-powered backup systems, and usage of widespread and low-cost mobile technology. Suggested measures found with regard to inadequate infrastructure include:

- Mobile telephone technology (12, 21, 26, 40, 43);
- Solar and wind powered systems (17, 43);
- Computerized-manual and connected-disconnected solutions (30);
- Software infrastructure level (24), Mesh network infrastructure (24);
- Network-connected and disconnected solutions (30);
- Offline data entry (12), Offsite data system (12), Security tools (32).

9.3.6   CHALLENGES: LACK OF ACCEPTANCE

Third of the six themes of challenges is lack of acceptance. Low levels of comfort with the use of technology may have to do with, for example, a fear of computers or technological concerns as there could be a perception of risk, including concerns about safety, validity and reliability of the technology, but especially privacy, security and confidentiality concerns. A key potential barrier might not necessarily be technological complexity, but could be job security concerns (8). Also, enforcing e-health legislation in developing countries could be difficult in developing countries and acceptance by the community for the transformation of any system can be hard (3). Below are the details:

- Cultural and ethical challenges (1, 3, 5, 7, 13, 20, 21, 23, 26, 43–45);
- Costs (1, 3, 7, 9, 16, 21, 22, 26, 31, 33, 34);
- Resistance to technology or change (2, 3, 5, 7, 13, 18, 19, 23, 25, 42);
- Benefits not perceived (3, 20, 22, 25, 31, 32, 42, 44, 46);

- Ignorance, or lack of incentive (3, 20–22, 41, 46);
- Low levels of comfort with use of technology (4, 10, 23, 32–34);
- Privacy, security, confidentiality concerns (1, 2, 8, 18, 32, 42);
- General lack of acceptance (3, 13, 21, 45);
- Safety, validity and reliability concerns (7, 26, 37);
- Job security concerns (8).

9.3.7 SOLUTIONS: ACHIEVING ACCEPTANCE

As benefits are often not perceived, risk-benefit analysis, impact analysis and monitoring and evaluation are considered to be important. Note that monitoring and evaluation may not only be considered to be important for acceptance, but for e-health in general. For example, whenever possible, systems should be based on evidence of effectiveness of previous systems (30). Developing countries need to make proper cost evaluations, including a comparison of current and expected costs. This would then also aid in understanding the benefits of a system (32). Limited user participation appears to be a great contributor to failure of e-health projects (39), but getting users involved can also be a key to successful e-health implementations.

Some essential guidelines for these problems are: systems standards and guidelines need to be in place and systems should be assessed for suitability prior to widespread implementations (44). The compatibility of IT solutions to work practices is a key determinant to acceptance and because social influence is important, the support and backing of senior decision makers should be sought, as support technology transfer is likely to be unsuccessful without direct senior management (19).

E-health should be integrated into health care service delivery in order to be accepted by clients and health caregivers (40). Early involvement of users puts them at the leading end, instead of just forcing them to receive a developed system and imposing it on them (22). Change management with clients, as well as staff should be sensitized (18). A local champion who can be taught in depth about the system and can be the connection between developers and staff is a key success factor (12). This person could then communicate issues to the developers and align the introduction of a system with organizational improvements can actually strengthen its acceptance (12). Without the support and influence from the top, transfer of technology is likely to be unsuccessful (19). Suggested measures found with regard to a lack of acceptance include:

- Stakeholder participation (7, 10, 12, 22, 36, 37, 40, 42, 46);
- Continuous risk-benefit and impact analysis, monitoring and evaluation (6, 12, 25, 26, 30, 32, 46);
- Change Management (18, 22, 46), Mechanisms for feedback (3, 25);
- Suitability assessment (19, 44), Use of a local champion (12, 46);
- Align with organizational improvements (12);
- Continuous improvement approach (25);
- Integration in health care service delivery (40), Support from top (19).

9.3.8 CHALLENGES: LIMITED RESOURCES

Fourth of the six themes of challenges is limited resources. It is worth noting here that some concepts within this theme may be overlapping with the themes based on infrastructures. For example, technical resources such as on-site computers and computer systems can be seen as part of the broader ICT infrastructure. However, it was considered worth listing such resource constraints in a separate theme, along with more general resource constraints such as time and funding. Maybe the most important concept in this theme is the concept of limited or unsustainable funds. In the long-term, donor funding may pose a challenge for sustainability (20). Also, in attempting to scale-up e-health services, programs may be hampered by reliance on donor funding, highlighting the need for a transition to alternative and diversified resources (21). Below are the details:

- Limited or unsustainable funds (1, 2, 3, 5, 7, 13, 20–23, 30–32, 40, 42, 43, 46);
- Lack of technical resources (3, 9, 14, 20, 22, 32, 33, 37, 39, 43);
- Limited resources in general (3, 16, 20, 23, 30, 42, 43);
- Time constraint (1, 3).

9.3.9 SOLUTIONS: BUILDING A RESOURCE BASE

Even though Open Source software has its own barriers such as contracting, documentation and language, it may prove to have great potential in developing countries due to its open methodology and low costs (5). Moreover, developing countries could build on current already existing and working technologies, while adding their own relevant modules (42 Governments may play a big role here as well as they could control computer prices, systems prices, and internet access costs, such that more people can have access to such resources (32). The alternative revenue sources may include government contracts, payment from consumers, or insurance) (21). Suggested measures found with regard to a lack of resources include:

- Use of Open source (5, 12, 30, 32, 33, 34, 43);
- Use of existing, validated systems (12, 42);
- Pool and reuse resources (30), Control computer and system prices (32);
- Alternative revenue sources (21).

9.3.10   CHALLENGES: INADEQUATE INFORMATION COMMUNICATION

Fifth of the six themes of challenges is inadequate information communication. The term "information communication" is used to describe challenges having to do with information exchange and handling by both systems and stakeholders. It comes down to the fact that, even though adequate technology may be in place, it does not necessarily mean that it can be used for appropriate information communication. Also, there may be a general lack of information exchange between stakeholders. Therefore, this theme is concerned with information and data.

Some of the challenges to e-health implementations in developing countries could be the lack of information sharing between agencies (23), lack of relevant content of applications (39), lack of information in a culturally and linguistically appropriate format, as to inform about e-health (44), problems with delivering information from a multitude of sources in a secure way. Information exchange is hindered by non-interoperability of heterogeneous database (28). Related to this topic is the lack of a Patient Unique Identifier, with which patient records can be distinguished (32). Technological challenges include ensuring interoperability, integration of information systems, but also securing the privacy of information (20). Below are the details:

- Interoperability challenges (3, 11, 20, 25, 28, 30, 42);
- Lack of information (23, 28, 33, 34, 39, 44);

- Lack of integration (2, 3, 20, 28, 30, 39);
- Lack of standards (2, 7, 11, 30, 32, 39);
- Language barriers (2, 5, 13, 32, 37, 42);
- Data processing or management challenges (2, 23);
- Duplication of systems or work (2, 25);
- Lack of patient unique identifier (32, 39);
- Lack of specifications or requirements (37);
- Limited attention for reusability (30);
- Difficulties securing privacy of information (20);
- Lack of operational compatibility (28), Lack of information infra-structure (2);
- Communication gap health-it (2).

9.3.11 SOLUTIONS: BUILDING AN INFORMATION COMMUNICATION BASE

Key to e-health implementations is not only the provision of the right information, at the right place, at the right time. Keeping a core dataset as a point of comparison for data correctness and user traceability – audit logs of user activity, page viewing and editing (12). A straightforward solution to address interoperability issues is the setting of national or global standards (32). Standards such as Health Level Seven (HL7) – for the transmission of e-health care information, or Logical Observation Identifiers Names and Code (LOINC) (35). To overcome interoperability and maintenance problems, simple, local technology and simple, locally adopted user-friendly software should be adopted (40). For example, using birth certificate numbers to identify patient records (32), using a smartcard that allows for information to be kept in one place (39). Suggested measures found with regard to inadequate information communication base include:

- Standards such as HL7 (24, 25, 30, 32, 35), Audit logs (12);
- Birth certificate numbers for patient unique identifier (32);
- Core data set (12), Smartcard (39), Simple, local technology (40).

9.3.12 CHALLENGES: INADEQUATE PROCESS GUIDANCE

The last theme of challenges is that of inadequate process guidance. The theme process guidance relates to those aspects and processes that are needed to guide the entire process of implementing e-health in developing countries. As will be discussed later, this can be seen as a critical theme encompassing aspects of all of the other themes. With regard to global health, it is argued that the developing countries may become even further isolated from e-health benefits because of isolated policies that are formulated in a parochial manner (23). For example in Pakistan, there is actually a mismatch between policies and ground realities (36), while in Africa, telemedicine projects fail because of insufficient coordination and evaluation (46). This is apart from the common project implementation issue that if larger goals are not clearly defined, development processes tend to be slow and inefficient and the product will likely be inadequate (12). Below are the details:

- Limited, inadequate or lack of policies (2, 10, 12, 18, 20, 23, 32, 36, 39);
- Absence of impact measurement, research and evaluation (2, 6, 38, 40, 44, 46);
- Inadequate legal framework (2, 3, 7, 12, 18, 46);
- Lack of political or managerial commitment or support (2, 9, 20, 22, 31, 40);
- Unstable social, economic or political environment (7, 11, 30, 44);
- e-health not part of strategy (7, 23, 46), Lack of vision (2, 12, 46);
- Insufficient leadership (20, 23, 32), Ownership ambiguity (12, 40);
- Insufficient coordination (46), Lack of governing body for guidance (9);
- Poor public-private partnerships (2), Policy reforming challenges (7).

9.3.13 SOLUTIONS: IMPROVING PROCESS GUIDANCE

Measures to improving process guidance are outlined below, merged into a set of steps for implementations of e-health by Mechael (26). Note that the steps themselves can be seen as too straightforward. However, the order of these recommended steps may be seen as an approach to process guidance itself and may therefore be seen as a measure in its entirety. Therefore, other measures found are embedded in this order of steps:

1. Assess current state of e-health;
2. Identify health conditions;
3. Examine workflow and priorities;
4. Identify e-health capacity and business cases for service;
   - Business model and research (38, 40, 46)

5. Develop strategic plans;
   - Architectural patterns (30)
   - Enterprise Architecture (11, 30)
6. Define roles, responsibilities and accountability;
   - Partnerships (4, 7, 18, 20, 26, 33, 34, 40, 43, 44, 46)
7. Develop guidelines and policies;
   - Globallocal policy (23, 40)
   - e-Government policy (11, 29)
8. Establish targets and measures;
9. Monitoring and evaluation (40).

9.3.14 SOLUTIONS: PLANNING AND EXECUTION

Detailed, careful, and well-researched Business model for e-health is needed. For example, taking a long-term view of usage (25) and setting up a taskforce (29). The factors that need to be taken into account when planning and executing early stage of an e-health project are primary health benefits, sustainability, scalability, costs, careful consideration of local needs and customs, safety, effectiveness, business model (46).

9.3.15 SOLUTIONS: POLICIES

Health policies should address e-health maintenance and support (39), regulations for privacy benefits, cultural differences, interoperability, and capacity building (32). These policies and strategies should "protect citizens, promote equity, observe cultural and linguistic issues in cyberspace, ensure interoperability (the ability of different technology systems to work together), and allow for capacity development so that all citizens can access e-health solutions" (40). e-health policy and enabling policy-environment are essential to the success of e-health initiatives (40).

9.3.16 SOLUTIONS: APPROACH – GLOCAL, PARTNERSHIP, COLLABORATIVE PARTNERSHIP

The Glocal approach is engaging wisdom and experience of stakeholders at both global and local levels (39). Partnerships among institutions in

Developed and Developing Countries that are putting in place e-health tools, may help train people to use them, and are likely to result in the development of human capacity and commitment to e-health in developing countries. Partnerships can increase the speed and breadth of improvement, as individual health care organizations, academic institutions, government agencies, private companies, private donors and philanthropic organizations alone are incapable of implementing e-health solutions and local capabilities should be built, not dependencies (43). Partnerships would include public-private partnerships, and partnerships between all kinds of stakeholders.

Collaborative partnership may entail all stakeholders to work as a team before, during, and after the implementation of the system. Working with a team then would entail understanding cultural differences (44).

9.4   DISCUSSIONS

This chapter has given some insight into challenges and solutions of e-health implementations in Developing Countries and possible measures to overcome them. Six major themes emerged with regard to challenges to e-health implementations in developing countries: (i) lack of skilled stakeholders, (ii) lack of acceptance, (iii) limited resources, (iv) inadequate information communication, (v) inadequate process guidance, and (vi) inadequate infrastructure. For each of these, a number of measures were proposed. The themes that have been elaborated in the results may actually be influencing each other. For example, a lack of awareness of e-health and a lack of information communication could be reason for ignorance or a lack of acceptance. Also, a lack of technical resources obviously has influence on a lack of ICT infrastructures.

Based on the analysis of the result and the assumption of interrelations among the themes, we attempted to develop a theoretical model. Firstly, stakeholder acceptance, based on our study, is central to e-health implementations in developing countries. Even when all technology requirements are in place, when it is not accepted by the ones who have to use or implement it, it is of no use; Secondly, an appropriate skill base should be built for these same stakeholders; Thirdly, a resource base should be built, not just around technical equipment, but also with regard to funding; Fourthly, infrastructure base should be built, encompassing both supporting infrastructure as ICT infrastructure; Fifthly, an information communication base should be built, which includes setting standards for interoperability and integration and making sure that patients can be identified, and is about the communication of information between systems as well as stakeholders; and Finally, process guidance is a theme that should be active through the entire process, from planning, to implementing, to monitoring implementations of e-health in developing countries. Strategic direction should be given, policies should be made, and adequate leadership is necessary through responsibilities and accountabilities. Even after the implementation is complete, performance measurement should be a guiding process. Moreover, for successful e-health implementations in developing countries, such process guidance should address each of the themes of challenges found. This discussion allowed us to create the proposed model shown below.

In this model (Figure 9.1), we have three major focus points drawn from the articles reviewed: (i) process guidance, (ii) acceptance, and (iii) building a base.

- Process guidance as an overarching theme, process is a recurring theme within the subsets of our model. We think that process orientation is essential in successfully managing the e-health implementations in Developing Countries. Guidance in trying to adopt and managing the e-health or when building a skill base, information communication base, infrastructure base, and resource base through strategy, planning, leadership, policies and necessary processes.
- Stakeholder acceptance, central to this model is the stakeholder as they will in the end carry on the task of the initiators of this project. Stakeholders would include hospitals, clinics, users, government and others. In the process guidance category special focus should be given to stakeholder acceptance.
- Base, although it is considered as a subset concern it is an important block that make the proposed model. These are: (a) skill, (b) information communication, (c) infrastructure, and (d) resource.

9.4.1   LIMITATIONS AND FUTURE RESEARCH

Note that the themes and concepts describing challenges are not necessarily true for every developing country and would differ in circumstances and environments. There was no distinction between rural and urban areas

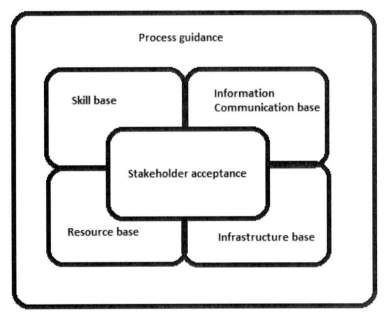

FIGURE 9.1   E-health adoption model.

within developing countries. However, the results of the literature review as displayed in the proposed model show themes to which attention should be paid, as those appear to be the areas in which problems occur for e-health implementations in developing countries in general. Moreover, these are themes describing challenges to e-health implementations in general, not to specific e-health applications such as Telemedicine or EHR's. Finally, the theme process guidance is quite broad and, in future research, may be split into smaller process steps and assessed on how it differs from e-health Governance. Therefore, we provide a comprehensive overview of these challenges and measures in a way that we found suitable.

Future research can be done on validating the model together with the unified theory of acceptance and use of technology.

9.5   CONCLUSIONS

The role of e-health in resolving some of the societal challenges in Health is still unraveling and much work is needed to successfully implement solutions from developed countries to developing countries. Through the systematic literature review conducted by the researchers they were able to consolidate the previously identified problems and link this with proposed solutions. These problems and solutions were then used as basis for a proposed theoretical model that is catered to developing countries. The model is held together by the principles of system of inter-relations and interactions for a successful implementation.

APPENDIX 1: DATA SET ANALYZED

[1–46]

1. Abodunrin, O., Akande, T. "Knowledge and perception of e-health and telemedicine among health professionals in LAUTECH Teaching Hospital, Osogbo, Nigeria," International Journal of Health Research (2:1) 2009.
2. Al-Shorbaji, N. "e-health in the Eastern Mediterranean Region: a decade of challenges and achievements," Eastern Mediterranean Health Journal (14) 2008, S157–S173.
3. Anwar, F., Shamim, A. "Barriers in adoption of health information technology in developing societies," International Journal of Advanced Computer Science and Applications (2:8) 2011, 40–45.
4. Arora, P. "Is the doctor on? In search of users for medical software in rural Himalayas," Development in Practice (22:2) 2012, 180–189.
5. Bagayoko, C.-O., Dufour, J.-C., Chaacho, S., Bouhaddou, O., Fieschi, M. "Open source challenges for hospital information system (HIS) in developing countries: a pilot project in Mali," BMC medical informatics and decision making (10:1) 2010, p 22.
6. Blaya, J. A., Fraser, H. S., Holt, B. "E-health technologies show promise in developing countries," Health Affairs (29:2) 2010, 244–251.
7. Burney, A., Mahmood, N., Abbas, Z. "Information and communication technology in healthcare management systems: prospects for developing countries," International Journal of Computer Applications (4:2) 2010, 27–32.

8. Chang, L. W., Njie-Carr, V., Kalenge, S., Kelly, J. F., Bollinger, R. C., Alamo-Talisuna, S. "Perceptions and acceptability of mHealth interventions for improving patient care at a community-based HIV/AIDS clinic in Uganda: A mixed methods study," AIDS care:ahead-of-print) 2013, 1–7.

9. Dasgupta, A., Deb, S. "Telemedicine: A new horizon in public health in India," Indian journal of community medicine: official publication of Indian Association of Preventive & Social Medicine (33:1) 2008, p 3.

10. Durrani, H., Khoja, S., Naseem, A., Scott, R., Gul, A., Jan, R. "Health needs and e-health readiness assessment of health care organizations in Kabul and Bamyan, Afghanistan," EMHJ (18:6) 2012.

11. Foster, R. "Review of developing country health information systems: A high level review to identify health enterprise architecture assets in ten african countries," Jembi Health Systems.

12. Fraser, H., Blaya, J. "Implementing medical information systems in developing countries, what works and what doesn't," AMIA Annual Symposium Proceedings, American Medical Informatics Association, 2010, p. 232.

13. Ganapathy, K., Ravindra, A. "Telemedicine in India: the apollo story," Telemedicine and e-Health (15:6) 2009, 576–585.

14. Gour, N., Srivastava, D. "Knowledge of computer among health-care professionals of India: a key toward e-health," Telemedicine and e-Health (16:9) 2010, 957–962.

15. Hersh, W., Margolis, A., Quirós, F., Otero, P. "Building a health informatics workforce in developing countries," Health Affairs (29:2) 2010, 274–277.

16. Hitchcock, C. L. "The future of telepathology for the developing world," Archives of pathology & laboratory medicine (135:2) 2011, 211–214.

17. Idachaba, F., Idachaba, E. "Robust e-health communication architecture for rural communities in developing countries," Engineering, Technology & Applied Science Research (2:3) 2012, pp. 237–240.

18. Isabalija, S., Mayka, K., Rwashana, A., Mbarika, V. "Factors affecting adoption, implementation and sustainability of telemedicine

information systems in Uganda," Journal of Health Informatics in Developing Countries (5:2) 2011.

19. Kifle, M., Payton, F. C., Mbarika, V., Meso, P. "Transfer and adoption of advanced information technology solutions in resource-poor environments: the case of telemedicine systems adoption in Ethiopia," Telemedicine and e-Health (16:3) 2010, 327–343.

20. Leon, N., Schneider, H., Daviaud, E. "Applying a framework for assessing the health system challenges to scaling up mHealth in South Africa," BMC medical informatics and decision making (12:1) 2012, p 123.

21. Lewis, T., Synowiec, C., Lagomarsino, G., Schweitzer, J. "E-health in low-and middle-income countries: findings from the Center for Health Market Innovations," Bulletin of the World Health Organization (90:5) 2012, 332–340.

22. Malik, M. A., Khan, H. R. "Understanding the implementation of an electronic hospital information system in a developing country: a case study from Pakistan," Proceedings of the Third Australasian Workshop on Health Informatics and Knowledge Management-Volume 97, Australian Computer Society, Inc., 2009, pp. 31–36.

23. Mars, M., Scott, R. E. "Global e-health policy: a work in progress," Health Affairs (29:2) 2010, 237–243.

24. Masi, M., Pugliese, R., Tiezzi, F. "e-Health for Rural Areas in Developing Countries: Lessons from the Sebokeng Experience," in: e-Infrastructure and e-Services for Developing Countries, Springer, 2012, pp. 187–196.

25. Matheson, A. I., Baseman, J. G., Wagner, S. H., O'Malley, G. E., Puttkammer, N. H., Emmanuel, E., Zamor, G., Frédéric, R., Coq, N. R., Lober, W. B. "Implementation and expansion of an electronic medical record for HIV care and treatment in Haiti: An assessment of system use and the impact of large-scale disruptions," International Journal of Medical Informatics (81:4) 2012, 244–256.

26. Mechael, P. N. "The case for mHealth in developing countries," innovations (4:1) 2009, 103–118.

27. Mishra, S., Ganapathy, K., Bedi, B. S. "The current status of e-Health initiatives in India," Making the e-health Connection: Global Partnerships, Local Solutions. Bellagio, Italy) 2008.

28. Mishra, S., Singh, I. P. "mHealth: A developing country perspective," Making the e-health connection. Bellagio, Italy) 2008, 1–9.

29. Mishra, S. K., Kapoor, L., Singh, I. P. "Telemedicine in India: current scenario and the future," Telemedicine and e-Health (15:6) 2009, 568–575.

30. Moodley, D., Pillay, A. W., Seebregts, C. J. "Position paper: researching and developing open architectures for national health information systems in developing african countries," in: Foundations of Health Informatics Engineering and Systems, Springer, 2012, pp. 129–139.

31. Mupela, E., Mustard, P., Jones, H. Telemedicine and primary health: The virtual doctor project Zambia UNU-MERIT, Maastricht Economic and Social Research and Training Centre on Innovation and Technology, 2011.

32. Omary, Z., Lupiana, D., Mtenzi, F., Wu, B. "Analysis of the challenges affecting e-healthcare adoption in developing countries: a case of Tanzania," International Journal of Information Studies (2:1) 2010, 38–50.

33. Ouma, S., Herselman, M. "E-health in rural areas: Case of developing countries," International Journal of Biological and Life Sciences (4:4) 2008, 194–200.

34. Ouma, S., Herselman, M. E., Van Greunen, D. "Implementing successful e-health implementations within developing countries, ") 2009.

35. Piette, J. D., Lun, K., Moura Jr, L. A., Fraser, H. S., Mechael, P. N., Powell, J., Khoja, S. R. "Impacts of e-health on the outcomes of care in low-and middle-income countries: where do we go from here?," Bulletin of the World Health Organization (90:5) 2012, 365–372.

36. Qureshi, Q. A., Ahmad, I., Nawaz, A. "Readiness for e-health in the developing countries like Pakistan," Gomal Journal of Medical Sciences (10:1) 2012.

37. Qureshi, Q. A., Shah, B., Najeebullah, N., Kundi, G. M., Nawaz, A., Miankhel, A. K., Chishti, K. A., Qureshi, N. A. "Infrastructural

barriers to e-health implementation in developing countries," European Journal of Sustainable Development (2:1) 2013, 163–170.

38. Rey-Moreno, C., Reigadas, J. S., Villalba, E. E., Vinagre, J. J., Fernández, A. M. "A systematic review of telemedicine projects in Colombia," Journal of Telemedicine and Telecare (16:3) 2010, 114–119.

39. Ruxwana, N. L., Herselman, M. E., Conradie, D. P. "ICT applications as e-health solutions in rural healthcare in the Eastern Cape Province of South Africa," Health information management journal (39:1) 2010, 17–26.

40. Shiferaw, F., Zolfo, M. "The role of information communication technology (ICT) towards universal health coverage: the first steps of a telemedicine project in Ethiopia," Global Health Action (5) 2012.

41. Shoaib, S. F., Mirza, S., Murad, F., Malik, A. Z. "Current status of e-health awareness among healthcare professionals in teaching hospitals of Rawalpindi: a survey," Telemedicine and e-Health (15:4) 2009, 347–352.

42. Sood, S. P., Nwabueze, S. N., Mbarika, V. W. A., Prakash, N., Chatterjee, S., Ray, P., Mishra, S. "Electronic medical records: a review comparing the challenges in developed and developing countries," Hawaii International Conference on System Sciences, Proceedings of the 41st Annual, IEEE, 2008, pp. 248–248.

43. Tierney, W. M., Kanter, A. S., Fraser, H. S. F., Bailey, C. "A Toolkit For E-Health Partnerships In Low-Income Nations," Health Affairs (29:2), February 1, 2010–2010, 268–273.

44. Were, M. C., Meslin, E. M. "Ethics of Implementing Electronic Health Records in Developing Countries: Points to Consider," AMIA Annual Symposium Proceedings, American Medical Informatics Association, 2011, p. 1499.

45. Wootton, R. "Telemedicine support for the developing world," Journal of Telemedicine and Telecare (14:3) 2008, 109–114.

46. Wynchank, S., Fortuin, J. "Developing Nations' e-health and Telemedicine: Lessons Learned, Especially for Africa,"

eTELEMED 2012, The Fourth International Conference on e-health, Telemedicine, and Social Medicine, 2012, pp. 49–54.

KEYWORDS

- **developing countries**
- **e-health**
- **stakeholders**
- **systematic literature review**
- **theoretical model**

REFERENCES

1. ISI, ISI Developing Countries (in 2013), http://www.isi-web.org/component/content/article/5-root/root/577-developing2012, Accessed 20 January, 2014.
2. Jamison, D. T. Mosley, W. H. *American Journal of Public Health*, 1991, 81, 15–22.
3. European Commission, Societal challenges, http://ec.europa.eu/programs/horizon2020/en/h2020-section/societal-challenges.
4. Kwankam, Y. in *International Hospital Federation Reference Book*, 2008, pp. 56–58.
5. Blaya, J. A., Fraser, H. S., Holt, B. *Health Affairs*, 2010, 29, 244–251.
6. Eysenbach, G. *Journal of Medical Internet Research*, 2001, 3.
7. Oh, H., Rizo, C., Enkin, M., Jadad, A. *Journal of Medical Internet Research*, 2005, 7.
8. Rodrigues, R. J. *International Journal of Healthcare Technology and Management*, 2003, 5, 335–358.
9. Bath, P. A. *Journal of Information Science*, 2008, 34, 501–518.
10. Kirigia, J., Seddoh, A., Gatwiri, D., Muthuri, L., Seddoh, J. *BMC Public Health*, 2005, 5, 137.
11. Drury, P. *World Hospitals and Health Services*, 2005, 41, 38.
12. Venkatesh, V., Morris, M. G., Davis, G. B., Davis, F. D. *MIS quarterly*, 2003, 425–478.
13. Jesson, J., Matheson, L., Lacey, F. M. *Doing Your Literature Review: Traditional and Systematic Techniques*, Sage Publications, 2011.
14. AMS, AMS low dues rate for developing countries, http://www.ams.org/membership/individual/mem-develop, Accessed 1 July 2013.

PART 5:

SOCIAL NETWORKS

CHAPTER 10

HOW MEDITERRANEAN LOCAL GOVERNMENTS USE FACEBOOK TO ENHANCE STAKEHOLDER ENGAGEMENT

ALEJANDRO SÁEZ-MARTÍN, MARÍA DEL MAR GÁLVEZ-RODRÍGUEZ, and MARÍA DEL CARMEN CABA-PÉREZ

Department of Economics and Business, University of Almería, Almería, Spain

CONTENTS

ABSTRACT

Social networking sites are increasingly considered to be key elements in enhancing relationships between governments and citizens, and with this in mind, Mediterranean governments, among others, have implemented initiatives to encourage citizens to seek the full benefits that these technologies offer and to enhance citizens' engagement in democratic processes.

Taking these considerations into account, this paper reports a comparative analysis of the use of Facebook as a communication strategy for encouraging citizens' engagement with local governments in Spain, Italy and France. In general, we find that in all three cases local governments need to make greater effort in promoting the use of these technologies among their citizens. Focusing on the statistical differences identified, France has obtained a higher degree of online citizens' engagement than Spain and Italy. Although the level of commitment is generally low, Italian citizens show the least interest in giving their opinions about the activities reported by their local governments. Moreover, Spanish and Italian citizens show the least concern in making use of Facebook pages to boost the virality of the messages posted by their local governments.

10.1   INTRODUCTION

In recent years there has been growing concern about the limited extent to which governments are implementing social networking sites (SNS) to improve their relationships with citizens [28, 44, 55]. These tools are Web 2.0 technologies that facilitate collaboration, the rapid exchange of information and the sharing of created contents [31]. In view of the many benefits they provide as channels for one and two-way communication, SNS are considered a very important means of increasing governmental transparency, promoting citizens' engagement with their local administration and fostering public participation in democratic processes [13, 38].

The adoption of SNS, thus, is viewed as a significant aspect in the modernisation and democratization of European governments [10]. In this regard, a major EU initiative is the European Digital Agenda,[1] which

---

[1] http://ec.europa.eu/digital-agenda/

encourages society to take full advantage of the economic and social potential offered by Information and Communication Technologies (ICT), and encourages European governments to work toward this goal. And an essential element in this program is the use of SNS by local governments [49]. In this respect, too, with particular regard to Mediterranean countries, and in line with the European Digital Agenda, the HOMER[2] project aims to foster the use of web technologies by governments in order to enhance access to information. The countries participating in this latter project are Spain, Italy, France, Malta, Greece, Slovenia, Cyprus and Montenegro.

With respect to the implementation of SNS by local governments, Akadwani [2] noted their increasing use of Facebook, observing that its low cost and ease of use makes it an effective and efficient strategy to ensure good communication between local governments and society and to foster citizens' participation. In Europe, the populations of Spain, Italy and France are among those making most use of Facebook [12]. In this respect, Bonsón et al. [10] noted that Spain, France and Italy are among the top ten countries with the highest numbers of visits to local government Facebook pages.

Despite the undeniable impact of SNS as a new model of communication between governments and citizens, relatively few studies have been undertaken to investigate the role played by these technologies in promoting citizens' engagement [60]. In this regard, some studies have examined perceptions among public servants of SNS as a tool for citizen engagement, in the USA [38] and in Central Mexico [44]. Other authors have addressed the question by analyzing the local government use of SNS in a specific region, such as western Europe [10]. However, little is known as to whether among different countries there are significant differences in the use of SNS as a mechanism to encourage citizens' participation.

Taking into consideration the importance of SNS in local government and the growing popularity of these technologies in Mediterranean countries, we conducted a comparative analysis of the use of Facebook as a communication strategy for encouraging citizens' engagement, among local governments in Spain, Italy and France. This study is intended to contribute both to theoretical knowledge and to practical use. Thus, it

---

[2] http://www.iam-project.eu/?page_id = 2

advances our understanding of the acceptance of SNS as a new means of facilitating relations between local governments and their citizens. In addition, we identify differences in the degree of online citizens' engagement in three Mediterranean countries. In practical terms, the study aims to identify convergences and divergences in the use of Facebook as a channel to boost citizens' engagement and to provide relevant information about what should be done in this respect.

The remaining sections of this chapter are structured as follows. Section 10.2 provides an overview of social networks and their importance in public administration. Section 10.3 presents the theoretical framework and a review of the literature. The next section describes the research methodology applied and explains how the study sample was selected. Section 10.5 presents the empirical study, and discusses the results obtained. The final section summarizes the main conclusions drawn.

10.2   ONLINE RELATIONS BETWEEN CITIZENS AND LOCAL GOVERNMENT THROUGH SOCIAL NETWORKS

SNS are considered among the most useful means of conducting dialogic communication, thanks to the possibilities they offer for interaction and communication [67]. Accordingly, at diverse levels of government SNS are viewed as a means of achieving direct and straightforward communication with society [22, 50, 53]. SNS are relatively new, and are being used as a strategy to reach citizens who might not otherwise relate to the public authorities through traditional forms of interaction [23, 41, 52].

This development has produced a change in relations between government and society, as SNS facilitate communication and interaction, encouraging active participation by society in general and by citizens in particular [27, 52]. Thus, SNS have become one of the main tools used in the application of e-government between the local administration and citizens [7, 55, 60]. In this respect, it is noteworthy that over 43% of European citizens who access the internet make use of SNS [5], and consider it to be one of the most useful and simple ways of keeping in touch with government activities; SNS provide an alternative mechanism for citizens to have their say, thus awakening the interest of society in the affairs of government [20]. Although governments have used other tools of e-government, such as web pages, to establish relations with the public, SNS are gaining wide-ranging acceptance as a communication channel, because of the considerable interactivity offered and the possibility for governments and citizens to create and share content [43]. Accordingly, most public administrations now make use of social networks as a channel to represent artifacts of their core mission, to engage the public, to participate in conversations on issues and to network with stakeholders [38].

Among the different levels of government involved, authors such as Bonsón et al. [9, 10] and Sáez et al. [51] have noted that it is important to understand how municipalities use SNS to promote citizens' participation and engagement. Due to the greater proximity between local governments and citizens, these public entities are well placed to ascertain and to satisfy citizens' information needs [62]. Moreover, citizens have a greater interest in participating directly in local affairs than in those of a more general scope [9]. Therefore, authors such as Sæbø et al., [49] have emphasized the need for a greater use of SNS by local governments, citing the following benefits they provide:

- Greater accessibility to local government: authors like Song and Lee [56] argue that SNS contribute to making government more accessible to citizens and to keeping the public informed about government policies and results.
- Increased legitimacy and democratization of local government: if SNS are used to meet the needs and interests of citizens, they can become a valuable tool for increasing the democratization of local government [26].
- Greater transparency, trust and participation: the use of SNS promotes more deliberative and dialogic relations between government and citizens [9, 13].
- Increased citizens' engagement: SNS encourage and strengthen public participation, heightening citizens' interest in becoming involved in government issues via the social networks [11].

Despite the benefits that accrue to society from SNS, research suggests that organizations, including local governments, fail to make optimum use of the dialogic strategies offered by SNS, and so do not obtain all the benefits offered [12, 14, 36, 37, 48, 63]. In this regard, authors such as Burson and Marsteller [14], Kietzmann et al., [32] and Wright and Hinson [66] have argued that the way in which SNS content is shared, created and modified may determine the success or failure of organizations in developing online relationships with their stakeholders.

Among all SNS, the use of Facebook is considered to be one of the most valuable instruments for involving society [57], and it is very widely used by local governments, due to the increased interaction with citizens that can thus be obtained [19, 61]. In this line, Chun and Warner [16] argued that many citizens are active Facebook members, which has led it to become one of the SNS most commonly used by local governments to reach large numbers of citizens. In this regard, it has been argued that through the use of Facebook local governments have the opportunity to meet citizens "personally" [11]. Furthermore, Facebook is the most popular SNS among citizens, according to the Alexa ranking [3], and therefore it is necessary to learn more about its use.

10.3 THEORETICAL FRAMEWORK AND LITERATURE REVIEW

Numerous theories have been used to create a theoretical framework to explain the strategies and actions implemented by local governments through social networks [51], including agency theory [64], the theory of information technology [17], Buckner's theory on rumor transmission [29], the theory of legitimacy [8] and stakeholder theory [8, 9]. However, during the last decade, the dialogic communication theory of Kent and Taylor [30] has been the most widely employed to explain the importance of using SNS as a strategic tool to promote communication and dialog [8, 12, 37, 48, 51, 63, 66]. This theory emphasizes the use of Web technologies as a key element in improving interactivity and communication by local governments and in enhancing citizens' satisfaction and commitment with this process, via SNS, achieving greater transparency and participation and strengthening trust between organizations and their stakeholders [8, 9].

Most previous studies on the use of SNS by governments have focused on aspects such as identifying the type of SNS most commonly used, the number of followers, the number of messages sent by public administrations, their content, etc. [4, 6, 7, 12, 15, 21, 33–35, 46–48, 55, 59]. However, very few have examined the use of SNS as a communication channel to strengthen the involvement and participation of citizens in government affairs [1, 26, 51, 52].

In this field, however, a significant study has been conducted by Bonsón et al., [10], who analyzed citizens' engagement with local governments in western European cities. These authors implemented the metric developed by Bonsón and Ratkai [8] for use in the corporate sector, adapting it to the context of public administration. In their approach, citizens' engagement is defined by three dimensions: popularity, commitment and virality. "Popularity" measures engagement in terms of citizens' interest in a local government's affairs via its posts to the Facebook page; if the issue arouses interest, people will "like" the post. The "commitment" dimension, on the other hand, reflects a more interactive engagement, in which citizens participate more actively, by commenting on the post made on the local government's Facebook page. Finally, "virality" demonstrates citizens' involvement in the active disclosure of matters concerning government activity, through the action of "sharing" a Facebook post. Their engagement, thus, is measured by the action taken in divulging the messages and activities of the local government.

However, although Bonsón et al., [11] performed a comparative study of citizens' engagement in the context of European local governments, their sample was composed of just five local governments in each country, which might be considered insufficiently representative to determine the real degree of citizens' engagement through SNS in each country. Therefore, in line with the proposals for future research made by Bonsón et al., [11], and in view of the scant number of studies made to analyze citizens' engagement with local governments through their Facebook pages, this paper considers the question in greater depth, covering a larger sample of local governments to obtain a better understanding of the use of this tool in three European countries: Italy, France and Spain.

10.4 METHOD

10.4.1 COMPARATIVE ANALYSIS

In this comparative analysis of the use of Facebook as a communication strategy to encourage citizens' engagement with local government, in Spain, Italy and France, we reviewed the official Facebook pages of local governments in each of these countries. This level of government was chosen because it is the one in which citizens are most strongly involved (Oakerson 1999), which most directly affects them and their interests, in which most interest is taken in the administration's functioning and decisions (Gaventa and Valderrama, 1999), and where the current process of strengthening participatory democracy is most evident (Licha, 2002).

Our review examined the popularity, commitment and virality of local governments' Facebook pages following the metric developed by Bonsón and Ratkai (2013), as adapted by Bonsón et al. (2014a) to measure the level of citizens' engagement in public sector Facebook pages. As indicated in Table 10.1, popularity is measured by the "likes" on Facebook, commitment refers to the number of comments made and virality is measured by the number of "shares."

To determine whether there are significant differences among the three sample groups (Spain, Italy and France) the *Kruskal-Wallis* test was applied, following previous studies related to online information disclosure [11, 58]. This nonparametric technique compares the medians of two or more independent samples to determine whether the samples come from different populations. Therefore, it tests whether the mean ranks of

TABLE 10.1 Metrics Used to Measure Citizens' Engagement with Local Government via Facebook

Name	Sign	Formula	Measures
Popularity	P1	Posts with likes/total posts	Percentage of total posts liked
	P2	Total likes/total posts	Average number of likes per post
	P3	(P2/number of fans) * 1000	Popularity of messages among fans
Commitment	C1	Posts with comments/total posts	Percentage of total posts that have been commented on
	C2	Total comments/total posts	Average number of comments per post
	C3	(C2/number of fans) * 1000	Commitment of fans
Virality	V1	Posts with shares/total posts	Percentage of the total posts that have been shared
	V2	Posts with shares/total posts	Average number of shares per post
	V3	(V2/number of fans) * 1000	Virality of messages among fans

Source: Bonsón and Ratkai [8].

the measurement variable are the same in all the groups. This technique is implemented when the samples do not present a normal distribution, in which case one-way ANOVA cannot be applied, as is the present case [54].

10.4.2 SAMPLE

For the sample, we selected the 50 largest local governments in each country (Italy, France and Spain), an approach that is in line with numerous previous studies that have focused on the largest cities in terms of population [25, 45]. Moon [39], Ho [24], Bonsón et al. [10, 12] and Sáez et al. [51], among others, have analyzed large cities, arguing that they are more likely to adopt e-government innovations and other new technologies, to reduce implementation costs and to cover their greater needs for information and disclosure. Therefore, our initial sample consisted of 150 local governments, 50 in each country.

The search of Facebook pages was initially carried out through the official websites of each local government, seeking to identify the link to the official Facebook profile. If it was not found, the word "Facebook" was used in the search box on the local government website. If no links to a Facebook profile were found on the official website, we then searched on Google, combining Facebook with the name of the local government [19].

Of the 150 local governments analyzed, only 120 were found to have an official Facebook page. These municipalities represent 14.3% of the French population, 20.9% of the Italian population and 20.7% of the Spanish population.

These data were examined for a period of one month (September 2014), which was considered sufficient time in which to analyze the information presented on the social media [40]. Altogether, we obtained and analyzed 6418 posts by local governments, 11,613 comments, 316,051 "likes" and 69,793 "shares."

10.5 RESULTS

Table 10.2 presents the results obtained from the comparative analysis of the use of Facebook as a communication channel for promoting citizens' engagement. It shows significant differences among degrees of citizens'

TABLE 10.2   Citizens' Engagement: Mean Values and the Krustall-Wallis Test

		Spain Mean	France Mean	Italy Mean	Krustall-Wallis test (2-tailed) p-value
Popularity	P1	78%	96%	75%	0.000
	P2	21.62	86.91	29.15	0.000
	P3	4.08	5.85	3.36	0.000
Commitment	C1	27%	54%	36%	0.000
	C2	1.20	3.02	1.08	0.002
	C3	0.54	0.30	0.20	0.000
Virality	V1	48%	71%	53%	0.000
	V2	5.78	18.86	5.84	0.000
	V3	0.99	1.31	0.77	0.001

Significant at: $p < 0.01$.

participation in the local governments examined, in Spain, France and Italy. These results are in line with Bonsón et al., [10], for example, that there is considerable heterogeneity in the use of Facebook by local governments and citizens in western Europe. In general, the highest levels of citizens' engagement with local government are found in France, followed by Spain and Italy.

In all cases, the items most commonly used to participate in information disclosure by local governments on their Facebook pages are those related to the popularity of the pages ("liking" the information posted), followed by the items related to the virality ("sharing" the information posted), while "commitment" (making comments on the posts) is the aspect that is least often used. These data coincide with those of previous studies, which have reported low levels of citizens' engagement; most people prefer to take the easiest option, clicking the "like" button, rather than invest time in giving productive feedback to their local governments [69]. These results also seem to be in line with the trend in the United States, as Norris and Reddick [42] indicate that U.S. citizens present little demand for online mechanisms to enable interaction with their local governments.

In view of the significant heterogeneity observed among the results obtained, we decided to implement the Kolmogorov-Smirnov test to identify possible divergences between pairs of countries (Spain-France, Spain-Italy, France-Italy) (see Annexes 1, 2 and 3). This test revealed a clear difference between France, on the one hand, and Spain and Italy, on the other. However, fewer differences were found between Spain and Italy. These results indicate that Italian and Spanish citizens use Facebook in a similar way as a means of interacting with their local governments, and in a different way from French populations.

Focusing on each of the aspects that describe the degree of citizens' engagement, with respect to popularity, "likes" are commonly used to indicate agreement with the information posted by local governments. Thus, nearly all the information posted by French local governments receives "likes" (96%), while Spanish and Italian local governments receive fewer "likes," with 78% and 75% respectively (P1). Moreover, the level of support for each post published by French local governments is quite high (86.91), in contrast to the Italian local governments and even more so the Spanish ones (P2). As measured by the number of fans, the degree of popularity of the information published is low; however, the messages posted by French local governments obtain the highest popularity, and Italian ones, the lowest (P3).

"Commitment" is the aspect that received least citizens' engagement. These results are in line with Pina et al. [45], who concluded that few European local governments implement web technologies to promote dialog with their citizens. As regards the total information published, slightly over half of the French local governments analyzed in our study received comments on the information published, whereas fewer than 40% and 30% comments were recorded in the corresponding local governments in Spain and Italy, respectively (C1). With respect to the number of comments per post, the French local governments received most, with 3.02 comments per post on their Facebook pages, while those in Spain obtained 1.20 comments, and the Italians 1.08 (C2). However, Spanish local governments present a higher level of commitment (0.54) than the French ones (0.30) and a much better result than is obtained in Italy (0.20) (C3).

In the "virality" aspect, the majority (71%) of the posts published by French local governments are shared by their citizens, whereas slightly over half of the information disclosed by the Italian local governments is shared, and only 20% of that posted by Spanish local governments (V1).

Regarding the average number of "shares" per message posted, the citizens from France made 18.86 "shares" on average, in contrast to just 5.78 and 5.84 "shares" by those in Spain and Italy, respectively (V2). The same pattern was observed with respect to the virality of messages (V3). Therefore, French citizens appear to be the most interested in spreading the information provided by their local governments on their Facebook pages.

10.6   CONCLUSIONS

In recent years SNS have acquired increasing importance in the communication strategies of local governments, due to the possibilities they offer to foster dialogic communication and thus increase the degree of participation by citizens in municipal affairs.

Although EU countries have undertaken various initiatives to encourage the use of SNS by local governments and citizens, it is not clear whether these initiatives are being accepted to the same degree in all the countries involved. This paper examines whether there are significant divergences in the use of Facebook as a communication strategy to foster citizens' engagement with local government, in Spain, Italy and France. This study enhances our understanding of the question by showing how citizens in these countries react to the opportunity of participating and extending the democratic process via the use of Facebook. A further contribution made is our analysis of whether greater effort is needed by local governments in promoting the use of information technologies and SNS.

The following main findings were obtained: in general, citizens from all three countries are more interested in expressing approval of the information posted by local government than in participating in its diffusion. The least common reaction is that of responding to the information published on local governments' Facebook walls. Thus, there is little interest in providing feedback regarding municipal actions.

Focusing on the differences among the three cases analyzed, France obtains a higher degree of online citizens' engagement than do Spain and Italy. In this regard, French local governments are more popular than those in Spain and Italy, with a notably high level of "likes" per information item posted, indicating that French citizens are more willing to support the policies being implemented by their local governments. Although the level of commitment is generally low, Italian citizens show the least interest in giving their opinions about the activities reported by their local governments. Moreover, Spanish and Italian citizens present the lowest level of awareness regarding the use of Facebook pages as a means of boosting the virality of the messages posted by their local governments.

According to the dialogic communication theory, SNS constitute an appropriate instrument to create conditions for dialog. However, this condition depends on the process in which the dialog is generated. Therefore, a low degree of citizens' engagement could be due to the fact that the local government needs to make a greater effort in stimulating such a dialog. In view of the results obtained, we conclude that local governments should improve the way in which they make use of Facebook to foster citizens' engagement. Moreover, initiatives should be taken to make citizens more aware of the need to use SNS as a tool to create a better and more democratic society. Similarly, local governments should take steps to encourage the use of these technologies, by measures such as training programs.

This paper focuses on France, Spain and Italy because these Mediterranean countries receive the highest numbers of visits to their Facebook pages in the region. However, to enrich this line of research, it would be useful to go further and examine a larger sample. Another important question would be to inquire into the opinions of officials within local governments, who could provide fresh insights into the obstacles perceived regarding the use of Facebook as a communication tool to enhance citizens' engagement.

KEYWORDS

- **citizens' engagement**
- **Facebook pages**
- **local governments**
- **Mediterranean countries**
- **social networking sites**

REFERENCES

1. Agostino, D. (2013). Using social media to engage citizens: A study of Italian municipalities. *Public Relations Review*, 39, 232–234.
2. Akadwani, A. M. (2014). Gravitating towards Facebook (GoToFB): What it is? and How can it be measured? Computers in Human Behavior, 33, pp. 270–278.
3. Alexa.com. (2014). Retrieved from http://www.alexa.com/siteinfo/facebook.com
4. Bannister, F., Connolly, R. (2011). Trust and transformational government: A proposed framework for research. *Government Information Quarterly*, 28(2), 137–147.
5. Eurostat 2013. http://epp.eurostat.ec.europa.eu
6. Belanger, F., Carter, L. (2008). Trust and risk in e-government adoption. *The Journal of Strategic Information Systems*, 17(2), 165–176.
7. Bonsón, E., Torres, L., Royo, S., Flores, F. (2012). "Local e-government 2.0: Social media and corporate transparency in municipalities." *Government Information Quarterly*, 29, 123–32.
8. Bonsón, E., Ratkai, M. (2013). A set of metrics to assess stakeholder engagement and social legitimacy on a corporate Facebook page. Online Information Review, 37(5), 787–803.
9. Bonsón, E., Royo, S., Ratkai, M. (2013). *Analysis of european municipalities' facebook channels activity and citizens' engagement*. XVII Congreso AECA 'Ética y emprendimiento: valores para un nuevo desarrollo,' Pamplona (Spain).
10. Bonsón, E., Royo, S., Ratkai, M. (2014a). Facebook Practices in Western European Municipalities: An Empirical Analysis of Activity and Citizens' Engagement, Administration and Society published, available at: http://aas.sagepub.com/content/early/2014/09/04/0095399714544945.abstract.
11. Bonsón, E., Bednarova, M., Escobar-Rodríguez, T. (2014b), Corporate YouTube practices of Eurozone companies," Online Information Review, 38 (4), 484–501.
12. Bortree, D., Seltzer, T. (2009). Dialogic strategies and outcomes: An analysis of environmental advocacy groups' Facebook profiles. *Public Relations Review*, 35(3), 317–319.
13. Brainard, L. A., McNutt, J. G. 2010. "Virtual government-citizen relations: informational, transactional, or collaborative?" *Administration & Society*, 42(7), 836–858.
14. Burson and Marsteller. (2011). 2011 Fortune Global 100 Social Media Study. Retrieved from http://www.bursonmarsteller.com/Innovation_and_insights/blogs_and_podcasts/BM_Blog/Lists/Posts/Post.aspx?List = 75c7a224–05a3–4f25–9ce5–2a90a7c0c761&ID = 254
15. Chi, F., Yang, N. 2010. Twitter in Congress: Outreach vs Transparency. *Social Sciences*, 1–20.
16. Chun, S. A., Warner, J. (2010). Finding information in an era of abundance: Towards a collaborative tagging environment in government. Information Polity, 15, 89–103.

17. Curtis. L, Edwards, C., Fraser, K. L., Gudelsky, S., Holmquist, J., Thornton, K., Sweetser, K. D. (2010). Adoption of social media for public relations by nonprofit Organizations, *Public Relations Review*, 36, 90–92.
18. Digital Agenda Scoreboard 2012. Life on line, European Commission. Available at: http://ec.europa.eu/digital-agenda/sites/digital-agenda/files/scoreboard_life_online.pdf
19. Ellison, N., Hardey, M. (2013). Developing political conversations? Information, Communication & Society, 16, 878–889.
20. European Commission (2012). *Smart Cities and Communities – European Innovation Partnership*. Retrieved June 1, 2013, from http://ec.europa.eu/eip/smartcities/
21. Golbeck, J., Grimes, J. M., Rogers, A. (2010). Twitter Use by the U. S. Congress. *Journal of the American Society for Information Science*, 61, 1612–1621.
22. Grimmelikhuijsen, S. G. (2010). Transparency of Public Decision-Making: Towards Trust in Local Government? *Policy and Internet*, 2(1).
23. Harfoush, R. (2009). Yes We Did! An inside look at how social media built the Obama brand 1st Aufl.: New Riders Press.
24. Ho, A. T. K. (2002). Reinventing local governments and the e-government initiative. Public Administration Review, 62(4), 434–444.
25. Holzer, M., Kim, S. T. 2005. *"Digital Governance in Municipalities Worldwide, A Longitudinal Assessment of Municipal Web Sites Throughout the World."* E-Governance Institute.
26. Hong, H. (2013). Government websites and social media's influence on government-public relationships. *Public Relations Review*, 39, 346–356.
27. Hui, G. Y., Hayllar, M. R. (2010). Creating Public Value in E-Government: A Public-Private-Citizen Collaboration Framework in Web 2.0. *Australian Journal of Public Administration*, 69(S1), 120–131.
28. Kavanaugh, A. L., Fox, E. A., Sheetz, S. D., Yang, S., Li, L. T.. Shoemaker D. J. Natsev, A., Xie, L. (2012). Social media use by government: From the routine to the critical, Government Information Quarterly, 29, 480–491.
29. Kazienko, P., Musial, K., Kajdanowicz, T. (2011). Multidimensional Social Network and Its Application to the Social Recommender System. *IEEE Transactions on Systems, Man and Cybernetics-Part A: Systems and Humans*. In Press.
30. Kent, M., Taylor, M. (1998). Building dialogic relationships through World Wide Web, *Public Relations Review*, 24, 321–334.
31. Kim, W. Jeong, O.-K., Lee, S.-W. (2010): On social Web sites, Information Systems, 35(2) 215–236.
32. Kietzmann, J. H., Hermkens, K., McCarthy, I. P., Silvestre, B. S. (2011). Social media? Get serious! Understanding the functional building blocks of social media. Business Horizons, 54, 241–251.
33. Linders, D. (2011). *We-Government: an anatomy of citizen coproduction in the information age*. Paper presented at the Proceedings of the 12th Annual International Digital Government Research Conference: Digital Government Innovation in Challenging Times, College Park, Maryland.
34. Maciel, C., Roque, L., Bicharra, A. C. (2009). *Democratic citizenship community: a social network to promote e-deliberative process*. 10th Annual International

Conference on Digital Government Research: Social Networks: Making Connections between Citizens, Data and Government.

35. Maciel, C., Roque, L., Bicharra, A. C. (2010). Interaction and communication resources in collaborative e-democratic environments: The democratic citizenship community. *Information Polity*, 15(12), 73–88.

36. McAllister, S. M., Kent, M. (2009). Dialogic public relations and resource dependency: New Jersey community colleges as models for website effectiveness. *Atlantic Journal of Communication*, 17(4), 220–239.

37. McAllister, S. M. (2012). How the world's top universities provide dialogic forums for marginalized voices. *Public Relations Review*, 38, 319–327.

38. Mergel, I. (2013). Social media adoption and resulting tactics in the U. S. federal government, Government Information Quarterly, 30, 123–130.

39. Moon, M. J. (2002). The evolution of e-government among municipalities: Rhetoric or reality? Public Administration Review, 62(4), 424–433.

40. Nah, S., Saxton, G. D. (2013). Modeling the adoption and use of social media by nonprofit organizations. New Media and Society, 15, 294–313.

41. Nakki, P., Back, A., Ropponen, T., Kronqvist, J.; Hintikka, K. A., Harju, A. (2011). *Social Media for Citizen Participation: Report of the Somus Project*. Publication 755, VTT Technical Research Center: Espoo, Finland, 2011. Available online: http://www.vtt.fi/inf/pdf/publications/2011/P75–5.pdf (accessed on 22 November 2011).

42. Norris, D. F., Reddick, C. G. (2013). E-Participation among American Local Governments, Lecture Notes in Computer Science 8075, 37–48.

43. O'Reilly, T. (2005). *What is Web 2.0: Design Patterns and Business Models for the next generation of software*. O'Reilly website, 30th September 2005. O'Reilly Media Inc. Retrieved May 1, 2013, from http://www.oreillynet.com/pub/a/oreilly/tim/news/20–05/09/30/wha t-is-web-20.html

44. Picazo-Vela, S., Gutiérrez-Martínez, I., Luna-Reyes, L. F. (2012). Understanding risks, benefits, and strategic alternatives of social media, Government Information Quarterly, 29(4), 504–511.

45. Pina, V., Torres, L., Royo, S. (2010). "Is E-Government promoting convergence towards more accountable local governments?" *International Public Management Journal*, 13(4), 350–380.

46. Purser, K. (2012). *Using social media in local government: 2011 survey report*. Sydney: Australian Centre of Excellence for Local Government, University of Technology.

47. Rodríguez, L., Garcia, I. S., Gallego, I. (2011). Determining factors of e-Government development: A worldwide national approach. *International Public Management Journal*, 14(2), 218–248.

48. Rybalko, S. Y., Seltzer, T. (2010). Dialogic communication in 140 characters or less: How Fortune 500 companies engage stakeholders using Twitter. *Public Relations Review*, 36, 336–341.

49. Sæbø, Ø., Rose, J., Nyvang, T. (2009). The role of social networking services in eParticipation. In Macintosh, A., Tambouris, E. (Eds.), Electronic Participation, Lecture Notes in Computer Science (46–55). Springer, Linz (Austria).

50. Sadeghi, L. (2012). *Web 2.0*. In M. Lee, G. Neeley, and K. Stewart (Eds.), The practice of government public relations, 25–140, CRC Press.

51. Sáez-Martín, A., Haro-de-Rosario, A., Caba-Perez, C. (2014). "A vision of social media in the Spanish smartest cities," Transforming Government: People, Process and Policy, Vol. 8, Iss: 4, pp. 521–544.

52. Sandoval, R., Gil, J. R. (2012). Government-Citizen Interactions Using Web 2.0 Tools: The Case of Twitter in Mexico. *Public Administration and Information Technology*, 1, 233–248.

53. Sandoval, R. (2013). Open Government Success Factors in Government Websites: The Mexican Experience, in Gil, J. R. (ed.), *E-Government Success Factors and Measures: Theories, Concepts, and Methodologies*. USA: IGI Global.

54. Sheskin, D. (2000). Handbook of parametric and nonparametric statistical procedure, 2nd Edition. Florida.

55. Snead, J. T. (2013). Social media use in the U. S. Executive branch, Government Information Quarterly, 30(1), 56–63.

56. Song, C., Lee, J. (2013). *Can Social Media Restore Citizen Trust in Government?* Public Management Research Conference, Madison, Wisconsin, June 20–22, 2013.

57. Strecker, A. (2011). Flocking to Facebook: How local governments can build citizen engagement. Paper presented at the annual Capstone Conference for the University of North Carolina at Chapel Hill Public Administration Program, Chapel Hill, North Carolina, April 20. http://www.mpa.unc.edu/sites/www.mpa.unc.edu/files/AmyStrec ker.pdf. (accessed August 2012).

58. Syhu, H S; Kapoor, S. (2010). Corporate Social Responsibility Initiatives: An Analysis of Voluntary Corporate Disclosure, South Asian Journal of Management; 17(2), 47–80.

59. Towner, T. L., Dulio, D. A. (2012). New media and political marketing in the United States: 2012 and beyond. *Journal of Political Marketing*, 11(1/2), 95–119.

60. Zavattaro, S. M. (2013). Social media in public administration's future: A response to Farazmand. Administration and Society, 45, 242–255.

61. U. S. Census Bureau (2011). U.S. and World Population Clocks. Retrieved 2011, 20-February from http://www.census.gov/main/www/popclock.html

62. Watt, P. 2004. "Financing local government." *Local Government Studies*, 30(4), 609–623.

63. Waters, R. D., Burnett, E., Lamm, A., Lucas, J. (2009). Engaging stakeholders through social networking: How non-profit organizations are using Facebook. *Public Relations Review*, 35(2), 102–106.

64. Wattal, S., Schuff, D., Mandviwalla, M. y Williams, C. B. (2010). Web 2.0 and politics: The 2008 U. S. Presidential Election and E-Politics Research Agenda. *MIS Quarterly*, 34(4), 669–688.

65. Wright, S. (2009). Political blogs, representation and the public sphere. Aslib Proceedings, 61, 155–169.

66. Wright, D., Hinson, M. (2012, March). A four-year longitudinal analysis measuring social and emerging use in public relations practice. Paper presented at the International Public Relations Research Conference, Miami, FL, USA.

67. Smith, B. G. (2010). Socially distributing public relations: Twitter, Haiti, and interactivity in social media. Public Relations Review, 36, 329–335.

APPENDIX

APPENDIX 1   Spain and Italy: Comparative Analysis of Citizens' Engagement. Mean Values and Results of the Kolmogorov-Smirnov Test

		Spain Mean	Italy Mean	Kolmogorov Smirnov Z test
Popularity	P1	78%	75%	−1.188
	P2	21.62	29.15	−1.115
	P3	4.08	3.36	−2.789
Commitment	C1	27%	36%	−2.373**
	C2	1.20	1.08	−1.165
	C3	0.54	0.20	−1.878*
Virality	V1	48%	53%	−0.914
	V2	5.78	5.84	−0.500
	V3	0.99	0.77	−1.780*

APPENDIX 2   France and Italy: Comparative Analysis of Citizens' Engagement. Mean Values and Results of the Kolmogorov-Smirnov Test

		France Mean	Italy Mean	Kolmogorov Smirnov Z test
Popularity	P1	96%	75%	2.460
	P2	86.91	29.15	2.795
	P3	5.85	3.36	2.348
Commitment	C1	54%	36%	1.453
	C2	3.02	1.08	1.789
	C3	0.30	0.20	2.012
Virality	V1	71%	53%	1.789
	V2	18.86	5.84	1.901
	V3	1.31	0.77	1.901

APPENDIX 3 Spain and France: Comparative Analysis of Citizens' Engagement. Mean Values and Results of the Kolmogorov-Smirnov Test

		Spain Mean	France Mean	Kolmogorov Smirnov Z test
Popularity	P1	78%	0.96	−2.855
	P2	21.62	86.91	−5.225
	P3	4.08	5.85	−2.473**
Commitment	C1	27%	54%	−4.832
	C2	1.20	3.02	−4.360
	C3	54%	30%	−1.738*
Virality	V1	0.48	0.71	−3.653
	V2	5.78	18.86	−4.306
	V3	0.99	1.31	−1.824*

CHAPTER 11

UTILIZING SOCIAL NETWORKS FOR EDUCATIONAL PURPOSES

ABOUZAR SADEGHZADEH,[1] MARYAM HAGHSHENAS,[2] MOJTABA NASSIRIYAR,[3] and ROGHAYEH SHAHBAZI[4]

[1]MSc Graduate, Electronics and Telecoms Engineering, University of Bradford, UK, E-mail: Abouzar_10@homail.com

[2]PhD Student, Media Management, University of Tehran, Iran, E-mail: M_haghshenas@ut.ac.ir,

[3]MSc Graduate, IT Management, University of Tehran, Iran, E-mail: M.nassiriyar@gmail.com

[4]MSc Student, IT Management, Alzahra University, Tehran, Iran, E-mail: Shr.shahbazi@gmail.com

CONTENTS

ABSTRACT

This chapter aims to explain the efficacy of social media in aiding teaching and learning. Using digital technology via social network websites such as Facebook, MySpace, Linked In, etc. outside traditional classrooms enhances group learning. By taking Facebook as an example, it is clear how it is perceived as an innovative tool in a student based learning environment, which advances students learning experience and encourages student cooperation with their peers and members of teaching staff. In terms of teaching, using Facebook in a group-based pedagogy augments the teaching and learning process considerably and allows teachers to utilize student's digital learning methods and presents new ways of motivating students to learn.

11.1   INTRODUCTION

Starting with emergence of the Internet as a public sphere, many unprecedented changes have occurred in communication types and formats in daily life. Face-to-face communication in interpersonal relationships has been gradually replaced with communications via technological devices. This change has also been associated with new types of relationships (Murray, 2008). Social networks within the scope of social media are almost at the heart of these virtual communication forms. Social networks are platforms for virtual social lives created by people over the Internet. Individuals define themselves in such networks so that they communicate with other people sharing same or different cultural backgrounds/ dimensions through powerful communication opportunities provided by the Internet. The first known network in this area is considered the SixDegrees, which was constructed in 1997. Particularly after 2003, rapid and important developments were experienced in social networks; consequently the number of users has increased quickly. The worldwide growth of social communication networks gained incredible pace and popularity. Among these networks, the most common one is known as Facebook (Toprak et al., 2009).

As of February 2011, Facebook has more than 500 millions of users around the world. Fifty percent of total users actively login the site every day. Users spend a total of 700 billion minutes per month on Facebook. More than 200 million users have mobile connection to Facebook.

There are about a billion of locations (pages, groups, activities etc.) in Facebook where users interact with each other. An ordinary user is connected to 80 groups, activities or society pages and shares an average of 90 contents per month. More than 30 billions of contents per month are shared by users.

When over 500 millions of people worldwide started to spend a substantial part of their daily lives in Facebook, this social network also attracted interest of sociologists and psychologists. According to researchers, social networks improve communication skills, enhance participation as well as social commitment, reinforce peer support, and ensure realization of education based on collaboration strategies. Furthermore, social networking sites can be easily and inexpensively used without substantial support from universities. That is also to say that they can be successfully integrated into educational processes. In appears that this type of use rapidly becomes widespread all around the world (Gulbahar et al., 2010).

It is a fact that Facebook is the largest social network with the biggest audience compared to similar media, particularly because it enables people communicate with their friends and exchange multimedia-based information conveniently. Hundreds of people, unaware of each others, may gather around a particular purpose via social network and they may even decide to act together. This chapter discusses the ways social media in particular social websites can be utilized for educational purposes.

11.2   SOCIAL MEDIA DEFINITIONS

Social networking covers a wide range of online environments, with many formal definitions broad enough to encompass almost any Web 2.0 collaborative environment (Alexander, 2006). While various public social collaborative environments existed on the Internet as early as the 1980s, the emergence of social networking as it is best understood today arose with the large commercially-supported sites such as Friendster (2002), LinkedIn and MySpace (2003), and Facebook (2004), along with content-sharing focused sites with limited social network features such as Flikr (2004) and YouTube (2005). Other social networking sites were developing which have higher usage outside the U.S. including Orkut (2005), popular in South America and Asia/Pacific areas, Bebo (2005) in Europe and Australia, and QQ (2006) in China.

With the development of Twitter in 2006, social networking took a new twist that increased immediacy and incorporated mobile phones into the social mix.

Boyd and Ellison (2007) include three criteria in their definition of social network sites (SNSs) which are web-based services that allow individuals to (1) construct a public or semipublic profile within a bounded system, (2) articulate a list of other users with whom they share a connection, and (3) view and traverse their list of connections and those made by others within the system. Further, they note that many SNSs commonly allow users to leave persistent comments on "friend's" profiles and send private messages although these are not universal features. In particular, rather than communities organized by topic, SNSs are "structured as personal (or 'egocentric') networks, with the individual at the center of their own community" (Boyd & Ellison, 2007). Perhaps more critical from an educational viewpoint, many of the SNSs are enhanced with multiple collaborative tools that go beyond the personal profile and "friending" links, including the ability to post and share files (text, images, audio and video), participate in discussions or blogs, co-create and edit content with wiki-like tools, and link in and tag external resources from other web sites paralleling social bookmarking. Sites such as Flikr or YouTube are in fact more commonly seen as environments primarily for sharing content, digital pictures and video respectively, rather than SNSs despite meeting the Boyd and Ellison criteria.

11.3   MOBILE LEARNING NETWORKS

The emergence of Web 2.0 not only accelerates the development of diverse communities but also promotes socialization of the Internet (Huang et al., 2009). Lots of social software tools are created along with Web 2.0. The socialization of the Internet has become powerful and trendy. It inspires social networking websites such as Blog, Facebook, etc. The mechanism of socialized Internet improves close interpersonal relationships and provides nonverbal communication media such as multimedia audio-visual objects, images, pictures, and other diverse media. By communicating and sharing with others through resourceful media, interpersonal interaction becomes closer.

Nowadays many researches try to use the trend of Web 2.0 to push forward a new learning model, for example, applying Blogs in learning and conducting knowledge sharing through Blogs.

As a result, Huang et al. (2009) brought up the idea of supporting collaborative learning by Blogs. In view of knowledge sharing, Yang and Chen (2008) also put forward a learning model established by collaborative learning through knowledge sharing based on social networks. Furthermore, due to the progress of wireless Internet and mobile devices, the mobile learning environment has gradually become stable and mature, for example, the mechanism of providing learning services in the mobile learning environment. As a result, Huang et al. (2008) brought up an annotation service to support mobile collaborative learning. In addition, Kuo and Huang (2009) put forth an authoring tool to support the adaptive presentation of learning resources on mobile devices. These studies all aim at supporting collaborative learning, and also promoting sharing resources with each other through social networks. Further than that, these studies integrate with the idea of mobile learning, all together putting forward the learning service mechanisms needed in collaborative learning, such as annotation services or authoring tool. These mechanisms enable smooth learning, creating a new learning model. Therefore, the important research issue at present is how to provide efficient learning services to support learners' needs when learning.

Many researches are devoted to technical support and application (Yang, 2006; Yang et al., 2007, Huang et al., 2008), however, the online collaborative learning model not only needs technical support, but more importantly, is to conduct interaction with people. As a result, to interact with what sorts of people is the most concerned issue of researches at present. In resent researches, focusing on exploring the formation of new groups, Chen, Kinshuk, Wei, and Yang (2008) brought up the idea of using a self-contained group area network to support mobile collaborative learning. Another researcher Yang and Chen (2008) put forth the idea of knowledge sharing communities to be formed based on social networks. From the above, we know that it appears to be more important to focus on how to find suitable people to conduct interaction and how to form CoP (Communities of Practice) under social networks to support the collaborative learning model.

Furthermore, through the services of Social Software such as Blog, Wiki, Facebook, Del.icio.us, Flickr, etc., social networking between users is established. This social networking helps users locate people with shared interests and thus form CoP. Through these social platforms, collective intelligence is realized.

Afterwards, people can bring different CoPs together to form CoIs (Communities of interest) which can provide unique opportunities to bring social creativity alive by transcending individual perspectives. Accordingly, some researchers assert that social network mainly puts emphasis on building various CoPs so that users can share and exchange information with each other based on their similar interests (Rachel, 2008).

Indeed, products of knowledge sharing and creating by users are mostly on certain social platforms. For example, Flicker allows users to share pictures or images and Del.icio.us allows users to share bookmarks. These products are the aggregate of collective intelligence. However, real collective intelligence should not be limited to sharing and creating products. The most significant resources are producers of these products. As Diederich and Iofciu (2006) pointed out, "using tag-based profile can give more recommendations than standard object-based user profiles." It means that producers play an important part in forming collective intelligence. If users can find those who share the same interests with them and interact with each other, innovation of knowledge and new world would be inspired by collective intelligence.

On the other hands, although the Internet technology has made it possible for people to collaborate effectively without staying physically together, they have led to the unintended consequence of increasing isolation among people with respect to their academic peers. In bygone times, the inconvenience of having to share resource sites (for example, computer centers and unscheduled laboratory use) afforded opportunities to develop computer-oriented social groups for virtual collaboration. Computer Supported

Cooperative Work (CSCW) provides a virtual collaboration technology that offers participants a promising option of not being physically present at cooperation. Applied to collaborative learning, CSCW techniques allow students to study in a virtual team without physically staying at a common place (Weinberger, and Fischer, 2006). Computer-Supported Collaborative Learning (CSCL) was thus coined in 1996 (Koschmann, 1996) to refer to adopting CSCW technology to provide a computer and network supported collaborative learning platform for students to study cooperatively to acquire knowledge (Yang and Chen, 2008).

Many researchers have proven that social relationships and interactions have significant impacts on collaborative learning (Yang and Chen, 2008; Chen, Kinshuk, Wei, and Yang, 2008).

Fischer et al. (2002) also concluded that social relationships have an impact on knowledge acquisition in a collaboration mode. The technique of social network is thus used to represent a determinable networking structure of how people know each other (Yang and Chen, 2008). A social network can be formalized into a net structure comprising nodes and edges. In such a network, nodes represent individuals or organizations. Edges that connect nodes are called ties, which represent the relationships between individuals and organizations, either directly or indirectly. The strength of a tie (weight of an edge) indicates the strength of the relationship.

Many kinds of ties may exist between nodes in a social network. One popular tie is a social interaction tie which refers to the structural link created through social interactions between individuals in a social network (Yang and Chen, 2008). Ahuja et al. (2003) suggested that an individual's centrality in an electronic network of practice can be measured using the number of social ties an individual has with others in the social network. Tsai & Ghoshal (1998) reported that social interaction tie has positive impacts on the extent of inter-unit resource exchange. Wasko and Faraj (2005) discovered that the centrality established by social interaction ties significantly impacts the helpfulness and the volume of knowledge contribution. Kreijns et al. (2005) concluded that social interactions largely affect group forming and group dynamics.

Moreover, there are researches centering on "human." Artiles et al. (2005) used "people name" to locate people, so that in this way, they solved problems of naming ambiguity and the same name. They designed WePS

(Web People Search) as a test platform. Users inputted "people's names" to search for relevant people resulting in a ranking list. Although it is helpful to search people by their names, it is possible to lose people whose names seem to be similar to the users. Accordingly, researches of locating people of similar fields by analyzing documents were conducted. Some researchers (Wan, Gao, Li, and Ding, 2005) proposed a method to search for relevant "people" from relevant people-document and provided resolution for the above-mentioned problem to find out people of relevant fields. Diederich and Iofciu (2006) proposed another method to locate people in the website, using tag-based profiles to find people with similar interests. In terms of this, if people in similar fields can be located in a social network and thus gathered to form a CoP through community activities, collective intelligence is realized. It will highlight the efforts of the people search service.

Other researchers such as Mori et al. (2008) proposed an important finding. They discovered that finding relevant people in certain fields is important to collaborative system. Thus, they recommended a people search interface and tool which requires users to input information about those who are searched for before people search. In this way, search results of the target people can be more accurate. According to the above-mentioned relevant researches, the research of using "human" as a resource in the era of Web 2.0 becomes significant.

It is important to be able to find suitable learning partners for learners to conduct collaborative learning. As a result, a collaborative service to support learning activities becomes vital. In another respect, through this recommendation mechanism, learners can conduct social activities driven from the same interests and specialties; and further, to learn from each other based on the concept of knowledge sharing in this kind of social activities to form the so-called mobile learning knowledge networks.

11.4 SOCIAL MEDIA AND KNOWLEDGE MANAGEMENT

Social media has recently emerged as a promising technology for knowledge management (KM) (Levy, 2009; Yates and Paquette, 2011). It is defined as "a group of internet-based application that builds on the ideological and technological foundations of Web 2.0, and that allow the creation and exchange of user generated content" (Kaplan and Haenlein, 2010, p. 61).

Graphically, the processes of KM can be represented with KM cycles (e.g., Bukowitz and Williams (2000), McElroy (1999), Wiig (1993)). An integrated KM cycle comprising of three major stages of KM, organizational culture and KM technologies was put together by Dalkir (2011). The three stages of KM include: 1) knowledge capture and/or creation; 2) knowledge sharing and dissemination and 3) knowledge acquisition and application. The three stages of KM supported by technologies are facilitated by a favorable organizational culture that promotes information and knowledge sharing. Ruggles (1997) classified KM technologies as tools that intervene in the three knowledge-processing phases mentioned by Dalkir (2011). Recently, social media, with their distinctive features that allow proactive participation, social connectivity and user collaboration, have become important tools in facilitating knowledge management processes in business and education institutions (Dames, 2004; Lee, 2003).

McDermott (2000) defined knowledge as an output from active social construction. Owing to its powerfulness in disseminating information, soliciting comments and links, and classifying and archiving entries, blogs have gained vast recognition as a KM tool, especially in business organizations (Ojala, 2005). Ferdig and Trammell (2004) perceived blogs as a relatively more advanced platform for effective information and knowledge sharing when compared to the more traditional technologies such as emails and discussion forums. Research has found blogs to be effective in organizing, articulating, developing, and sharing ideas (Mortensen and Walker, 2002), as well as in developing and maintaining community relationships (Fiedler, 2003). Chu, Kwan and Warning (2012) reported that the participating university students have found blogs useful in helping them manage and share knowledge gained from their professional experiences.

Although Facebook is one of the most commonly used social media tools nowadays, there have been very few studies concerning the use of Facebook as a KM technology. A recent study conducted by Chu and Du (2013) examined the use of Facebook by academic and public libraries in English-speaking countries a tool for knowledge sharing, information dissemination and knowledge gathering. The large volume of postings related to knowledge sharing and information dissemination appears to suggest Facebook as a potential KM tool. Other research examined Facebook through the lens of community of practice (e.g., Wong, Kwan and Leung, 2011). Communities of Practice (CoP) are groups of people who share a common concern and seek to construct and share knowledge with others within the group (Lave and Wenger, 1991). CoP is a fundamental concept related to organizational culture as it encourages "collaboration and sharing resources in knowledge domain on the Web" (Wong, Kwan and Leung, 2011, p. 319). The same study has found that Facebook, which emphasizes interaction, sharing and collaboration, is a motivating tool that fosters social learning.

Facebook has advanced group features where all messages posted onto the group wall would generate group notifications and greater attention among the group members (De Villiers, 2010). When anything is posted in the group, everyone can view it. With notifications users can read the most updated comments from their friends easily and immediately. Hence, it facilitates knowledge sharing. Facebook also enhances collaboration and facilitates the exchange of ideas as a result of peer interaction.

By actively engaging individuals in online discussion, it could support communications and peer-to-peer feedback among individuals who face common dilemmas (Selwyn, 2009; Ziegler, 2007), constituting what could be a significant collaboration platform for knowledge construction (Kabilan, Ahman, and Abidin, 2010; Wang et al., 2011). Apart from its advanced group features, Facebook is also associated with greater offline social connection, which can promote knowledge dissemination and sharing. However, there are more functions in Facebook and we could include more people like colleagues. For instance, classmates can get in touch with their colleagues through Facebook. In this way, more is learnt. Unlike blogs, the primary function of Facebook is for "social searching,," e.g., find out more about people who they already knew (Lampe, Ellison, and Steinfield, 2006). Its users mainly use Facebook to connect and maintain relationships with existing offline contacts (Joinson, 2008). Therefore, it provides greater social motivation for its users to communicate and interact with others (classmates, colleagues and academic supervisors) on Facebook.

Facebook was found to enhance social support because its interactive element is strong. For instance, its Wall function enables the users to post items and leave comments easily on each other's Wall. It offers them a new form of communication that facilitates discussion and reciprocal interaction. In addition, Facebook users could read and respond to each other's postings by giving 'Like's. These kinds of interactive features therefore generate greater sense of social connection among Facebook users, which could be one indicative factor of the higher degree of knowledge management engagement found among Facebook users.

Facebook users are more likely to express social support on the platform. The perceived support from classmates and supervisors among the Facebook users is also significantly greater than that of blog users. Facebook is a more effective social media in building a culture of support, thereby facilitating the social-collaborative process of knowledge management. As a social networking platform, Facebook comprise more interactive and collaborative features (e.g., group notification, messaging), and provides more channels to demonstrate social support (e.g., giving "Like"). With the provision of a supportive social framework, individuals are therefore more willing to connect, communicate and reflect their knowledge and experiences on Facebook.

11.5 SOCIAL NETWORKS AND LEARNING

Typically, institutions use a range of various educational approaches in the classroom, tutorial, lab and lecture hall. Activities can take place face to face, but may also be mediated by social networking technologies include peer assessment, discussions, and collaborative work. Course designers have been quick to spot such opportunities by way of chat rooms, discussion forums and collaborative work support tools, which may be used in this way. The efficiency and effectiveness of such approaches is the subject of evaluation, analysis and debate. The study of social networks within a learning domain encompasses the processes of social learning that occurs when self-selecting groups of people who have a common interest in a subject collaborate to share ideas or find solutions. Observations of the processes and behaviors of self-selecting groups can be used to engineer interactions in groups orchestrated for specific educational purposes. Social networking applications which incorporate Web 2.0 technologies demonstrate affordances, which could be available to utilize within the classroom. These operate with paradigms which are different to those observed within conventional e-learning tools. However utilizing social networking tools with large student groups might present problems. An advantage of increased awareness or appreciation of the complexity of typical observed behaviors in a social learning environment may enhance the academic's ability to manage the tools. A recent study of the potential for semantic modeling of learners explores using Semantic Web-based social networks to facilitate the automatic and dynamic creation of students' networks within large online communities. Enriching the semantics of network and membership descriptions can provide valuable information. This can be used to assist in tuning group allocations, enabling the network to be used for specific educational objectives.

11.6   ADVANTAGES AND DISADVANTAGES OF SOCIAL MEDIA UTILIZATION

Since the web is arguably better linked than the "real world," finding people on-line with shared interests is easier, and more likely. Nowadays many applications allow users to keep in touch with long-term friends, family and to find new friends.

In addition, new relationships based on the links between friends, and friends of friends are created. These new relationships are not limited to people users already know. Indeed, links are created in the act of stating an interest, or joining a network; in this action, users find other people who share the same opinions, hobbies, or university. To maintain relationships, the computer-supported social network software provides various tools within the application (forums, tickets, online profiles, etc.). Thus users have more support options than when using one-to-one communications such as email. When a dedicated place is available for users to post specific comments, the opportunity to request information and gauge others' interactions creates available norms that can more easily be applied. Viewing others' comments and postings provides a double feedback to the user: first, they are using the right application at the correct place; secondly, other people have the same questions, interests, or ideas. This promotes a much-needed sense of community. This reciprocating interaction applied to the university environment offers not only benefits to students but in the long-term to the entire community.

Links among individuals based on trust, affinity, and expertise versus friendship are not as well defined as in the real world. As the definition of strong and weak ties are vague in their application to online relationships, social software struggles to model and implement real world relationships. Online profiles can be a source of deception. Indeed, when a person misstates their true identity or intentions, trust is broken, potentially negating the foundation of their online relationships. It is much easier to lose contact online since online interaction is asynchronous communication. One party is invariably waiting for the other to reply. Conversely, interacting in close physical proximity exploits the non-verbal cues inherent in face-to-face communication. Naive students over-reliant on online communication can be unwittingly cut off from the necessary communities with possibly disastrous academic consequences. In terms of trust, the security of personal information online is increasingly important. Whilst the authenticity of online identities may be questioned, conversely the vulnerability of personal information online is generating increasing numbers of 'horror stories' covered in the mass media. Many social networking sites work on a basis of presumed trust, with users' profiles being displayed and available to registered users and guests (meaning non-registered users) by default, even if they do not belong to the same network or do not share the same interests.

Progressively, many networks now give the users the facility to set their own level of disclosure, at a community level or to close friends only. However, this precaution is not yet available for all applications. Negative consequences of sharing personal data in this way include spam to private email accounts, 'phishing' attempts at fraud and identity theft and vulnerability to malicious real-world activity aimed at damaging individuals or their property. Standards and policy for protecting user information is a vastly researched area, and beyond the scope of this paper. However, for users to fully benefit from online applications in safety, it is clear that an evaluative understanding of online communities and online security issues is necessary.

11.7   SOCIAL MEDIA AND KNOWLEDGE SHARING

For organizations, social media offers a fast moving, fragmented and fun space for intelligent individuals to connect and share ideas. Business is keen to know how the spare mental capacity available in social interaction can be used in the service of solving business problems and improving performance. The potential knowledge benefits of implementing social media are enormous. The promise of greater breadth of input and faster access to deep expertise would vastly improve prospects for innovation, increase efficiency and enhance strategic reputation, customer intimacy and organizational responsiveness in a volatile business environment.

Nevertheless, this convivial social network world has a dark side that has to be managed. The platform is neutral; it's the users' behavior and whether their views and opinions are contagious that matters. Masses of self-interested individuals deciding together can lead everyone straight into the 'tragedy of the commons' trap. Overly coherent groups may indulge in unreflective group-think, blind to the consequences of collective stupidity. Depth can exclude diversity of perspective. Clay Shirky's caveat is to the point. You can have large groups, active groups and groups that pay attention to the same thing, but generally only two out of three are sustainable at the same time. Why? Because at some point, even the most generous volunteer involved in a large active group gets impatient with differences of opinion and wants coherence. Of course opinion can go viral. Large groups become enamored of a singular position, unleashing collective and infective madness.

But over time, either the less radical activists become disenchanted and the group becomes smaller, or groups divide into core and peripheral participants: the core tends to be those who feel rewarded (not necessarily with money – satisfaction and glory are valuable rewards too); the rest lose interest or become less active. At best, active smaller groups start to focus on something and collaborate.

Executives are inevitably cautious because no one can predict what and who will become a 'strange attractor,' that emerging force for change, which could be constructive or destructive. These challenges can become inhibiting pre-occupations if they are not addressed head on. Whilst the majority of people agree that usage in business will grow in the next few years, and a large proportion of executives see social media as a high priority topic and integral to strategy, 61% of respondents to a recent HBR survey considered that they had a significant learning curve to travel before they get involved in it. A recent research project carried out by the Henley KM Forum suggests the following conclusions:

- Although interest in social media is building, many managers are watching and waiting before getting involved, partly because they don't understand the real benefits for them and partly because of concerns over the risks and consequences of a new set of tools which are open to misuse.

- Senior executives are naturally nervous of engaging, because the openness of the social media pulls against all the established responsibilities of senior management for co-ordination in a purposeful direction. The risk to organizational reputation of irresponsible involvement looms large in senior managers' minds, particularly when anyone's sentiments can go viral, often based on shallow perception rather than deep founded knowledge.
- Middle management may be concerned about the risks to their traditional role as gatekeepers of standards.
- Knowledge workers and the younger generation may not feel sufficiently motivated or empowered to get involved.

In 2010, the PR consultancy Weber Shandwick conducted an audit of on-line communications by CEO's of the world's largest 50 companies. They found that:

- 64% were not engaging with people through the company web site or social media.
- Only 18% incorporated video/podcast on company websites or YouTube channels.
- Just 16% have a profile on Twitter (8%), Facebook (14%), MySpace (4%) or
- LinkedIn (4%).
- Only 12% were featured on their corporate video channel.

Even if executives see social media as a priority if CEO's are nervous about leading by example and prioritizing it as their communication channel, then it is less likely to be perceived as a critical to business activity and is more likely to deliver tactical rather than strategic benefits. Only a minority (11%) still sees social media as a passing fad for business, yet the responses highlight how insufficient appreciation of the potential of social media to make a difference and the challenge of convincing people of its value may contribute to an early inertia in organizational adoption. Probably some hard measure of effectiveness would be an incentive to get involved.

To overcome this early inertia in organizations, it is important to:

- Develop a purposeful strategy for where the organization could benefit from social media implementation. Often the best way to achieve

this is to learn from the lessons of those who have already made progress. However, some organizations have also experimented with social media to develop their strategy. A structured framework for examining success stories is helpful and our research has uncovered many broad examples of social media opportunities.
- Realistically acknowledge and address the risks at each level openly and in dialog with those affected. The key is finding the win-win approach that delivers benefits for both the individual and the organization.

For the organization: at best it can enhance and at minimum, protect the organizational reputation.

For the individual: at best, it can enhance personal reputation as a knowledgeable individual, at minimum it offers the freedom to influence events and have their say. The win-win approach happens when co-ordination aligns individual and organizational interests. Individuals build their reputation and have the freedom to apply their knowledge to projects that advance their career. They will therefore be 'care-full' in their involvement to protect their personal reputation. Their care and willingness to apply their knowledge to organizational problems can build business reputation, which in turn becomes an attractor for the sort of talent and custom that fits with organizational strategic capability. This mutual contract based on shared interest builds a flexible organization that can thrive and learn in a volatile world.

11.8   SOCIAL MEDIA AND HIGHER EDUCATION

Learning on demand is becoming a type of lifestyle in modern society (McLoughlin and Lee, 2007). Learners constantly seek information to address a problem at work, school, or to just satisfy a curiosity. To do so, they take advantage of digital and networked technologies not only to seek information, but also to share information. Thus, learners should not be considered as passive information consumers; rather, they are active co-producers of content. Additionally, learning in the context of social media has become highly self-motivated, autonomous, and informal, as well as an integral part of the college experience (McGloughlin and Lee, 2010; Smith, Salaway, and Caruso, 2009; Solomon and Schrum, 2007). However, higher education institutions are still primarily relying on traditional platforms such as course and learning management systems (CMS/LMS) that do not capitalize on the pedagogical affordances of social media for example allowing learners to manage and maintain a learning space that facilitates their own learning activities and connections to peers and social networks across time and place (McGloughlin and Lee, 2010; Selwyn, 2007; Valjataga, Pata, and Tammets, 2011; van Harmelen, 2006).

The 2010 ECAR (EDUCAUSE Center for Applied Research) study of undergraduate students and information technology revealed that students' use of social media has steadily increased from 2007 to 2010 and that the gap between older and younger student use of social media is shrinking (Smith and Caruso, 2010). More specifically, the 2010 ECAR study showed that 33.1% of the participant undergraduate student sample (N = 36,950) reported using wikis; 29.4% used SNS; 24.3% used video-sharing websites; 17.4 used web-based calendars; 11.6% used blogs; 4.3% used micro-blogs; and 2.8% used social bookmarking tools. Additionally, the percentages of those using social media for coursework related collaboration was particularly noteworthy (30.7% of wiki use, 49.4% of SNS use, 33.4% of video-sharing use, 37.6% of blog use, 40.2% of micro-blog use, and 30.5% of social bookmarking use). These data reveal that college students are integrating social media in their academic experience both formally and informally. Furthermore, college faculty is increasingly using social media to support teaching and learning activities (EDUCAUSE Learning Initiative, 2007, 2007). For example, some are encouraging students to use blogging platforms (e.g., WordPress) for the development of e-portfolios, which have become an important authentic assessment tool in higher education (Rosen and Nelson, 2008). Others are using Twitter (a micro-blogging platform) to stimulate student engagement in the classroom (Rankin, 2009) and wiki software (e.g., PBworks) to engage students in collaborative projects that support the creation, editing, and management of content (Hazari, North, and Moreland, 2009).

These efforts by faculty and students are creating new ways of teaching and learning leading to the emergence of constructs such as e-learning 2.0, pedagogy 2.0, student 2.0, faculty 2.0, and classroom 2.0, with the suffix 2.0 characterizing themes such as openness, personalization, collaboration, social networking, social presence, user-generated content, the people's Web, and collective wisdom, and demarcating areas of higher education where a potentially significant transformation of practice is underway

(Alexander, 2006; Dabbagh and Reo, 2011b; Jones, 2008; Lindstrom, 2007; Norton and Hathaway, 2008; O'Reilly, 2005; Sessums, 2006). For example, Hilton (2009) believes that higher education is being challenged by perceptions that Web 2.0 technologies (social media in particular) are empowering students to take charge of their own learning resulting in what some interpret to mean that there is no arbiter of their knowledge, work, publication, or thinking. Others (e.g., Anderson, 2008; Cormier, 2008; Dede, 2006; Katz, 2008; Siemens, 2005; Siemens and Tittenberger, 2009; Weigel, 2002) argue that Web 2.0 technologies are inducing a pedagogical transformation where the community is the curriculum rather than the path to understanding or accessing the curriculum and that higher education institution should integrate social media platforms that enable the creation of personal and social learning spaces to support more learner-centered "personalized"" education systems (Dabbagh and Reo, 2011b; Dron, 2007; McGloughlin and Lee, 2010; Selwyn, 2007). As a result of these social media induced pedagogical challenges and practices, the concept of Personal Learning Environments or PLEs is listed in the 2011 Horizon Report as an emerging technology that is likely to have a large impact on teaching and learning within education around the globe and a time-to-adoption of four to five years (Johnson, Adams, & Haywood, 2011).

While there is growing evidence that social media is increasingly supporting informal learning at home and in the community and that informal learning is becoming a vital element of education for learners of all ages (Selwyn, 2007), research has also revealed that PLEs can help integrate formal and informal learning in higher education contexts (McGloughlin and Lee, 2010). Formal learning is described as learning that is institutionally sponsored or highly structured, for example, learning that happens in courses, classrooms, and schools, resulting in learners receiving grades, degrees, diplomas, and certificates, whereas informal learning is learning that rests primarily in the hands of the learner and happens through observation, trial and error, asking for help, conversing with others, listening to stories, reflecting on a day's events, or stimulated by general interests (Cross, 2007; Selwyn, 2007). Attwell (2007) reported that in the workplace, informal learning through asking questions, observing coworkers, and other uncoordinated and independent learning activities accounts for 80% of an individual's knowledge about this/her job.

Research studies should be designed to trace students' trajectories (paths) of social media use across the levels of the framework with the goal of documenting how students transition through the levels and examining the degree to which self-regulated learning strategies (e.g., goal setting, time management, self-monitoring, and self-evaluation) influence the design and advancement of their PLE. Such studies would also need to consider students' motivational beliefs such as self-efficacy beliefs as well as learning styles since PLEs are individualized by design and will differ from student to student. Results of such studies would inform higher education faculty whether PLEs can be used as an effective pedagogical and educational tool.

Faculties are well aware of social media, and the majority uses the sites for both personal and professional reasons. It appears those faculties do make considerable use of social media in their teaching: nearly two-thirds of all teaching faculty have used social media in their class sessions, and 30% have posted content for students to view outside class. Not all social media sites are used equally within a given course. Online video is by far the most common type of social media used in class and posted outside class for student use. Podcasts and blogs are next in popularity, but at rates far below the rate of online video.

Several of the sites commonly used for personal purposes, such as Facebook and Twitter, are rarely used as part of a course. There are some differences in the selection of social media sites for professional uses (within a professional career or within a course) versus use for personal purposes. The top two sites for personal use; YouTube and Facebook remain the top two for professional use, but research shows their order is usually reversed and the overall level of use for professional purposes is much lower than for personal use. Given the concerns about the amount of time it takes to use social media and about issues of privacy and integrity, it might be expected that faculty do not see a great deal of potential for social media use in classes. That's not the case, however. Faculty sees considerable value in many social media sites for use in class. Faculty responses from various questionnaires around the world show that online video from either YouTube or other online video sites are seen as having the greatest value for use in classes. After online video, faculties report that podcasts are next in value for class use, followed by wikis and blogs. Not all social media sites are seen as being valuable for teaching; Facebook and Twitter are not seen as having value for class use. A large proportion of faculties believe that Facebook and Twitter have negative value for use in class.

11.9   SOCIAL MEDIA AND NEW LEARNING METHODS

Social media constitute an increasingly important context wherein individuals live their everyday lives. The most immediate significance of social media for higher education is the apparently changing nature of the students who are entering university.

In a practical sense, the highly connected, collective and creative qualities of social media applications are seen to reflect (and to some extent drive) more flexible, fluid and accelerated ways of being. Social media are therefore associated with an increased tendency for young people to multitask, to rely on a 'digital juggling' of daily activities and commitments (Subrahmanyam and Šmahel, 2011). More subtly, these technologies are also associated with an enhanced social autonomy with young people now used to having increased control over the nature and form of what they do, as well as where, when and how they do it. Indeed, social media users are described as having an enhanced capacity to self-organize. As Tapscott and Williams (2007: 52) continue, these young people 'are not content to be passive consumers, and increasingly satisfy their desire for choice, convenience, customization, and control by designing, producing, and distributing products themselves.'

While these descriptions could be said to apply to users of all ages, often the changes associated with social media are seen in distinct generational terms. There will soon be cohorts of students who know nothing other than a life with the internet, having been 'born into a world woven from cabled, wired or wireless connectivity' (Bauman, 2010: 7). For many educators, therefore, the increased presence of social media in higher education settings is essential if universities are to connect with these students.

Social networking sites are now being used by universities as alternative spaces wherein students can adapt to the university lifestyle through interacting online with peers and faculty (Yu et al., 2010). Indeed, many universities now maintain profiles and groups on social networking sites such as Facebook, where students and faculty can interact, share resources and express 'learner voice.' As Mason and Rennie (2007:199) reason, 'shared community spaces and inter-group communications are a massive part of what excites young people and therefore should contribute to their persistence and motivation to learn.' Of course, it could be argued that the top-down, mass institution of 'the university' is poorly placed to be a meaningful part of students' hyper-individualized social media use. Some critics are pointing towards an ever-growing digital disconnect between students and their education institutions. It is also argued that even the best-intentioned universities are able only to offer their students an artificially regulated and constrained engagement with social media. Thus, alongside other institutions such as schools, libraries and museums, universities are seen to face distrust and a growing loss of faith amongst younger generations (Downes, 2010). This clash is particularly evident in terms of the linear and hierarchical ways in which universities set out to structure communication, learning and access to knowledge.

There are clear disparities between the educational rhetoric and educational realities of social media use, which should come across as a little surprising.

There has been a long standing tendency in education for digital technologies to eventually fall short of the exaggerated expectations that initially surrounded them which can be described as a cycle of 'hype, hope and disappointment' (Gouseti, 2010). In this sense, perhaps the most pressing challenge for the higher education community at present is to engage in considered and realistic debates over how best to utilize social media in appropriate ways that hopefully reduce this eventual disappointment.

Of course, clear lines need to be drawn between the immediate practical tasks of developing forms of social media use that better fit within the current grammar of formal higher education systems, and addressing the rather more difficult longer-term issues of system-wide reform and redesign. In terms of this latter point, there is a clear need to thoroughly consider and discuss what higher education is, and what forms it should take in a 21st-century digital age. Indeed, many of the controversies and tensions concerning the use of social media in higher education have little to do with the technology itself. Instead, these are issues that are driven by personal belief and opinion about 'the essentially ethical question of what counts' as worthwhile learning and worthwhile education (Standish, 2008:351). In this sense, social media are socially disruptive technologies that prompt a range of deeply ideological (rather than purely technical) questions about the nature of institutionalized education.

Yet these wider debates notwithstanding, higher educators also face the immediate task of integrating social media into their current provision and practice. As such, universities clearly need to continue to consider, for example, the practical challenges of how to assess students' collaboratively authored work or how best to design blended curricula (Gray et al., 2010; Buckley et al., 2010). Further thought certainly needs to be given to how best to support staff and students alike in making sustained and meaningful use of these internet technologies. In these ways, universities need to play an important role in supporting students' supposedly self-directed activities providing students with a good core and governance in 'arranging the furniture' of technology-based learning (Crook, 2008).

All these issues will certainly become clearer as the 2010s progress, and as society's expectations and general understanding of social media become less exaggerated and more realistic and objective. Indeed, many technological commentators have now moved on to enthuse about the next set of 'next big things' within the world of digital technology such as the semantic web, cloud computing and the internet of things. In this sense, there is now room for the higher education community itself to assume a greater role in shaping the development of social media on the ground in higher education settings. After all, social media technology is something that is supposed to be created by its users including higher education institutions and educators.

11.10   ENTERPRISE SOCIAL NETWORK SITES

Social network sites (SNSs) such as Facebook are being used by individuals today to accomplish a wide range of goals – asking questions of their network (Gray, Ellison, Vitak, & Lampe, 2013), getting support after losing a job (Burke and Kraut, 2013), even grieving for loved ones (Marwick and Ellison, 2012). They are also increasingly adopted to enhance organizational performance, especially in the context of knowledge sharing. Organizations today are increasingly distributed and networked, making it more challenging to share knowledge across time and space (Cross, Parker, Prusak, and Borgatti, 2001). Indeed, many large organizations today are turning to networked forms of organizing as a key organizational structure and relying on technology to facilitate coordination and support interdependent groups (Espinosa, Slaughter, Kraut, and Herbsleb, 2007). In particular, multinational organizations are turning to virtual teams and distributed work arrangements in order to coordinate and enable knowledge flow; however, team members face challenges identifying experts in distant parts of the organization (Faraj and Sproull, 2000), developing trusting relationships that encourage information sharing (Gibson and Gibbs, 2006), and sharing knowledge that is situated in local contexts and often taken for granted Cramton, 2001; Sole & Edmondson, 2002). Employees working in distributed arrangements must negotiate the tensions associated with working across geographic and other structural boundaries (Gibbs, 2009). Enterprise social network sites (ESNSs), a form of SNS used within organizations can help address these challenges because they provide affordances that enable large-scale knowledge sharing. ESNSs include the foundational features associated with SNSs but are implemented within organizations, are typically formally sanctioned by management, and have the ability to restrict membership or interaction to members of a specific enterprise, thus enabling the flow of information that would be inappropriate for public, commercial social media tools.

Large, distributed multinational organizations have led the adoption of enterprise social network technology, and thus most of the extant research speaks to this particular type of organization, although ESNSs can also serve an important role for smaller and co-located companies. Large multinational organizations are increasingly dependent on successful knowledge sharing among individuals, teams, and units because of their high degree of geographical dispersion across locations and time zones. Prior research from organizational communication and management scholars often treats knowledge sharing as a mechanical process of information transfer or transmission (Argote and Ingram, 2000; Szulanski, 1996, 2000), without adequately considering social dynamics or interpersonal processes.

Knowledge is treated as distinct from (though encompassing) data or information (Fulk, Monge, and Hollingshead, 2005) and recognize knowledge sharing as an equivocal communicative process involving sense making and interpretation (Weick, 1995; Zorn and Taylor, 2004).

Knowledge sharing may range from dyadic exchanges of information between individuals (Cummings, 2004; Gupta and Govindarajan, 2000), to ongoing problem solving and coordination in formal project teams (Ancona and Caldwell, 1992; Tsai, 2001), to large-scale organizational brainstorming to generate solutions to global problems or issues (Levin & Cross, 2004). In large multinational organizations, knowledge sharing is a complex process due to the need to negotiate meaning among diverse individuals as well as larger groups and collectives.

Distributed organizations face challenges above and beyond those that are co-located: knowledge sharing may be stymied by the fact that employees may not recognize who has relevant expertise (Farak and Sproull, 2001), may be reluctant to request or share information with strangers in other organizational units or job functions, may not be motivated or incentivized to contribute more than task-related information, or may be uncomfortable asking questions publically. Thus, organizational knowledge sharing is defined as the process of providing and receiving information, advice or feedback (Cummings, 2004; Hansen, 1999), acknowledging that individuals will interpret and process knowledge to co-create individual and shared meanings (Weick, 1995; Zorn and Taylor, 2004).

With an increased focus on virtual teams and distributed work arrangements, ESNSs are increasingly being adopted for internal knowledge sharing in large distributed organizations (Treem and Leonardi, 2012). Previous research has demonstrated that collaborative technology may enable knowledge sharing within and between organizational teams (Cummings and Kiesler, 2005; Hinds and Kiesler, 2002) but has not yet considered the unique affordances of ESNSs in this process. In addition, prior work has often emphasized the task-related dimensions of knowledge sharing as a process of information transfer (e.g., Szulanski, 2000; Hansen, 1999) without fully accounting for the role of social dynamics at the interpersonal and organizational levels.

Organizationally, ESNSs are utilized for sharing knowledge, engaging in organizational politics, understanding the work environment, and collaborating in the everyday work of teams, among other tasks. ESNSs are also used for social purposes, such as establishing ties, finding common ground and maintaining relationships with co-workers. Managers may use informal social networks to enable learning and gain access to information about new processes (Hansen, 1999). Furthermore, social cues received from peers and supervisors within a communication network will have a direct effect on how team members respond to new information (Fulk, 1993; Tsai, 2001). Thus, network position will affect a team member's acceptance and use of collaborative technology, and ultimately will affect knowledge sharing practices.

Employees' relationships with one another also have a direct impact on the ability of organizations to coordinate work and discover new knowledge. When organizations are faced with situations involving complex knowledge, strong ties are needed to facilitate the effective transfer of knowledge (Hansen, 1999). Weak ties, on the other hand, may provide an individual with access to non-redundant information sources, bolstering bridging social capital (Burt, 1992; Burt, 2005). ESNSs can support relationship maintenance activities with existing ties both strong and weak and can help individuals both identify relevant latent ties with valuable information and determine one's shared common ground with them (Ellison et al., 2007). There are limitations however; Aral and Van Alstyne (2010) note that gains in network diversity often result in a decrease in the communication bandwidth, and the increased communication flow may limit individuals' ability to locate useful knowledge.

In summary, ESNSs provide affordances that aid in the distribution of information and the sharing of knowledge at the individual and organizational level. Importantly, ESNSs support the socialization and interpersonal interaction that provides a foundation for many knowledge-sharing processes. This section has highlighted the value of an integrated approach to knowledge sharing in modern organizations that considers both social and task dimensions, especially in relation to the roles played by social capital dynamics, identity information, context collapse, and networked organizational structures in constraining, enabling, and reshaping knowledge sharing within the organization.

11.11   RECOMMENDATIONS

Anyone can register in social media networks, and because these are growing in popularity, UNESCO's objective should be to raise awareness of their educational relevance and to find ways to integrate them into the teaching process. Networking systems like Facebook and LinkedIn; micro blogging like Twitter; Wikis, MSN and Flickr are free to subscribe to. Similarly to the ICT applications that penetrated education the last four decades, the best recommendation is to allow social media in the classroom to explore the benefits and limits of the new level of social networking that they offer. Recent projects, such as web-based communities for teachers, show that social media can be used not only for the exchange of didactic methods and ideas. They also allow sharing more subtle personal concerns, such as legal conflicts and emotional states related to career development, which can be discussed with colleagues from other institutions and even other countries in order to avoid affecting hierarchic relationships within the school.

Social media use in educational institutions is not yet a sustainable solution for the traditional problems of education. However, they provide opportunities that are changing the way we learn.

For example, in the realm of life-long learning during one's professional career, they facilitate the sharing of practical solutions and make colleagues aware of new trends and topics. The first step needed is to let teachers explore the potentials of social media and gradually test-drive some of their benefits in classroom exercises. This will illustrate for teachers the effects of social media on learning, rather than illustrate for students how they may benefit from them. Inherent to social media is the tendency to improve the learning atmosphere rather than the direct instruction. The current trend towards integrating social media with gaming is not encouraged in the scope of social media. Complex issues like growing social awareness are not easily covered by direct instruction. They need a curious mind and experimentation by the learner, as well as an evidence-based analysis by the teacher. Ethical issues like intertwining recent experiences with social media need to be rubricated and enlisted in best practices of social media for education. As school reports are coming out now, we see social media as bridges between individual education and mass education. The implications for teacher education need to be clearly defined. Freshly trained teachers forget their ICT specialties as soon as they start working in an actual school. The reason is that students under a high-pressure regime in demanding courses tend to "escape" and "chill" in learning conditions that allow more freedom. Young teachers immediately feel this threat, especially compatible with the test-driven regime, and they forget about the more subtle advantages of learning with cognitive tools like simulations and social media. The prospects of social media for gender and cultural fairness, as well as the advantages they present for disabled learners, are still unexplored.

Before attempting to consolidate the new practices with social media in institutional policies, at least two more years of co-evolution between social media and contemporary school innovation are necessary. The future trend is to privatize schools and let other stakeholders like parents and enterprises enter this process. Categories like lower vocational training, grammar schools and gymnasium are being introduced again. How far this trend continues will depend on socio-political factors. In this context, social media present a platform for educators and parents to express their opinions and priorities in this regard.

11.12   CLOSING THOUGHTS

Social networks have affected the modern society positively and have changed some of people's habits. Effective use of features and opportunities of social networks supports instructors' empowering of the educational process with active learning, creativity, problem-solving, cooperation, and multifaceted interactions as well as students' using and improving their academic performance, inquiry, and alternative thinking skills.

It develops communication skills, extends participation and social commitment, strengthens peer support, and enables the realization of cooperation-based learning. In addition, social network sites can be used easily and conveniently, they can be integrated into the educational practices successfully and such utilizations are becoming widespread rapidly. Social networks provide the users a communication environment which is not limited with space and time because they are online tools that generate interaction by allowing new opportunities for more information, interest, and data sharing. Social networks also provide opportunities for the academicians because they are flexible and user friendly, they are used more easily than the other educational management systems. Social websites are used in the delivery of information, reference books, group assignments, and course sessions. Instructors and students can send materials, addresses of web sites, and videos regarding courses on Social websites and presentations, assignments, and other products of the students that can be shared by forming links to Google documents. Social websites can be used to share materials (video files, audio files, pictures, spreadsheet, presentation, database, web sites etc.) effectively, follow up current events, news, people or groups, and get involved in discussion environments to support the skills of learning for cooperation, research, discussion, critical thinking, problem solving and so on. Those who have social media website membership amongst the instructors participating in this study are using these features actively. Social networking sites are becoming more involved in our daily life day by day. As of today, instructors can neither conduct a course completely through social media nor can they ignore this development comfortably. The younger faculty members use social media and similar technologies more in their courses compared to senior faculty members. This is due to their familiarity and mastery regarding the use of these technologies.

Social networking is a tool, with both its advantages and problems for usage in teaching and learning. When used in a learning context where affordances of the technology are carefully evaluated in terms of pedagogical requirements and student learning outcomes, including those elements that result in a supportive and collaborative learning environment, these tools offer significant advantages for distance learning. Among the positive attributes are impacts on student engagement, motivation, personal interaction, and affective aspects of the learning environment. The direct contribution to student achievement remains to be proven, but when technology supports an affirmative, constructivist-learning environment and contributes to successful pedagogical strategies without distracting from essential objectives for development of knowledge and skills, the result of formative evaluation of social networking potentials for learning is positive.

KEYWORDS

- **distance learning**
- **higher education**
- **Knowledge Management**
- **mobile learning**
- **social media**
- **social networks**

REFERENCES

1. Murray, C. (2008). Schools and social networking: Fear or education? Synergy Perspectives: Local, 6(1), 8–12.
2. Toprak, A. (2009). Toplumsal paylasim agi Facebook. Istanbul: Kalkedon.
3. Gulbahar, Y., Kalelioglu, F., Madran, O. (2010). Sosyal aglarin egitim amacli kullanimi [Educational use of social networks]. XV. Turkiye'de Internet kullanimi konferansi. Istanbul: Istanbul Teknik Universitesi.
4. Alexander, B. (2006). Web 2.0: A new wave of innovation for teaching and learning? EDUCAUSE Review, 41, 32–44.
5. Boyd, D. M., Ellison, N. B. (2007). Social network sites: Definition, history, and scholarship. Journal of Computer-Mediated Communication.
6. Huang, Y. M., Jeng, Y. L., Huang, T. C. (2009). An Educational Mobile Blogging System for Supporting Collaborative Learning. Educational Technology & Society, 12 (2), 163–175.
7. Yang, S. J. H., Chen, I. Y. L. (2008). A Social Network-based System for Supporting Interactive Collaboration in Knowledge Sharing over Peer-to-Peer Network. International Journal of Human-Computer Studies, 66 (1), 36–50.
8. Huang, Y. M., Huang, T. C., Hsieh, M. Y. (2008). Using annotation services in a ubiquitous Jigsaw cooperative learning environment. Educational Technology & Society, 11 (2), 3–15.
9. Yang, S. J. H., Chen, I. Y. L. (2008). A Social Network-based System for Supporting Interactive Collaboration in Knowledge Sharing over Peer-to-Peer Network. International Journal of Human-Computer Studies, 66 (1), 36–50.
10. Rachel, P. (2008). Communities of practice: using the open web as a collaborative learning platform. iLearning Forum 2008, Proceedings, Paris.
11. Diederich, J., Iofciu, T. (2006). Finding communities of practice from user profiles based on folksonomies. Proceedings of the 1st International Workshop on Building Technology Enhanced Learning Solutions for Communities of Practice, Crete, Greece.
12. Weinberger, A., Fischer, F. (2006). A Framework to Analyze Argumentative Knowledge Construction in Computer-Supported Collaborative Learning. Computers & Education, 46, 71–95.
13. Koschmann, T. (1996). Paradigm Shifts and Instructional Technology: An Introduction. In T. Koschmann (Ed.), CSCL: Theory and Practice of An Emerging Paradigm, Mahwah, NJ: Lawrence Erlbaum, 268–305.
14. Tsai, W., Ghoshal, S. (1998). Social Capital and Value Creation: The Role of Intrafirm Networks. Academy of Management Journal, 41 (4), 464–476.
15. Artiles, J., Gonzalo, J., Verdejo, F. (2005). A testbed for people searching strategies in the WWW. Proceedings of the 28th
16. Annual International ACM SIGIR conference on Research and Development in Information Retrieval (SIGIR'05), 569–570.
17. Wan, X., Gao, J., Li, M., Ding, B. (2005). Person resolution in person search results: Webhawk. Proceedings of the 14th ACM international conference on Information and knowledge management, New York: ACM, 163–170.

18. Mori, J., Basselin, N., Kroner, A., Jameson, A. (2008). Find me if you can: designing interfaces for people search. Proceedings of Intelligent User Interfaces Conference, 377–380.

19. Levy, M. (2009). Web 2.0 implications on knowledge management. Journal of Knowledge Management, 13(1), 120–134.

20. Bukowitz, W., and Williams, R. (2000). The Knowledge Management Fieldbook. UK, London: Prentice-Hall.

21. Kaplan, A. M., Haenlein, M. (2010). Users of the world, unite! The challenges and opportunities of social media. Business Horizons, 53(1), 59–68.

22. McElroy, M. W. (2003). The New Knowledge Management: Complexity, Learning and Sustainable Innovation. Burlington, MA: KMCI Press/Butterworth-Heinemann.

23. Dalkir, K. (2011). Knowledge Management in Theory and Practice. England, Cambridge: The MIT Press.

24. Ruggles, R. (1997). Knowledge tools: Using technology to manage knowledge better. Boston: Butterworth-Heinemann.

25. Dames, M. (2004). Social software in the library. Accessed on April 13, 2013 at http://www.llrx.com/features/socialsoftware.htm

26. Ojala, M. (2005). Blogging for knowledge sharing, management and dissemination. Business Information Review, 22, 269–276.

27. Ferdig, R. E., Trammell, K. D. (2004). Content delivery in the 'Blogosphere.' Technological Horizons in Education Journal, 31, 7. Retrieved from http://www.thejournal.com/magazine/vault/articleprintversion.cfm?aid = 4677

28. Mortensen, T., Walker, J. (2002). Blogging thoughts: personal publications as an online research tool. In A. Murray-Harvey, R. (2001). How teacher education students cope with practicum concerns. The Teacher Educator, 37, 117–132.

29. Chu, S. K. W., Kwan, A. C. M., Warning, P. (2012). Blogging for information management, learning, and social support during internship. Journal of Educational Technology & Society, 15(2), 168–178.

30. Chu, S. K. W., Du, H. (2013). Social Networking Tools for Academic Libraries. Journal of Librarianship & Information Science, 45(1), 64–75.

31. Wong, K., Kwan, R., Leung, K. (2011). An Exploration of Using Facebook to Build a Virtual Community of Practice. Proceedings of the 4th International Conference, ICHL 2011 (pp. 316–324), Hong Kong, China.

32. Lave, J., Wenger, E. (1991). Situated Learning: Legitimate Peripheral Participation. Cambridge: Cambridge University Press.

33. De Villiers, M. R. (2010). Academic use of a group on Facebook: Initial findings and perceptions. Proceedings of Informing Science & IT Education Conference (InSITE) 2010, 173–190. Retrieved from http://proceedings.informingscience.org/InSITE2010/InSITE10p173–190Villiers742.pdf

34. Selwyn, N. (2009). Faceworking: exploring students' education-related use of Facebook, Learning, Media and Technology, 34(2), 157–174.

35. Kabilan, M. K., Ahman, N., Abidin, M. J. Z. (2010). Facebook: An online environment for learning of English in institutions of higher education? The Internet and Higher Education, 13(4) 179–187.

36. Lampe, C., Ellison, N., Steinfield, C. (2006). A Face(book) in the crowd: Social searching vs. social browsing. Proceedings of the 2006–20th Anniversary Conference on Computer Supported Cooperative Work (pp. 167–170). New York: ACM Press.

37. Joinson, A. N. (2008). Looking at, looking up or keeping up with people?: Motives and use of Facebook. Proceeding of the Twenty-Sixth Annual SIGCHI Conference on Human factors in Proceedings of the 44th Hawaii International Conference on System Sciences. 1–10.

38. Al-Khalifa, H. S., and Davis, H. C. Harnessing the wisdom of crowds: How to semantically annotate web resource using folksonomies. In Proceedings of IADIS Web Applications and Research (WAR2006) (2006).

39. Alani, H., Dasmahapatra, S., O'hara, K., and Shadbolt, N. Ontocopi – using ontology based network analysis to identify communities of practice. IEEE Intelligent Systems 18 (2003), 18–25.

40. Andriessend, E., Soekijad, M., Veld, M. H. i. t., and Poot, J. J. Dynamics of knowledge sharing communities. Tech. rep., Delft University of Technology, 2001.

41. Angeli, C., Bonk, C., and Hara, N. Content analysis of online discussion in an applied educational psychology course. [On-line] (1998).

42. Archee, R. Using computer-mediated communication in an educational context: Educational outcomes and pedagogical lessons of computer conferencing. The Electronic Journal of Communication 3, 2 (1993). http://www.cios.org/getfile/ARCHEE_V3N293

43. Bales, R. F. Interaction Process Analysis. A Method for the Study of Small Groups. Cambridge, MA: Addison–Wesley, 1950/1951.

44. Balsamo, A. Technologies of the Gendered Body: Reading Cyborg Women. Duke University Press, New York, December 1995.

45. Bandura, A. Self-efficacy: Towards a unifying theory of behavioral change. Psychological Review 84 (1977), 191–215.

46. Berglund, A., and Eckerdal, A. What do cs students strive to learn? Computer Science Education 16, 3 (2006), 185–195.

47. Berkson, L. Problem-based learning: Have the expectations been met? Academic Medicine 68, 10 (1993), S79–S88.

48. Bolin, M. End-User Programming for the Web. PhD thesis, MIT, http://groups.csail. mit.edu/uid/ projects/chickenfoot/mbolin-thesis.pdf, May 2005.

49. Booth, S. Learning computer science and engineering in context. Computer Science Education 11, 3 (2001), 169–188.

50. Castells, Manuel (1996, second edition, 2000). The Rise of the Network Society, The Information Age: Economy, Society and Culture Vol. I. Cambridge, MA. Oxford, UK: Blackwell. ISBN 978–0631221401.

51. Castells, Manuel (2009). Communication power. Oxford/New York: Oxford University Press. ISBN 9780199567041.

52. Dillenbourg, P., Tchounikine, P. (2007). Flexibility in Macro-Scripts for Computer- Supported Collaborative Learning. Journal of Computer Assisted Learning, 23(1), 1–13.

53. Festinger, L. (1954). A theory of social comparison processes. Human Relations, 7(2) 117–140.

54. Glasersfeld, Ernst von (1995), Radical Constructivism: A Way of Knowing and Learning, London: RoutledgeFalmer.

55. Palincsar, A. S. (1998). Social constructivist perspectives on teaching and learning.

56. Annual Review of Psychology, 49, 345–375.

57. Suls, J., Wheeler, L. (2000). A Selective history of classic and neo-social comparison theory. Handbook of Social Comparison. New York: Kluwer Academic/Plenum Publishers.

58. Gray, R., Ellison, N. B., Vitak, J., Lampe, C. (2013). "Who wants to know?" Question-asking and answering practices among Facebook users. In Proceedings of the 2013 Conference.

59. Burke, M., Kraut, R. (2013). Using Facebook after losing a job: Differential benefits of strong and weak ties. In Proceedings of the 2013 Conference on Computer Supported Cooperative Work (CSCW '13) (pp. 1419–1430), New York, NY: ACM Press.

60. Marwick, A., Ellison, N. (2012). "There isn't Wifi in heaven!" Negotiating visibility on

61. Facebook memorial pages. Journal of Broadcasting and Electronic Media, 56, 378–400.

62. Cross, R. L., Parker, A., Prusak, L., Borgatti, S. P. (2001). Knowing what we know:

63. Supporting knowledge creation and sharing in social networks. Organizational Dynamics, 30(2), 100–120.

64. Espinosa, J. A., Slaughter, S. A., Kraut, R. A., Herbsleb, J. D. (2007). Team knowledge and coordination in geographically distributed software development. Journal of Management Information Systems, 24, 135–169.

65. Faraj, S., Sproull, L. (2000). Coordinating expertise in software development teams. Management Science, 46, 1554–1568.

66. Gibson, C. B., Gibbs, J. L. (2006). Unpacking the concept of virtuality: The effects of geographic dispersion, electronic dependence, dynamic structure, and national diversity on team innovation. Administrative Science Quarterly, 51, 451–495.

67. Cramton, C. D., Orvis, K. L., Wilson, J. M. (2007). Situation invisibility and attribution in distributed collaborations. Journal of Management, 33, 525–546.

68. Argote, L., Ingram, P. (2000). Knowledge transfer: A basis for competitive advantage in firms. Organizational Behavior and Human Decision Processes, 82, 150–169.

69. Fulk, J., Monge, P., Hollingshead, A. B. (2005). Knowledge resource sharing in dispersed multinational teams: Three theoretical lenses. In D. L. Shapiro, M. A. Von Glinow, J. L. C. Cheng (Eds.), Managing multinational teams: Global perspectives (Advances in international management) (Vol. 18, pp. 155–188). San Diego, CA: Elsevier.

70. Weick, K. E. (1995). Sensemaking in organizations. Thousand Oaks, CA: Sage.

71. Cummings, J. (2004). Work groups, structural diversity, and knowledge sharing in a global organization. Management Science, 50, 352–364.

72. Hansen, M. T. (1999). The search-transfer problem: The role of weak ties in sharing knowledge across organization subunits. Administrative Science Quarterly, 44, 82–111.

73. Szulanski, G. (2000). The process of knowledge transfer: A diachronic analysis of stickiness. Organizational Behavior and Human Decision Processes, 82(1), 9–27.

74. Szulanski, G. (1996). Exploring internal stickiness: Impediments to the transfer of best practice within the firm. Strategic Management Journal, 17, 27–43.

75. Treem, J. W., Leonardi, P. M. (2012). Social media use in organizations: Exploring the affordances of visibility, editability, persistence, and association. Communication Yearbook, 36, 143–189.

76. Zorn, T. E., Taylor, J. R. (2004). Knowledge management and/as organizational communication. In D. Tourish & O. Hargie (Eds.), Key issues in organizational communication (pp. 96–112). London: Routledge.

77. Fulk, J. (1993). Social construction of communication technology. Academy of Management Journal, 36, 921–950.

78. Tsai, W. (2001). Knowledge transfer in intra-organizational networks: Effects of network position and absorptive capacity on business unit innovation. Academy of Management Journal, 44, 996–1004.

79. Burt, R. S. (1992). Structural holes: The social structure of competition. Cambridge, MA: Harvard University Press.

80. Burt, R. S. (2005). Brokerage and closure: An introduction to social capital. New York, NY: Oxford University Press.

81. Ellison, N. B., Steinfield, C., Lampe, C. (2007). The benefits of Facebook "friends": Social capital and college students' use of online social network sites. Journal of Computer-Mediated Communication, 12, 1143–1168.

82. Aral, S., van Alstyne, M. (2010). The diversity-bandwidth trade-off. American Journal of Sociology, 117, 90–171.

CHAPTER 12

SOCIAL NETWORKS AND INFORMATION DISSEMINATION FOR DISASTER RISK MANAGEMENT

MOHAMMED ZUHAIR AL-TAIE[1] and
SITI MARIYAM SHAMSUDDIN[2]

[1]*PhD Candidate, UTM Big Data Centre and Faculty of Computing, Universiti Teknologi Malaysia (UTM), Skudai, 81310 Johor, Malaysia, E-mail: mza004@live.aul.edu.lb, Tel: 07–5538793, Ext: 8793*

[2]*PhD, UTM Big Data Centre and Faculty of Computing, Universiti Teknologi Malaysia (UTM), Skudai, 81310 Johor, Malaysia, E-mail: mariyam@utm.my, Tel.: 07–5538793, Ext: 8793*

CONTENTS

Disasters happen, and no one is immune. Pandemic, industrial accidents, geological and meteorological hazards, airplane crashes, and bombings cause the death of tens or hundreds of people and lead to severe damages to economy and environment. Therefore, it became more urgent than before to put active emergency management policies and programs.

Dealing with damage brought about by accidents can be difficult and even daunting if the world countries do not adopt right procedures for disaster risk management. Therefore, knowing how to prevent disasters (or alternatively mitigate their effects), prepare the communities to cope with them, and coordinate the efforts and experiences across the world can provide a concrete foundation to manage risks.

Currently, as per the International Transportation Association, there are 900 airlines and 20,000 aircraft that transport yearly over 1.6 billion passengers. Furthermore, 40% of the world goods is transported by airplanes. Hence, investigating in transportation system failures such as airplane accidents has become a worldwide necessity, especially when such failures can directly or indirectly affect the global economy and our environment.

In this chapter, the readers first will take a look at disasters, what they mean and how communities can recover from them with minimal costs. Then, the authors address a number of accident analysis techniques that are both accepted and approved by experts. The aim of these methods is to investigate accidents through identifying causal factors, which are meant to prevent further accidents in the future.

In addition to the use of these techniques for sophisticated and complex mishap investigations; they can also be used for limited-size accident investigations (for example airplane crash or ship sinking). Besides, some of these tools have been applied successfully in transport means reliability and safety.

The authors focus on the disappearance of Flight MH370, which created a surge of public grief and distress across the world.

What makes this accident exceptional and different from other aviation accidents is its unknown destiny and the number of countries participated in search efforts. About 26 countries helped provide data, intelligence, assets, civilian as well as military support. This makes it one of the biggest and most costly search and rescue operations, of its kind, until today.

12.1   HAZARDS AND DISASTERS

Scientists usually define a "hazard" as a dangerous event that strikes places singly or collectively at different times of the year. However, they do not come about with the same strength: they vary in the degree of intensity and severity [1].

On the other hand, a "disaster" results from three components: hazard, condition of vulnerability and insufficient capacity to mitigate the adverse effects of it [2].

That is to say: a hazard will turn into a disaster when it causes massive human, material or environmental losses in a way that physical and psychological victim recovery is unlikely without external aid.

It is necessary, when a hazard strikes, that experts recognize the underlying factors that lead the hazard to occur (called hazard analysis) [3]. Such analysis gives information on the risk, how serious it is, and what are the possible ways to remove; intervene or reduce its consequences [4].

Several countries have arrived at awareness that they should organize their efforts to better deal with the significant risks. This attention came after a series of accidents that stroke worldwide during the fifty or sixty years ago. Examples of recent events include: the tsunami of the Indian Ocean 2004, the active hurricane season in 2004 that hit central Florida in the U.S., the 2005 Pakistan earthquake, the 2008 Sichuan-China earthquake, the 2010 Haiti and Chile earthquake, the 2009 Victoria wildfires in Australia, the massive floods in Pakistan in 2010 and in Brazil, South Africa and Australia in 2011, the 2011 Japan earthquakes and tsunami, the Deepwater Horizon oil spill in the Gulf of Mexico in 2010, 2001 attacks of New York and Washington, the bombings of Madrid in 2004 and of London in 2005 and the attempts to bombing aircrafts in the U.S. in recent years. All these events have called the attention not only to natural disasters but also to the other types of hazards.

EM-DAT, is a world disaster database that contains core data on hazards from 1900 to the present. A hazard should meet one or more conditions to be included in this database [5].

1. Ten or more people are confirmed or presumed dead;
2. 100 people or more are affected and require immediate assistance;

3. The hazard is associated with a declaration of a state of emergency;

4. The affected country (or coutries) call for international assistance.

When looking at the number of deaths and injuries, and the size of damage caused by giant disasters, some people may imagine that the disappearance of Flight MH370 is a "small" or "near miss" accident.

However, the attention paid by big countries and the vast amounts of money spent on search efforts, makes the disappearance one of the most extraordinary events in the modern age.

Ted Ferry, the accident investigation expert, stresses the significance of what apparently looks 'little event' when he said [6]:

"... to the contrary, the 'little' event may have all the implications and complexities of a catastrophe. It only appears less important because serious injury or property loss did not result. The same causes and modes of failure are present in both large and small mishaps."

12.1.1 CLASSIFICATION OF DISASTERS

Disasters vary in terms of: intensity, type, duration, locus and human impact (see Figure 12.1). Researchers have classified disasters based on certain criteria:

- **Cause**: some disasters happen due to natural causes, such as tidal waves, tsunamis or hurricanes, whereas others happen because of human reasons such as wars and chemical plant leaks.
- **Duration**: some disasters last for years, such as famines or wars, while others last for few seconds, such as earthquakes.
- **Size of Effect**: some disasters have very limited effect, like house burning or woman rape, while others like earthquakes or wars can have devastating effects over the entire community or country. For instance, the Chernobyl atomic meltdown on 26, April-1986 released large quantities of radioactive particles into the air. The clouds stretched out over much of the western part of the former Soviet Union as well as some parts of Europe [7].
- **Intention**: disasters can be initiated deliberately, such as personal assaults and ethnic cleansing, or unintentionally, such as some airplane crashes and industrial accidents.

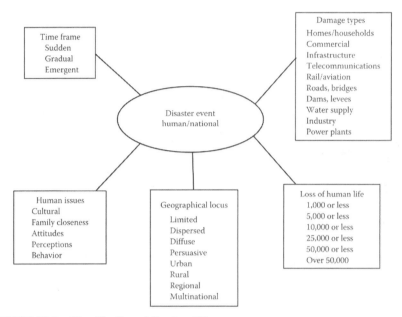

FIGURE 12.1   Classification of disasters [8].

- **Occurrence Pattern**: some disasters occur regularly (e.g., cyclones), some occur randomly (e.g., earthquakes) and some occur in tandem with another (e.g., tsunamis).

In addition to the classification of disasters based on the above criteria, we can as well classify them based on the cause of disaster into: natural disasters and technological disasters [5].

12.1.2 NATURAL DISASTERS

Some countries are more vulnerable to disasters than others; they lack the national resources that are critical to absorb and recover from the effects of a hazard.

Natural disasters cause the death of around 60,000 people every year and directly affect the life of about 250 other million people [3]. While floods are the most expensive to the economy, drought and famine are more devastating in human terms.

Many natural disasters are cyclical and regular in nature. For example, the following disasters periodically strike the same area or the same country [3]:

- **Tropical Storms**: for example, India, Bangladesh.
- **Earthquakes**: for example, Indonesia, Afghanistan, Haiti.
- **Landslides**: for example, Burundi, Pakistan.
- **Droughts**: for example, Kenya, Somalia, Sudan.
- **Floods**: for example, Cambodia, Mozambique, Ethiopia.

Researchers have noticed that the unprecedented effects of natural disasters worldwide (e.g., earthquakes, tropical cyclones, floods, and wild-fires), that escalated especially during the last five decades, are attributed to a number of factors [7]:

- **Climate Change**: It is accepted that the current trend in extreme weather events is the result of changes in the global mean temperature [9]. Such changes include major heat waves and droughts, floods in localized areas, tropical cyclones and others. Such changes have effects on precipitation rates, soil moisture, wind velocities and vegetation cover. For example, in 2007 alone, Mexico saw the worst floods in five decades. Sudan, Burkina Faso and Costa Rica suffered from the most severe flooding in years while China saw the heaviest snowfall in 56 years [2]. Furthermore, it is expected that extreme weather events will increase in severity and frequency, and sea levels will continue rising as glaciers and ice polar caps melt [3].
- **Environment**: man has participated in the destruction of critical elements of the ecological balance such as the depletion of natural resources, deforestation, and so on.
- **Poverty:** The increase in the world's population and poverty forced a growing number of people to stay in vulnerable or unsafe areas. For example, in some countries (especially in developing countries) people prefer, forced by economic factors, to live in simple houses built with cheap materials.

Natural disasters can take one of the following three forms:

- **Hydro-Meteorological Events:** for example, landslides, avalanches, droughts, famines, floods, and forest fires (although forest fires can also be ignited by man).

- **Geophysical Events:** for example, earthquakes, volcanic eruptions, and tsunamis.
- **Biological Events:** for example, locusts, pest infections, and epidemics.

In terms of scope of effect, natural disasters have the following three forms of effect [2]:

1. **Direct Effects**: these include physical damages to productive capital and stocks (e.g., industrial plants), economic structure (e.g., roads and electricity suppliers) and social infrastructure (e.g., homes and schools).
2. **Indirect Effects**: they include downstream disruption to the flow of services and goods (e.g., lower output from destroyed or damaged assets), and the disruption of the provision of essential services (e.g., telecommunication or water supply).
3. **Secondary Costs**: they include short- and long-term effects on the overall economy performance or socio-economic conditions such as fiscal and monetary performance or levels of national indebtedness.

12.1.3 TECHNOLOGICAL DISASTERS

While natural disasters are caused by forces of nature and thus governments and organizations can to some extent predict them, man-made disasters are unpredictable, and often malicious. However, not all man-made disasters are deliberate, especially in the era of high technology and expansion in world population.

The mismanagement in critical sectors such as transportation, chemical and nuclear power was the reason of a number of industrial accidents related to radioactive and chemical materials like the one happened in 1976, where an explosion at a pesticide plant in Italy led to the release of 2,3,7,8-Tetrachlorodibenzo-dioxin. Other examples include the explosion of a chemical plant in Bhopal (India), in 1984, which led to the release of methyl isocyanate, and the Switzerland-Sandoz chemical incident that happened in 1989.

Wars and armed conflicts made the issue much worse: during 1990s, about 48% of reported conflicts occurred in Asia while 42% of them occurred in Africa. The IFRC estimates that due to armed conflicts, nearly 230,000 people were killed every year in the 1990s, in addition to 31 million people, annually, being directly affected [3].

The reoccurrence of such accidents made some governments agree on a number of international regulations aimed at preventing similar accidents. The regulations emphasized the importance of safety in technology. They also forced some countries to quit or restrict their nuclear programs [9].

Technological disasters fall into three groups:

- Industrial Accidents: such as chemical spills, fires, gas leaks, explosions, and so on.
- Transport Accidents: accidents by transportation means using air, rail, or road.
- Miscellaneous Accidents: accidents that are not related to industry itself, such as the collapses, explosions and fires in domestic structures.

12.1.4   IMPACT OF DISASTER

There has been a high increase, in recent decades, in both impact and number of disasters, especially the natural. For example, in the period 1997–2006, the number of reported natural disasters grew from 4,241 to 6,806 [2].

Previous studies estimated that since the 1990s, nearly 60,000 people a year on average died, and the lives and livelihoods of other millions are affected. More than 200 million a year have been directly affected in addition to unaccountable number of indirectly affected people [2]. The number of affected people rose from 1.6 in the period (1984–1993) to almost 2.6 billion in the period (1994–2003).

Disasters affect people, environment and livelihoods. Based on type of effect, disasters have the following forms of losses [10]:

- **Death**: disasters can lead to tens or hundreds or even thousands of deaths. For example, in the period 1997–2006, the number of reported deaths jumped from 600,000 to over 1.2 million, compared to the previous decades [2]. Mortalities have shown to be very low in industrial countries and were rapidly decreasing in developing countries through economic growth and enhanced disaster risk management systems.

- **Economic Loss**: they also lead to losses in houses, buildings, workplaces, livestock and so on. In constant dollars: between 1990 and 1999, disasters costs were USD 652 billion in material losses, compared to the period between 1950 and 1959, where the costs were USD 38 billion [2].
- **Looting and Violence**: looting can be problematic during catastrophes such as war, rioting, political and military victory, or natural disasters. It is often associated with the use of power to take goods. Some forms of looting and violence were present during the Hurricane Katrina, 2005.

In addition to looting, disasters can be associated with activities such as price gouging and anti-social behavior (such as crime or rape).

Besides the adverse effects of large-scale disasters, the threat posed by small- and medium-scale disasters to the well-being of rural and marginal urban communities is increasing [2]. This risk adds more to the accumulative impact of disasters on the overall world poverty.

12.1.5 DISASTER RISK MANAGEMENT

Disasters are neither unpredictable nor unavoidable [2]. Therefore, and in order to prevent hazards from turning into disasters, it is crucial to developing a comprehensive emergency management plan that can minimize vulnerability and risk.

The implementation of these operations requires the use of administrative organizations, decisions, operational skills and capacities that aim to lessen the impacts of disasters and subsequent environmental and technological crises. These processes are usually three cyclical non-linear phases (see Figure 12.2) that influence each other:

- **Disaster Preparedness**: disaster preparedness (risk reduction) aims at preventing its occurrence or alternatively reducing its effects, if the disaster is must. It aims at strengthening the nation's capacity to withstand, respond and recover gradually from hazards. Preparedness

FIGURE 12.2 Disaster risk management processes.

covers activities such as gathering of information, training of responders, establishment of public awareness, maintenance of financial, human and material resources, as well as development of early warning systems. Countries can apply their plans at different levels, from villages and cities to local authorities and central government.

This phase is complex and incorporates the combination of political, participatory, resource mobilization and technical components.

- **Disaster Mitigation**: disaster mitigation aims at preventing, reducing the chance or lessening the size, scale, frequency, intensity and impact of emergencies. Such measures include fighting fires, providing medical aid, searching for survivors, strengthening houses and public buildings, raising river banks, checking dams, reforestation and others. They can take place before, during or aftermath of a disaster. A successful mitigation strategy is the one that is able to reduce hazard losses.

An example of such a successful strategy: after the September 2001 attacks, the U.S. authorities realized that the number of victims and injuries could be substantially in tens of thousands. Therefore, Metropolitan Medical Response System (MMRS) cities were encouraged to maintain prophylactic treatment capacity of 40,000 victims and exposed people [11].

The International Association of Fire Chiefs has defined the "Golden Hour" in disaster response, as the critical time when a patient should be stabilized [8]. The golden hour period is not firm, for example, it can increase or decrease based on the type of crisis being confronted (volcano, earthquake, airplane hijack/crash and so on). However, it is unexpected that patients should wait for 6 hours before outside help arrives. During this time, they may receive assistance from local community or administration to support the search and rescue efforts at the scene (See Figure 12.3).

- **Disaster Recovery**: disaster recovery (a.k.a. reconstruction or rehabilitation) aims to take the necessary decisions and actions to rebuild the community in the aftermath of a disaster. It also includes restoring or improving the pre-disaster conditions and encouraging necessary adjustments to minimize disaster risks.

It is worth noting that the severe effects of a disaster can last for a long time and affected countries deplete much of their resources during the post-impact phase. Successful and fast relief processes require the involvement of international parties.

International disaster management systems have benefited a lot from world crises. They summarized the experiences they had acquired in the following points [2]:

- Local community is the first to feel the effect of a disaster and the first to respond.

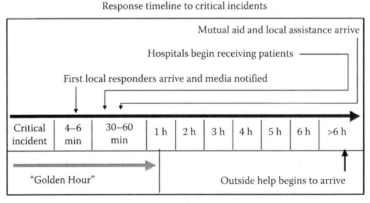

FIGURE 12.3   Response timeline to critical incidents [8].

- Direct people participation can achieve successful risk reduction measures.
- Not understanding the culture of communities risk behavior can badly affect disaster response.
- It is crucial for local people to get involved in emergency management plans.
- Community-based preparedness and early warning systems have a significant influence during disaster response.

12.1.6 PEOPLE COPING BEHAVIOR DURING DISASTER

Coping refers to the thoughts and acts adopted by disaster victims that aim to manage the internal and external demands posed by the traumatic event [12].

Researchers have divided coping strategies into two categories: "emotion-focused" and "problem-focused." The first type of which aims at changing or altering personal emotions associated with stress while the second type aims at changing the behavior of some aspect in the surrounding. Usually, a combination of the two is required.

During disasters and emergencies such as earthquakes and floods, people were noticed to show consistent patterns of coping behavior and psychological response as shown below [13].

12.1.6.1 Psychological Response

There is an incorrect image of how people react when confronted with disasters: victims get stunned, shocked, and cannot cope with the crisis. They were also pictured as dizzy and disoriented people who are waiting for outsiders to provide the necessary materials of living [14].

On the contrary, researchers found that in such cases, self- and kin-help relationships emerge, and people take actions that are adaptable for the situation. For example, most people who survived a disaster were saved by friends, family members or neighbors, and not by government responders, police or military. Besides, most of those who are not severely injured return to the disaster scene and participate in search and rescue activities [14].

The set of features that characterize communities during disaster time include social cohesion, feelings of trust, dependency and support as well as emotional and physical protection, even among those who never met before [15]. Typical psychological response patterns to disaster include:

1. **Stress**: The physical and psychological response to stress usually includes features such as the increase in neuroendocrine and cardiovascular measures, reflecting autonomic nervous system and hypothalamic-pituitary-adrenal responses [16].

Although stress can be maladaptive when people face threats, the "fight-or-flight" response scheme emerges in such conditions to control the allocation of physical energy and to better deal with the imminent danger. Moreover, stress was found to increase trust, trustworthiness and sharing behavior in social interactions, during emergency time. In other words, humans tend to provide and receive joint protection within groups.

2. **Mental State**: this includes symptoms such as pain, disability, fear, severe anxiety, or grief. Some of these symptoms vanish quickly, but others continue with some victims for longer periods. Feeling fearful, for example, is a typical response to extreme conditions. It is different from "panic" (discussed next) and can lead to positive coping behavior if consolidated by knowledge. It rarely results in inability to act, but can degenerate the ability to efficiently reason in complex situations.

3. **Panic**: A general misconception related to panic is that during disasters, panic is inherent. However, panic is not a primary characteristic in major disasters and is rarely noticed during natural or technological disasters. It has little significance and emergency organizations do not need to consider it in their plans.

BOX 12.1. Panic

Panic is an acute fear reaction, associated with loss of self-control and followed by non-social and non-rational behavior.

Panic is impossible without the following conditions:

- Belief of immediate and severe danger;
- Existence of only few escape routes;
- Belief that the escape routes are closed;
- Lack of communications about the situation.

12.1.6.2   Physical Response

In fact, citizens who experience disasters think and react normally, and usually do not tend to: freeze in fear, engage in irrational behavior or engage in panic flight. A typical example would be the United Airlines Flight 94: when people realized the significance of the threat, they organized and attacked the hijackers [11].

Information gathering behavior related to the change in environment is, a typical response pattern during disaster time [13]. People consider certain ways to gather information such as making phone calls, looking through windows, or switching TV channels to confirm a warning. Other behavioral patterns include preparations for evacuation (or alternatively strengthening house fortifications) and sticking to families.

Physical response during disaster is largely affected by a number of factors:

- **Physical Strength (readiness)**: In the situation of life or death, people with greater physical strength (like men or people in their prime age) have the advantage over more aged people, women and children. For example, in the case of "Titanic," where the ship needed almost three hours before ultimately sank, physical strength of individuals in a prime age gave 16% higher chance to survive than older persons [17]. The same study suggests that women passengers, compared to men with age and traveling class, had held constant a 53% greater chance of survival, whereas women who are both passengers and crew members had a survival probability of 64%. These results support the assumption that during disasters, people demonstrate pro-social actions. However, we should also put "time" into our consideration.
- **Time**: For analysts, time has proved to be significant during disasters since it helps explore behavior under extreme conditions (life and death). For example, Frey et al. [17]. studied people behavior during disasters through two use cases: The "Titanic," a British passenger ship that sank in the North Atlantic Ocean on April 14–1912, and the "Lusitania," another British ocean liner that sank three years later, for example, on 7 May 1915 during the World War I. While the Titanic needed two hours plus 40 minutes to sink, the Lusitania sank in 18 minutes. A more pro-social behavior (such as saving women

and children first) was shown in the first case, whereas selfish behavior and self-interested reactions were practiced in the second situation, when the ship was sinking rapidly.

- **Control over the Situation**: People, when experiencing a panic attack, believe that they cannot control the threatening agent such as fire, earthquake, aircraft hijacking, and so on [13].
- **Imminence of Danger**: Another important stimulus for panic behavior is the perception of imminence of danger. Control over the situation and imminence of danger are usually related to the physical strength and time, to mention but a few.

12.1.7   EMERGENT NORM THEORY

Emergency situations tend to disclose emergent group that had no existence prior to the disaster. Such groups often have transitory life but the collective work is crucial to the trans- and post-disaster response [18]. Members of such groups feel they are facing the same hazard and would meet the same destiny. Therefore, they feel they should work together.

Turner and Killian, who coined the concept of "Emergent Norm Theory," tried to explain the coping behavior of people in disasters by assuming that social behavior is driven by "Norms" established via two steps: crowd formation and keynoting.

BOX 12.2. Woman Response to Disasters

There are a number of characteristics that describe woman response to disasters:

- Women are more likely to evacuate. Besides, they have evacuation intentions;
- They are more likely to perceive a disaster as a serious hazard;
- Because of their social networks, women are more likely to receive communications about potential risks during emergencies;
- There are some indications that women prepare families to disaster.

Although the theory has collected a number of criticisms, it is still widely applauded and accepted, even by critics [19].

In case of extraordinary circumstances, such as civic emergencies, crowd members share a sense of uncertainty and urgency. These members should not necessarily know each other. Rather, the event brings these people together. The ultimate end of both the event and the crowd is unknown, and there is no beforehand agreement on what the crowd members should do and how they should do it. Such crowds exist only for a limited time and may vanish when the event is over.

Crowd members, according to Turner and Killian, begin asking questions like, "What is happening?" "Is this a terrorist attack?" or "What can we do about it." Nevertheless, the uncertainty collaps when new facts are discovered.

The next step of norm formation is what they called "keynoting." During this stage, crowd members start giving suggestions such as "let's surprise them," or "let's hide behind that building." Keynotes, to become group leaders, must be in harmony with what the large part of the crowd is thinking.

The combination of "crowd formation" and "keynoting" means that the "norm" emerges, and the group members fully understand the situation and agree to the actions decided. In other words, the crowd is ready to behave logically and purposely, and is capable of enforcing its norms.

12.1.8 PEOPLE COPING BEHAVIOR IN THE AFTERMATH OF DISASTER

Although the largest absolute financial losses occur in higher-income countries, lower-income nations suffer far more in relative terms.

Individually, people in the aftermath of the disaster experience a number of psychological symptoms such as sleep disruptions, nausea, anxiety, vomiting, bedwetting and irritability. Besides, serious psychological features such as depression, grief reactions and psychoses ensue, were present in some cases [11].

Scientists found that (i) community members in disaster time cooperate to provide support, and (ii) disasters generate broad consensus regarding a number of assets such as the value of life, property and community [15].

A real understanding of what happens during a disaster, supported by learning from previous lessons, provides appropriate mechanisms to deal with all likely effects.

Typically, a nation's response to disasters depends on two main factors:

- **Economic Development**: A series of connected actions aiming at promoting the standard of living and economic health. These activities include the development of critical infrastructure, human capital, regional competitiveness, health, safety, literacy, social inclusion and environmental sustainability.

Rich countries (e.g., Singapore, United States, Netherlands, Australia, Germany, etc.) have the human and financial resources that enable them to provide appropriate medical and food aids, efficient communication and transportation systems, as well as well-structured emergency plans. In contrast, poor countries (e.g., Zimbabwe, Liberia, Eritrea, Niger, Malawi and Nepal) cannot prepare such plans, owing to shortage vital facilities and resources. They suffer more significant losses relative to the gross domestic product (GDP) compared to richer countries. For example, more than half of the deaths that result from natural disasters occur in less developed countries [3]. Also, in developing countries, losses can reach 20 times higher, as a percentage of GDP, compared to industrial countries [2].

Regarding post-disaster coping behavior, poor countries, and instead of insurance, rely on informal insurance, microcredit and savings, or reciprocal exchanges (such as community self-help, kinship ties and remittances) [2].

- **Infrastructure**: it is the services and interconnected structural elements that are necessary for any economy to sustain. These elements can be either "Hard" such as transportation, solid waste, energy, water management, communications, management, earth monitoring and control networks, or "Soft" such as governance, economy, society, culture, sports and recreation.

The long-lasting problems that poor countries suffer from such as chronic malnutrition, poor health, substandard housing, inadequate dwellings, unstable hillsides, and mismanaged relief efforts; make them more vulnerable and less prepared to efficiently and effectively cope with the negative effects of disasters.

Poor countries are particularly more sensitive and more vulnerable to the ill effects of climate change, such that small changes can lead to devastating results on lives and livelihoods. On the other hand, rich countries have built strong infrastructures over the last century.

There are a number of steps that should be taken in the aftermath of disaster by governments, community volunteers, government organizations, agency aids and other responders [7]:

- **Emergency Response**: after survivors are taken to safe places and injured are transported to hospitals, people get engaged in search and rescue of missing and endangered from the perspective of being good citizens, good family members, or good neighbors and workmates. Trusting and reciprocal normative behavior in social relations can lead to more effective response when a disaster strikes.
- **Restoring Main Services**: in this stage, people work hard to rebuild their local facilities such as houses, roads, farms, buildings, power grids, communication networks, irrigation channels, et cetera. The significance of such participation is that it stimulates the soul of group work and the sense of one destiny, and helps people realize the benefits of involvement. The work consists of setting goals, preparing programs, implementing tasks and evaluating performance.
- **Evaluation Phase**: government organizations, aid agencies, technical experts and insurance companies examine the situation and estimate the requirements for the medium- and long-terms.
- **Assessing the Losses**: this phase incorporates the evaluation of losses and whether the nation needs urgent financial, food or medical aids to support implementing disaster recovery plans.

Based on the time the effects of a disaster take before recovery, disasters have two types of effect on local economy [5]:

- **Direct Consequences**: for example, damage to infrastructure, housing, crops.
- **Indirect Consequences**: for example, more unemployed personals, loss of revenues, destabilization of prices.

In the context of "social support" during disasters, Hurlbert et al. [20] studied the activation of network ties in the preparation and recovery phase of Hurricane Andrew 1992, to determine how network structures allocate resources for activation.

Based on the study results, people who exist in higher-density networks would activate network links for informal support much more than smaller systems. The same thing is true for networks with more gender diversity and networks with higher proportions of men, younger individuals and kin. A system consisting of men only would decrease the ties activated.

12.1.9 BELIEF ISSUES AFTER DISASTER

In the aftermath of a disaster, the notion of God and belief can be severely shaken, especially if the event leads to the death of some people [12]. All assumptions about faith, order, safety, goodness, fairness become at risk.

Although some of these impressions are present during the disaster, they are more persistent and more noticeable in the aftermath of a disaster.

People start looking for reasons of why they lost their beloved ones, usually from the perspective of religion or science. They ask people whom they trust, and religion men, about the rationale behind what happened: "Is this fair," "where was God when that happened," as well as other questions. If they do not receive good answers to their questions and there was no way to rationalize the matter in the shadow of God's will and wisdom, they may reconsider their current beliefs.

12.2 INFORMATION DURING DISASTERS

Just like food, water, medicine, shelter and other immediate requirements, disaster survivors need information about what is happening, where are their missing relatives and friends, how can they get medication, how can they tell government agencies and aid organizations about their urgent needs, how to involve in reconstruction works and so on. This behavior has noticeable impact on promoting transparency, trust and accountability.

Government organizations as well as aid agencies are also eager to obtain critical information so that they can begin search and rescue efforts.

The deterioration of communication infrastructures during the first several days after a disaster often results in: minimal or no electrical power, degraded or overwhelmed telephony services, minimal or no radio interoperability, overwhelmed satellite phone services, and limited Internet access [21].

The extensive use of Information and Communication Technologies (ICT), supported by high Internet speeds, paved the way to obtain critical information during mass emergencies. Phones, short messaging systems, radio, data sharing, email, social media and Geographical Information Systems (GIS) information, are some of the relevant ICT applications and tools that are important to responders and rescuers during disaster time.

Nevertheless, ICT devices for a typical use by responders should conform to a number of conditions: be small and lightweight, commercially available, non-military grade, energy independent and flexible [21].

Nevertheless, what is more important than having information is the dissemination of information throughout the communities at danger: before, during (if possible) and after the disaster. We should know how to put the information that we have into a practical form called "knowledge."

As expressed by the disaster expert "Ben Wisner," information in disasters has the following hierarchy of quality:[5]

- **Data**: raw information such as the statistical data about wind speed, rainfall intensity and death tolls. Information in this form is stored as-is, for example, without any processing.
- **Information**: organized data that are usually extracted from raw data and is meaningful.
- **Knowledge**: knowledge is accepted facts that come from the combination of acquired information and understanding. For example, when a medical doctor diagnoses a disease, he/she can prescribe the appropriate medicine based on the knowledge that he has.
- **Wisdom**: it is the ability to make judgments based on the exchange of information through dialog or multiple channels of communication.

Warning information must be timely enough and accurate to achieve their purposed objectives [5]:

First: Saving lives. The point here is to put people and technology at the center of warning systems. Real examples include:

- During the tsunami, and in the absence of official warnings, afflicted people used mobile phones to warn each other, assisted by civil society networks and online websites.
- In Singapore, when one of the people heard about the tsunami on the radio early in the morning, he gave a warning to 3,630 villagers on the eastern coast of India.

- During the 2004 Caribbean hurricane season, most countries were able to save lives by alerting people of the approaching storms.
- In Jamaica, Red Cross volunteers walk into streets and issue warnings, 48 hours before a periodic hurricane strikes.
- In Cuba, schools give lessons on disaster awareness. Also, evacuation drills are organized every year before the hurricane season.

Second: In addition to saving lives, information can reduce people suffering. Information about lost family members and lost friends, why the disaster happened, where people are going to stay or how much they are going to be paid under the provision of compensation act can greatly improve the morals of the homeless and traumatized.

For example, during Indonesia tsunami, Red Cross volunteers helped 3,400 survivors in Aceh district to meet their families.

Third: Information during disaster can help comply with the first two provisions by verifying that disaster relief is appropriate and well distributed. Assessing what is important against what is not can save precious time, money and resources.

12.2.1 EARLY WARNING SYSTEMS

Even though the number of yearly disasters has tripled since 1970s, Early Warning Systems (EWS) have been playing an important role decreasing death toll [3]. EWS aim at establishing community-based warning systems that operate during disasters to provide detailed local information.

These systems are not all-in-one products. Rather, they are being developed, separately, to provide information in situations like earthquakes, volcanoes, hurricanes, fires, dust storms, floods, drought and others [22].

EWS make use of today's widely-used technologies [3]:

- The Internet and mobile phones that opened up new potentialities to run early warning systems.
- Early warning messages that notify government organizations and aid agencies of possible hazards to take actions before a hazard deteriorates into a disaster.
- Radios. Although radios are old-fashioned method of communication, they are still important for women who work at homes or for

poor people. When powered with wind or solar power, listeners can use radios without the need to have electricity or batteries.

There is a need to establish these systems at a global level in order to avoid the disasters that traverse the borders of one country.

In fact, EWS are not just an assembly of technical networks (connections and devices) to deliver warnings before a disaster strikes. Rather, they incorporate other essential components such as political and cultural processes, ongoing crowd-sourced data and analysis, and broader public health strategies [22].

Although enough scientific information about the tsunami waves in the Indian Ocean was available, the absence of early warning systems was responsible for the rise in the number of deaths. Therefore, early warning information should be accurate, timely and credible.

However, the adoption of new technologies to use in disaster time faces a number of obstacles:

- Some technologies are still not broadly accepted, due to the high cost of development. However, aid agencies can sponsor the access to expensive technologies.
- Developing systems that link to a large portion of phones and radios is still far from reality for many regions.
- Language barrier is still one of the problems. One way to deal with it is to build websites using local languages.

12.2.2 SOCIAL MEDIA FOR DISASTER RISK MANAGEMENT

Without any doubt, social networking has become the most prominent activity for people working online, with a total consumption reaching 82% of the world's Internet use.

In 60 seconds: "168" million emails are sent, "694, 445" Google searches are performed, "695,000" Facebook statuses are updated, "98,000" tweets are added, "370,000" Skype calls are made, "6,600" Flicker pictures are uploaded, "1,500" new blog entries are published, and more than 600 YouTube videos are uploaded.

Social media refers to a set of internet-based tools that provide venues for social interactions to take place through many-to-many communications.

Examples of these tools include Twitter, Facebook, Google Maps and Flicker as well as social software packages like Ushahidi [23].

Social applications and services dramatically play a pivotal role in disaster response and recovery through providing response information before, during and after a disaster. For instance, social network applications have been used to gather information for the 2010 Haiti Earthquake, 2009 Oklahoma Grassfires and 2008 Sichuan earthquake in China. They also assisted during and the aftermath of the Japanese earthquake and subsequent tsunami.

In the aftermath of the Indian Ocean tsunamis, aid agencies used blogging websites more efficiently. They posted detailed information about the disaster like the death toll, the agencies that are ready to provide help, the names of the missing people, the types of assistance needed, the afflicted areas, and so on.

In the next two sub-sections, we will detail the talk on the use of Twitter, a microblogging service where people can tweet messages of no more than 140 characters. We will also detail on two social software applications and how they have been used in disaster time.

12.2.2.1   Social Software

Social software systems help establishing communication and collaboration between social entities for the sake of dissemination of information [24].

Different groups and organizations have been using social media to disseminate information to the public during disasters. The primary fields of implementation include:[23]

- **Fire**: Several tools (like Twitter) have been used by fire departments in the USA to disseminate and manage information during the different phases of fire emergencies. However, specific tools are more appropriate to use than others for specific phases. For example, the "Blog Talk Radio" tool is more helpful during the recovery stage.
- **Weather Disaster**: Blog posts have been used to increase public awareness against potentially severe weather conditions by sending alerts or warnings via Facebook and Twitter. The notifications make use of the services provided by the National Weather Service iNWS.

- **Earthquakes**: Real-time monitoring, estimation of the magnitude of ground shaking, rapid estimates of fatalities and economic damage and other measures are used by the National Earthquake Information Center (NEIC) to detect and alert against earthquakes. The process includes sending Twitter alerts to relief organizations, financial institutes, government agencies, media and general public.
- **Health Emergency**: Health-related organizations use social media applications during health crises to disseminate information. One of the tools is Twitter handle @CDCemergency, which became critical during the 2009 Salmonella Typhimurium outbreak.

Social network applications have been useful during emergency management and post-disaster recovery [25]. For example, Ushahidi Social Network Trends Map and Google Person Finder have assisted during and after the 2011 Japanese earthquake and tsunami by:

- provision of information about missing citizens;
- dissemination of information between governmental agencies and the citizenry.

"Ushahidi Social Network Trends Map" is a software developed by Japanese computer experts and Tufts University in USA. The victims who used the software in the disaster were able to upload critical information such as their locations and how can others find water, food and shelter by using phones or other mobile devices.

"Google Person Finder" is social network software developed by Google to help victims and families search for missing or displaced people. It was first used during the 2010 Haiti earthquake and was again successfully used in the aftermath of the Japanese earthquake to follow up with collation and record keeping of the missing, the displaced and the dead. The software is available in a number of languages including English, Japanese and Chinese.

Although some organizations have been using social media for disaster management, a full adoption is still facing a number of barriers [23]:

- **Limited Understanding**: Many emergency organizations are still unaware of the merit of social media as a tool for warning and alert. As a result, they prefer to use the traditional media tools (e.g., TV and radio).

- **Loss of Control**: Social media data is difficult to control. As a result, officials have the problem of evaluating the validity and reliability of these data if compared to traditional data sources.
- **Institutional Limitations**: Adoption of new tools and technologies or investing in the training of individuals requires significant budgets and lot of funding.
- **Standard Message Formats**: There is a need to put standard formats for the messages sent during disasters. The format should be machine-readable and compatible with the essential services. Successful examples include the recently adopted protocol: the Common Alerting Protocol (CAP).
- **Capability for Authorizing Users**: A distributed identity management system is required to ensure that only authorized individuals can issue official messages.
- **Authentication**: Digital signatures are necessary for the authenticity of messages.
- **Warning and Alerting Overload**: Limiting the transmission of messages to the area targeted protects the public from messaging overload and ensures that only intended people receive the messages.

12.2.2.2 Twitter

Launched in 2007, the service allows users to tweet about any topic with no more than 140 characters via the Internet-enabled devices.

Besides being a social network, Twitter is also a full-spectrum media ecosystem that can be used in disasters to provide services such as help coordinating rescue efforts, finding safe places, food, water, etc. For example, during the Haiti earthquake in January 2010, the Haitians who were trapped under debris were able to send text messages via social media channels like Facebook and Twitter, asking for help.

A number of reasons are behind the popular use of Twitter during disasters, even though there are many other tools [26]:

- Unlike Facebook, Myspace and other prominent social media tools, the relationships in Twitter are not compulsorily reciprocal. Therefore, Twitter messages (a.k.a. tweets) can be accessed by anyone (not only friends) via following hashtags. Moreover, being a follower means that the user receives all the messages of the user being followed.

- The simplicity of Twitter messaging system, which allows to send tweets with no more than 140 characters, makes the service suitable for use in different occasions.

12.3   AIRPLANE ACCIDENTS

Broadly, an accident is an unexpected, uncontrollable and unplanned event [4]. It usually results in personal injury or damage to property or equipment.

The airplane flight is divided into nine distinct phases [27]: taxi→ take-off→ initial climb→ climb→ cruise→ descent→ initial approach→ final approach→ and landing. Flight accidents can occur during any of the nine phases, from plane take-off to aircraft landing.

Takeoff and initial climb are probably the most dangerous stages of the flight because mechanical failures and malfunctions pose hazard when the airplane is moving fast or flying close to the ground because pilots do not have much time to deal with emerging problems [28].

Initial approach, final approach and landing are relatively hazardous phases. As with takeoff and initial climb, the airplane is also close to the ground leaving pilots with little time to react to mechanical errors or weather problems.

The climb, cruise and initial descent stages are substantially less risky, if compared to stages like takeoff and initial climb. Bombs, if placed inside the airplane, can explode during the cruise phase. The crashes of the Air India B-747 in 1985 and the Pan Am B-747 in 1988 were because of bombs placed inside the cargo hold and the luggage compartment of the two planes, respectively. The two accidents, together, led to the death of 599 people.

The last phase of flight, which is aircraft taxi, is relatively safe. In this phase, the airplane taxis to the gate where passengers deplane; and baggage and cargo are unloaded. A bomb explosion happened in 1985, when a baggage was being unloaded from one aircraft at the U.S. Dallas-Ff. The accident caused no deaths.

Although the average number of people who die every year due to plane crashes is 500, flying is still one of the safest transportation means, if we put in the mind the total number of daily flights and the total number of people traveling every day which is estimated by 3 million [29].

Two reasons were behind the gradual improvement in air accident rates [30]:

- The new regulations and rules imposed by airplane manufacturers which stipulated that (1) airplane designs must be approved by the local government, (2) pilots are needed to take further training, (3) accurate weather forecasts must be provided, and (4) new systems of traffic controllers, as well as radio beacons and radars, must be established.
- The efforts of aviation accident investigators over the past decades who sought to discover the reasons behind crashes and make sure no similar errors happen in the future.

During the period 1959–2001, there were 1,307 airplane accidents, resulted in 24,700 on board fatalities. They fall under the following three types of operations [27]:

- Passenger operations: 1,033 accidents (79%);
- Cargo operations: 169 accidents (13%);
- Testing, training, demonstration or ferrying: 105 accidents (8%).

Due to the importance of air transportation safety and trustability, various measures have been taken worldwide. For example, the U.S. Air Commerce Act, which became valid in 1926, emphasized the importance of examination and licensing of pilots and airplanes, proper accident investigations, and proper safety rules [27].

Most big aircraft manufacturers are located in USA, Canada, England, France, Russia and Brazil. Besides, each craft has its design, model, and idiosyncrasies that require special training and unique skills [30].

Boeing 777ER, the workhorse of modern commercial planes, is possibly one of the safest and most popular of all major airliners. The model was proposed to fill the niche between 300 and 500 seats, to respond to the increasing market demand, and to be an efficient market solution for any range. The 777 model offered cabin flexibility, range capability and multiple sizes of the same aircraft. This family offers a range of models that share the same flight crew type rating to help reduce costs of recurrence training, pilot pool size and transition training.

Apart from the September 11, 2001 attacks, where nearly 3,000 people died, there are nearly 20 aircraft incidents with a total number of deaths equals 250 or more. In addition, two other incidents have resulted in a total death of 500 or more.

12.3.1 CLASSIFICATION OF AIRPLANE ACCIDENTS

One of the principal sources of data related to aircraft accidents is the National Transportation Safety Board (NTSB) database. The database provides accident data for the three types of air operations (scheduled, non-scheduled and general) and gives statistics related to the number of fatalities, injuries, as well as descriptions of the aircraft involved and circumstances of accidents.

BOX 12.3. Notable Airplane Accidents

The first fatal airplane crash took place in Virginia during a demonstration flight by the USA Army, in 1908. The first sad event resulted in the death of one passenger, Thomas Selfridge, and the injury of another passenger, Orville Wright.

- **1937**: On July 24–1897, Amelia Earhart, who was the first female to fly alone across the Atlantic, disappeared in 1937, over the central Pacific Ocean when attempting to circumnavigate the globe. A campaign led by the Navy and Coast Guard was launched to find the plane that was supposed to reach Howland Island. After 19 days of continues search, no trace of Amelia, Fred Noonan or the plane wreckage was found. The campaign was the most costly in the USA history at that time. Several theories tried to explain the disappearance, but none of them was completely successful. The unresolved conditions of Earhart's disappearance tempted some people to produce a number of movies talking about Amelia's life and her last flight. The disappearance tempted also others to sing for her, write poems and books, and perform plays.
- **1977**: a Pan Am 747 collided with a KLM Boeing 747 on the runway at Tenerife Airport, Spain. A pilot's mistake (communication

misunderstanding) was considered the primary cause. The crash is regarded one of the worst accidents in the twentieth century as it led to the death of 583 people.

- **1985**: Boeing 747 crash into Mount Osutaka in Japan was the worst single-plane disaster, as it led to the death of 520 people. The airplane suffered from an explosive decomposition due to an incorrectly repaired aft pressure bulkhead.

- **2001**: on September 11, The Twin Tower of the World Trade Centre in New York City was targeted by two hijacked 767s airplanes: United Air Lines Flight 175 and American Airlines Flight 11 causing the two buildings to collapse. A third hijacked plane, American Flight 77, targeted the Pentagon in Washington, D.C. while a fourth hijacked aircraft, United Flight 93, crashed in a field in Pennsylvania. The four accidents led to the death of about 3,000.

- **2014**: Malaysian Airlines Flight MH370, that disappeared on March 8, 2014, remains one of the world puzzles.

Investigators have found that when using the sequence-initiating causal factor approach (described later), each accident should belong to one of the following categories [28]:

1. **Equipment**: mechanical or electrical failures that lead to accidents.
2. **Seatbelt Not Fastened**: A common source of passenger injuries like broken ankles, broken legs; and head and neck injuries.
3. **Environment**: A very common factor in many airline accidents. However, weather is not regarded as the cause of the accident. Rather, the cockpit crew when not sufficiently responding to weather conditions is considered the reason. Apart from weather conditions, animal strikes are common in general aviation. Such attacks often involve birds, deer, cattle, dogs, rabbits, water buffalo, buzzards, sea gulls and geese.
4. **Pilot**: Errors such as trying to land without lowering the landing gear or taxing into a stationary object.
5. **Air Traffic Control**: standard actions that should have been taken by air traffic control to prevent an accident.

6. **Ground Crew**: mistakes made by ground crews and lead to accidents, such as the one happened in October 1978 at O'Hare, when a service truck collided with a parking American Airlines airplane.
7. **Other Airplane**: accident that happen when an airplane collides with another. This definition does not necessarily imply that the other plane was at fault. Rather, the presence in that place initiated the accident or alternatively the accident would not have happened.
8. **Company Operations**: A consistently incorrect operation or procedure on the part of the company owning or operating an aircraft, or repeated improper actions by company personnel.

12.3.2 AIRPLANE ACCIDENT INVESTIGATION

Usually, when an airplane gets in trouble while in the sky, pilots communicate with air traffic controller and issue "Mayday" signals, which means that the airplane is showing one of the following features: engine failure, fire, severe weather, hijack, loss of control or any other symptom showing anomaly in plane behavior [30].

Airplane accident investigation is the examination of possible causal factors. Such investigations can help determine accountability, put plans for better pilot training and prevent similar accidents in the future [31].

The following are some typical materials (factual data) that experts need to examine, occasionally associated with security reports, during investigations [32]:

- Cockpit voice recorder (CRB);
- Communications Photographs of the accident site and the wreckage;
- Visual flight rules (VFR) charts or instrument flight rules (IFR) charts, if appropriate;
- Aircraft information;
- Pilot Operating Handbook (POH);
- Weight and balance information;
- Airframe, engine, propeller (if possible);
- Logbooks, pilot/crew information;
- FAA pilot certificates and medical records;
- Pilot training records;
- Weather information;

- Weather reports and forecasts;
- Testimony;
- Statements of eye or ear witness;
- Opposing expert reports;
- Depositions taken by attorneys on all sides.

Airplane accident investigation goes through a number of steps: collection of data, data analysis, knowledge building, and conclusions.

Studying the accidents of other pilots has become an important part of aviation education, especially in the case of inexperienced aviators. For example, in Boeing aircraft accidents, the airline sends a team of engineers and specialists in structures, operations, systems and data recorder. Depending on the severity of the accident, the company may even call other skills. Their aim is to arrive at a speedy conclusion of the crash reason(s) and to make a number of suggestions that can help prevent similar crashes in the future [30].

When investigators conduct accident investigation: they strive to know exactly what had happened and why it happened that way. They probe the possible causal factors (with one of them being the primary), the sequence of events, what must be done to prevent similar accidents, etc. The major cause is the one most responsible for the accident or alternatively the accident would not have happened. It can be a man, material, machine, methods, environment or management system.

After collecting data, investigators establish a detailed systematic search approach that has the following five major steps [6]:

1. Build the hypothesis, which is a tentative assumption made to draw out and test logical and empirical results;
2. Perform the analysis, which is the use of methods and techniques to arrange facts. It has the following benefits:
   - assist making decisions about what additional facts are necessary for the investigation;
   - build consistency, logic and validity;
   - develop necessary causal events;
   - guide inferences and judgments.
3. Draw inferences by moving from one fact to another
4. Make judgment, which is the process of forming an opinion or evaluating a proposition

5. Make recommendations that include specific methods and corrective actions that are believed to be correct, feasible, practical and sufficient. More elaboration on the investigation steps are found in Figure 12.4:

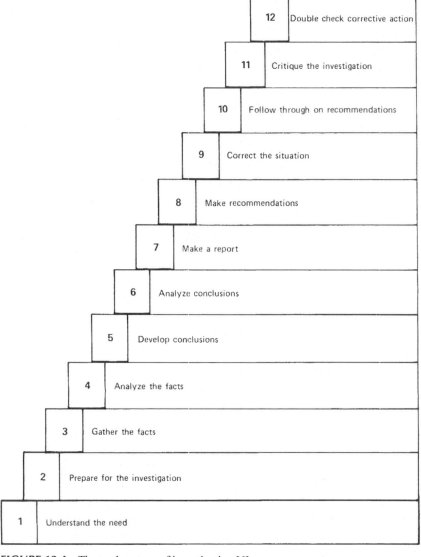

FIGURE 12.4 The twelve steps of investigation [6].

Accidents ought to be investigated in a timely manner, immediately after the event. However, it is not always a straightforward process. For example, in 1994, U.S. Air Flight 427, Boeing 737–3B7, which was supposed to land at Pittsburgh International Airport, crashed in West Palm Beach, Florida, and killed 132 on board. The investigations followed the accident were among the longest in aviation history: they lasted for four and a half years.

A noticed pattern in aviation accidents is that there is no single reason behind the accident. Rather, it is a chain of reasons that cause the accident [31]. In other words: if the chain is broken (by eliminating one or more causes) the accident may not take place.

Flight and cockpit voice recorders (a.k.a. black boxes, even though they are painted bright orange for better recognition during recovery) carried in the tail of the aircraft, provide critical information (e.g., the crew's voices, communications with air traffic control or other communications in the plane, weather briefings, etc.) to help solving the mysteries of airplane crashes. They have assisted in tens of Boeing airplane crashes throughout the world.

But if the aircraft wreckage is not found (like the case of Flight MH370), investigators cannot make use of the information stored in the black box and need to build speculations.

It was found that approximately 70% of air carrier accidents are due to the flight crew errors such as loss of situational awareness, fatigue or lack of training. Maintenance error accounts for 5%, while air traffic control and other airport-related issues are responsible for 4%. The remaining 20% of total accidents are due to mechanical failures, weather factors and miscellaneous causes [31].

Broadly, plane crashes occur because of the following reasons [30]:

- **Human Mistakes**: Actions unintentionally undertaken by pilots (due to fatigue, lack of training or incorrect airplane computer programming), and are shown later to be mistakes [33]. Pilot error has been the main reason for many airplane crashes (see Figure 12.5 for types of human error). Other human mistakes that lead to air accidents include: the mechanics forget fixing problems, airport controllers give wrong directions, two airplanes collide in mid-air, etc.

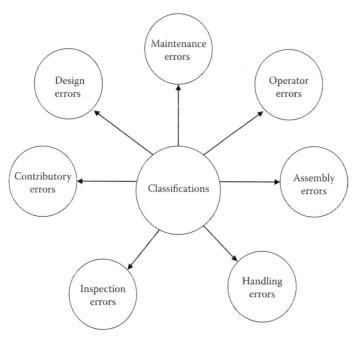

FIGURE 12.5 Common human error classifications [27].

A report issued by the National Aeronautics and Space Administration (NASA) stated that since the introduction of turbojet aircraft late 1950s, over 70% or airline accidents involved, at different levels, human factors [27]. Similar results emerged from the Boeing Company, the Daily Science journal, NASA and other bodies [27].

An example of pilot mistake leading to airplane accident: In 1979, the Allegheny Flight 561 that departed Benedum Airport, Clarksburg, West Virginia, with 25 persons on board, crashed. The accident led to the death of two among 25 persons on board in addition to eight serious injuries. An aircraft accident report issued in 1979 by the National Transportation Safety Board stated the following in regard to the crash of Flight 561:

"The Captain's decision to take off with snow and ice on the aircraft's wing and empennage surfaces which resulted in a loss of lift as the plane ascended out of ground effect. The pilot was not motivated to act as a professional and take the proper measures to ensure that the wings,

stabilizing surface, and control surfaces are clean and free of ice, snow, or frost before he attempts a take-off." [34].

- **Non-Human Mistakes**: airplanes also crash because of tank explosions, lightning strikes, ice forming on the airplane wings, fires, design errors, engine explosions, severe weather conditions, birds, empty fuel tanks or system failures (the malfunction of any part of aircraft system used to store or process information). Bombs, suicidal attempts, plane hijack and so on are also behind crashes.

12.3.3   CAUSAL FACTORS DISTRIBUTION ANALYSIS

Investigators in this type of analysis first classify accidents according to their cause and then compare the distribution of causes both over time and across segments of industry where safety performance is worse than elsewhere [28]. This approach builds upon the sequence-initiating cause in which data classification problems are avoidable.

As mentioned earlier, a single accident can have several causal factors. A major problem is how to assign a single cause to one accident? In this regard, researchers have been applying the following three principal approaches:

1.   Select the object that initiated the sequence of events.
2.   Select as the cause the object the last point at which the accident could not have happened.
3.   Select as the cause the two causes from the two previous points.

Let's take an example to see how each of the three approaches is applied in a different way [28]. Suppose an airplane engine fails during takeoff and the pilot was unable to take the proper action and make the plane land safely: according to second and third approaches, pilot error would show up as the cause, because in many of the accidents that are initiated by equipment failure, the pilot is able to avoid the accident by the flying skills. However, if we consider the first approach as the accident cause, then pilot error will show up as the cause.

The availability of accident databases (such as the NTSB) provides investigators with a baseline to assess weather subsequent occurrences are the result of known problems or they were just circumstantial.

12.3.4   MULTI-LINEAR EVENTS SEQUENCING ANALYSIS

To better show the sequence of events of any accident, we can use the Multi-linear Events Sequencing (MES) approach that takes into account the order of events on a time-line basis [6]. The model comprehends all the factors that may have role in the happening of an accident. Hence, it provides investigators with a way to evaluate the clues and understand the reasoning process.

The model was developed by Benner and is used by the National Transportation Safety Board (NTSB) as one of the analytical tools in mishap investigations, especially these involving hazardous cargoes.

The NTSB, which is one of two air-safety regulatory bodies in the U.S. besides the Federal Aviation Administration (FAA), is an independent agency with the aim of investigating every civil aviation accident in the U.S., as well as other significant transportation accidents. This Federal body focuses on carrying out studies related to transportation safety and assistance coordination during major transportation disasters.

A successful accident investigation process relies on a careful realization of when the accident began and ended up. Otherwise, investigators cannot develop the required corrective actions to prevent similar accidents.

Timeline is a significant scale that parallels the sequence of events and shows a time relationship between the events of an accident. It helps investigators to identify all possible causal factors. In Figure 12.6 below,

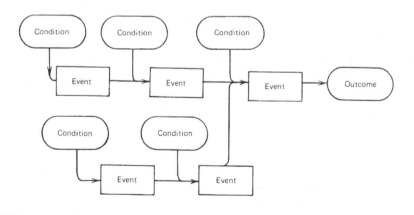

FIGURE 12.6   Activity events and outcomes for two actors and including conditions [6].

the conditions that influence the events can be inserted in the time flow in a logical order to show the flow of relationships.

When an accident happens, it is usually a series of events that are either sequential or parallel. The MES ensures the breaking of each sequence of events into single and precisely described events. Within MES processes, investigators consider the actions of each actor who (or which) bring about a change of state in the course. Events are posted in sequence so that each action leads to the next one.

If there is a gap in the understanding of an event, then there is a high probability that the event to the right has not been properly broken down and explained. In other words, we have not delved far enough into the sequence of events, actors or actions involved.

12.3.5   FAULT TREE ANALYSIS (FTA)

The fault tree analysis method was developed first at Bell Telephone Laboratories and refined later by the Boeing Company. Finally, it was adopted by the U.S. Department of Defense.

The tool was used to analyze the problems with the Minuteman missile launch control system [6]. It was also used as a preventive or failure analysis device (most probably this technique or one of its variations is currently being used in the investigations of Flight MH370 disappearance). The method is widely used in industry (e.g., in nuclear power facilities) to evaluate the reliability of engineering systems during the design and the development phases. Developers have also used the system for sophisticated aerospace systems, since it can deal with complex processes through models and charts.

A fault tree is a logic-based diagram that shows the relationships of cause and effect between likely events. The tool allows understanding where the problems are and how to produce a logical sequence of causal events.

The use of Boolean algebra and logic gates (e.g., AND or OR) within computer programs allows for applying the fault tree to process large amount of information and calculate the probability of occurrence of events accounting for the final event (see Figure 12.7). Investigators may need to build several fault trees to deal with for complex investigations.

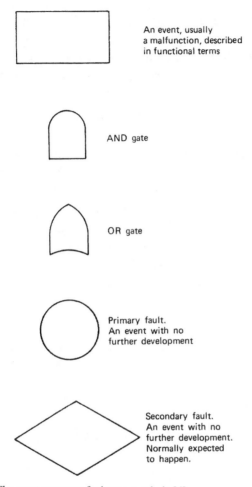

FIGURE 12.7   The most common fault tree symbols [6].

As mentioned earlier, the functionality of the tree incorporates the use of AND and OR. If all the events listed under some event are needed to make the next event possible, then the AND gate is used. Otherwise, the OR gate is used.

To find the occurrence probability of the OR gate output fault event [27]:

$$P(A) = 1 - \prod_{i=1}^{m} \{1 - P(A_i)\} \qquad (1.1)$$

where $P(A)$ = occurrence probability of the OR gate output fault event A; m = number of OR gate input fault events; $P(A_i)$ = probability of occurrence of the OR gate input fault event A_i, for $i = 1,2,3, ..., m$.

Also, to measure the occurrence probability of the AND gate output fault event:

$$P(B) = \prod_{i=1}^{k} P(B_i)$$ (1.2)

where $P(B)$ = occurrence probability of the AND gate output fault event B; k = number of AND gate input fault events; $P(B_i)$ = probability of the occurrence of the AND gate input fault event B_i, for $i = 1,2,3, ..., k$.

The system failure investigation approach understands that it does not deal merely with mechanical, electrical or hydraulic failures. Rather, any failure in environment, personnel or equipment impacts other components and therefore the three elements should be combined and structured to carry out certain functions.

Nevertheless, a number of conditions must be satisfied before FTA can be applied successfully [27]:

- deep understanding of all aspects of the system under consideration;
- clearly defined objectives and scope of analysis;
- clear definition of what constitutes system/item failure;
- clearly defined system and item physical bounds; and system and item interfaces;
- precise identification of associated assumptions;
- a comprehensive review of system and item operational experience.

12.3.6   SOCIAL NETWORK ANALYSIS

While the previous accident analysis techniques focus on collecting and studying tangible materials to draw conclusions; social network analysis SNA, on the other hand, focuses on investigating the nontangible person-person and group-group relationships and the analysis of the structural properties of social networks [35].

Social network analysis is a research approach that aims at studying the structures and processes of networks. Hence, it focuses on the relationships between people and the syntheses established through interactions, instead of the attributes of individuals such as age, gender, occupation etc.

SNA can quantify these relationships through the use of graphical representations, where nodes represent individual units and ties represent relationships between individuals [36].

It emerged from the efforts of three scientific groups: sociometric analysts, who worked on small groups and introduced important improvements to the field of graph theory; Harvard researchers of the 1930s, who discovered patterns of interpersonal relationships and clique formations; and Manchester anthropologists, who investigated the structure of community relationships in tribal and village societies.

It is a powerful technique in the study of: crime and cybercrime, politics, social movements, religious networks, group formation, consulting management, public health, disease spread, business applications, etc.

"Graph Theory" forms the backbone of all social network analysis applications. Therefore, it is common to encounter expressions like actors, ties, centrality measures, density, reachability, cohesion, brokerage roles, structural holes and others, which are basically borrowed from discrete mathematics and combined with notations from sociology.

12.4 FLIGHT MH370 PROFILE

Malaysian Airlines has a bright history of 40 years of flight [29]. During the 1990s, it contracted to buy '11' Boeing 777–200 planes and '4' planes from the larger size, Boeing 777–300 [37]. The 200 planes come with twin Rolls-Royce engines and can fly more than 11000 kilometers.

Based on the contract signed with the U.S.-based Boeing Company, Malaysian Airlines received in April 23–1997, the first plane of that batch, which was the 777–200ER IGW (IGW stands for Increased Gross Weight), a variant of the 777–200 that has a greater range and payload capability.

On March 8, 2014, Flight MH370 left Kuala Lumpur International Airport at 12.41am on a scheduled flight to Beijing Capital International Airport. The line of flight passes through a number of countries: Vietnam, Cambodia, Thailand, and Laos. There was nothing unusual during the flight but after 40 minutes, the airplane disappeared while over the South China Sea.

At the airport, relatives and friends gathered in the arrival hall to receive the beloved ones, who would never be there again.

The passengers came from a range of places: China (153), Malaysia (38), Indonesia (7), Australia (6), India (5), France (4), United States 3 (including two toddlers), Ukraine (2), New Zealand (2), Canada (2), Italy (1), Russia (1), Holland (1), Taiwan (1) and Austria (1). Although there were no Britons among the passengers, two travelers from China had been studying in England: Yue Wenchao, who was studying at Hull University and Li Yuchen, a PhD holder who recently graduated from Cambridge University.

No one knows what happened to the airplane or the 239 passengers: Are they still alive but captured? Have they died due to an explosion, suffocation, or crashing into the sea, mountain or flat ground? However, what is known is that the crew was well-trained and the pilot, Zaharie Shah, had 33 years of service with Malaysian Airlines and more than 18,000 flying hours. The co-pilot, Fariq bin Abdul Hamid, had worked with Malaysian Airlines since 2007 with about 2,800 flying hours.

Figure 12.8 shows the flight deck of Boeing 777–200, which is very similar to 777–200ER (the Malaysian airplane that went missing) in terms of basic dimensions, passenger capacity, cargo and typical cruise speed.

Trying to find the missing plane body, several countries have participated in the search and rescue efforts including Malaysia, Australia, U.S.,

FIGURE 12.8   Flight deck features color flat panel liquid crystal displays for the two pilot crew. electronic checklist increase pilot efficiency [37].

China, Singapore, Vietnam, Indonesia and the Philippines. Despite the use of modern technology and the heroic efforts: experts, scientists and engineers are still unable to solve the puzzle and perhaps it will take a longer time before they can find their missing body.

12.4.1 FLIGHT MH370 DISAPPEARANCE: THEORIES

On 17 July 2014, when the Malaysian Airlines Flight MH-17 was shot down near the Russian-Ukrainian borders, en route from Amsterdam to Kuala Lumpur, speculations emerged that Flight MH370, which disappeared earlier that same year, was also downed by a missile.

Malaysian Airlines posted on Twitter, during the very first days, that they were working hard with all parties to locate the aircraft, and that they understood peoples' concerns [29]. On the other side, some users on Twitter began to give false or hasty conclusions in terms of where the missing plane ended up. Some tweets said that it landed at Nanning Wuxu International Airport in southern China. Others claimed that the airplane went out of fuel or it was shot by a missile.

Conspiratorial thinking increases noticeably during war, earthquakes, outbreak of epidemics or when there are natural disasters such as tsunamis or earthquakes, which generates a social sense of uncertainties [19].

Many theories strived to explain the demise of Flight MH370: Some experts believe that the crew and passengers died from suffocation before crashing into the ocean. Some believe that the airplane was taken by force to a U.S. remote military base while others think that the plane was hit during a military operation near Thailand or was hijacked and flown to Afghanistan. Some tried to stick the disappearance to Bermuda Triangle, a place in North Atlantic where ships and planes were thought to disappear in the past, although recent science refuted the Bermuda Triangle claims.

12.4.2 FLIGHT MH370 DISASTER: CAUSAL FACTORS DISTRIBUTION ANALYSIS

Even though the "causal factors distribution analysis" explained in Section "12.3.3." is simple and promising, counting all contributing factors makes the interpretation of the distribution of causes difficult, especially in the case of the missing Flight, MH370, where the accident has many possible causes. Figure 12.9 below shows the assembly breakdown of Boeing 777.

12.4.3 FLIGHT MH370 DISASTER: MULTI-LINEAR EVENTS SEQUENCING ANALYSIS

If we want to apply the Multi-linear Events Sequencing analysis technique to the missing 777, we will end up with many unrealized events that leave us with huge gaps in terms of:

- **Events**: It is unknown exactly the number and the sequence of the events, and whether they occurred sequentially or in parallel. It is difficult to know exactly the first event and the last event in that sequence (examples of managerial causal factors are in Table 12.1)

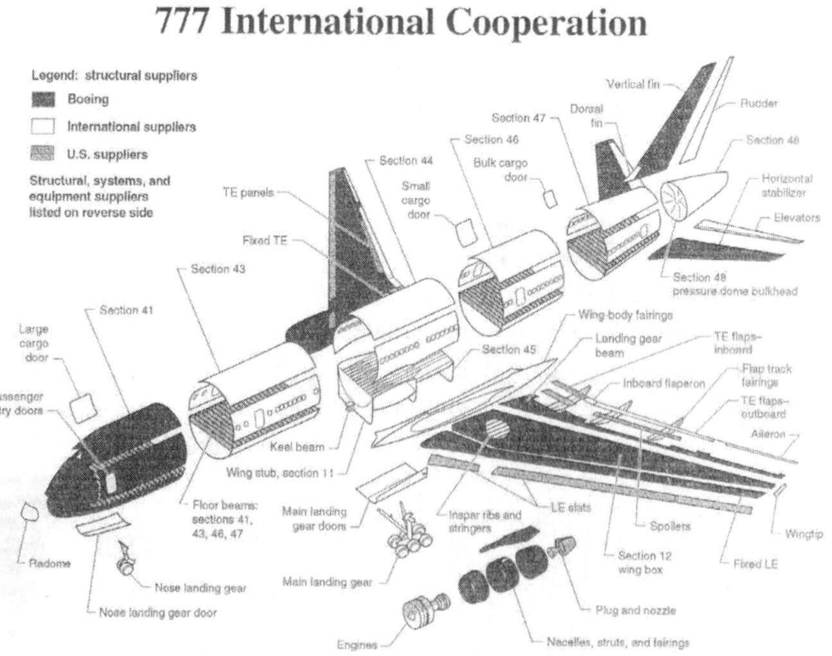

FIGURE 12.9   Assembly breakdown illustration of Boeing 777; structural suppliers are shown by the shading legend [37].

TABLE 12.1 Examples of Managerial Causal Factors

Item	Possible Underling Cause	Expected Consequences
1.	• Lack of skill and knowledge • Lack of motivation • Lack of proper procedures	Improper use of equipment, tools or facilities
2.	• Equipment and facilities are not recognized as unsafe/defective • Poor design/selection • Inadequate equipment	Unsafe use of equipment and facilities
3.	• Omissions • Error by designer or supervisor	Lack of proper procedure
4.	• Inadequate pilot training • Inadequate Airliner	Improvising unsafe procedures
5.	• Procedure unclear • Need not emphasized	Failure to following prescribed procedures from Air Traffic Controller
6.	• Complex instructions • Inadequate comprehension	Task not understood
7.	• Inadequate instructions • Inadequate warnings	Lack of awareness of expected involved hazards
8.	• Need not recognized • Deliberate act • Inadequate supply	• Lack of proper equipment, tools or facilities • Lack of guards and safety equipment • Lack of protective equipment
9.	• Inadequate warnings • Improper procedures • Inadequate instructions • Lack of comprehension • Deliberate act	Exceeding prescribed limits, strength, load and speed and so on
10.	• Inadequate comprehension • Lack of motivation	Neglecting of apparent safe practices
11.	• Excessive physical or mental needs	Fatigue, hypnosis or reduced alertness
12.	• Low morale and poor attitude • Misassignment	Deliberate failure to use protective materials

or the actors who (or which) brought about a change of state in the sequence (for example, the pilots, one or more of the passengers, system failure, weather conditions, etc.) Each of these events must be identified separately to explain the accident sequence.

- **Time**: the timeline of the events is incomplete. So far, we only know a few facts such as the time the airplane took off, the time it made its last call with the air traffic controller, the time it changed its flight direction, the time of its last location, etc.

Additionally, the idea of "event sequence" is not entirely appropriate (although it is used in most analytical techniques) for sophisticated investigations like the disappearance of Flight MH370, since many events and conditions may appear randomly over a considerable time span and do not lend themselves to exact sequencing [6].

The vagueness of the information obtained until now, together with the difficulty of having meaningful data to project the possible motivations and event genesis, impedes the structuring of the conceptual model.

Our MES chart can become very big and very complicated, if we want to incorporate all the details, because of the many events (the what-if chains) and the many conditions that we should process. Besides, we have an open-ended timeline.

12.4.4   FLIGHT MH370 DISASTER: FAULT TREE ANALYSIS

As said earlier, a fault tree is a diagram showing the relationships of cause and effect. A proper use of the fault tree for accident analysis requires extensive knowledge of the aircraft design, construction and how to make use of them during the analysis. It also requires an established knowledge of aviation methods and rules, which are currently unavailable and lie behind the scope of our experience and the objectives of this chapter.

Further, one limitation of the fault tree technique is that it is unable to handle human factors. Therefore, other approaches can be combined with this model, to consider the human factor as well. The key point here is that human beings are the subject to many more errors and more varied types of errors.

12.4.5   FLIGHT MH370 DISASTER: CROWD COPING BEHAVIOR ANALYSIS

Understanding interpersonal ties among group members of Flight MH370 as well as the sizes of the sub-networks are crucial to measure disaster response.

The afflicted 777 crowd included several groups of people. Some of them had, like in most flights, established rules (respect or obedience), predetermined leaders and leadership structure (elderly family members, family parents or group leaders) and formal, prearranged mechanisms for making decisions. Some of these groups (the Chinese artists and the Semiconductor engineers) gathered to participate in predetermined events (an art exhibition and a technical course, respectively) that attracted the crowd members.

As we stated earlier, many theories tried to explain the disappearance of the giant aircraft Boeing 777, such as the plane was hijacked by armed people, captured in a remote U.S. military base, downed by a missile, or kidnapped by aliens. In any case, the passengers either had enough time to show some counteraction (e.g., in the case of hijacking) or hardly had enough time to take any action (e.g., the passengers died because of suffocation, the assumption that is more acceptable until today or were forced to swallow a toxic or lethal drug which passes them at once into the narcotic state without any likely reaction).

12.4.6   FLIGHT MH370 DISASTER: CROWD SOCIAL NETWORK ANALYSIS

The passengers hailed from 14 countries. However, it was learned later that two people on the manifest, an Italian and an Austrian, were not onboard. Most of the 227 passengers were Chinese, while the 12 missing crew members were Malaysian. The rest of passengers were from France, United States, the Netherlands, Indonesia, Australia, the New Zealand, Ukraine, Canada, Russia and India. Some of those onboard were heading home while others just making a stopover. To some, it was only their first trip abroad.

News websites covered the disappearance of the Boeing 777 from different sides: telling the story of the missing aircraft, how it disappeared from radar, its possible location, background of the passengers, how the relatives of the passengers were dealing with the crisis and the efforts made by the international society to locate the missing plane.

The dataset of the passenger relationships contains 241 vertices and 1563 edges (excluding labels). It represents the known relationships between the flight passengers [38].

The dataset was collected during March to April 2014 from news websites (e.g., Yahoo! News Malaysia, CNN.com, The New Indian Express, the Economic Times, India Today and the Daily Express). It covers the period from when it was reported missing until end of April, when the big countries that participated in the search and rescue efforts declared that they were unable to find the airplane's body [38]. The data are in '.xls' format and is ready to download from the first author's website.

News agencies used three ways to collecting information about the missing airplane passengers:

- meeting families and friends, and visiting the places where they worked or lived;
- following up with information released by official channels;
- using archived data.

As shown in a previous study, there were three major groups on board: the "Artist Group," consisting of 29 members; the "Freescale Semiconductor," group consisting of 15 members; as well as the "Aircraft Crew" group, which consisted of 12 members [38].

In addition to the three larger groups, there was a number of smaller units: a family of six and a family of five from China, a family of four from Malaysia, a family from French, family from Australian and so on (Figure 12.10).

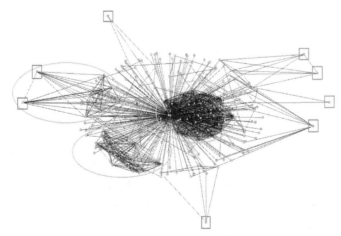

FIGURE 12.10   Visual representation of flight MH370 passenger network [38].

A visual representation of the passenger network can help imagine the different parts of the flight network. In the previous figure, we see three large parts (in circles): the largest one (giant component) is the Artists component (29 vertices). The other two significant parts are the Freescale Semiconductor group (14 nodes) and the Aircraft Crew (12 nodes). We also see a large number of groups with small size compared to the few big groups. In fact, this property was noticed previously in other types of networks (e.g., information, communication and biological networks.)

Apart from the large and small groups, there are eight peripheral nodes enclosed in boxes. These nodes are only connected to the main node 'Flight' and the rest of nodes in the network through the sub-network nodes. The peripheral nodes are:

- **Pilots**. Usually, pilots can only be reached through the rest of other crew members. In other words, they can be reached only through the ten other nodes in the network.
- **Toddlers**. Five toddlers were on board that can only be interacted with through family members.
- **Disabled**. On board, there was a 54 years old visually-impaired woman. The only people who were able to interact with the woman were her husband and two other people, friends to the family (See Table 12.2 to have a look at some of the overall Flight network statistics).

TABLE 12.2 Features of Flight MH370 Passenger Network

Metric	Value
Graph type	Undirected
Number of vertices (Dimension)	241
No. of edges	1563
No. of loops	5
Network density	0.05373530
Average degree	12.97095436
Connected components	1
Single-vertex connected components	0
Maximum vertices in a connected component	241
Longest shortest path	Between nodes 11 and 194

The number of components equals one, which means there are zero disconnected parts. This is because we are dealing with the aircraft system as one connected component, and hence all people can interact with all others, with some exceptions such as pilots, children, disabled, etc.

The network is undirected, which means that the links between people are symmetric. In this way, if a person 'A' interacts with person 'B,' then person 'B' also will communicate with a person 'A.'

The total number of edges is 1563 and of nodes (a.k.a. network dimension) is 241. We added two additional nodes to the initial number of nodes, which was 230: one node for connecting all passengers to a single network (because they are flying in the same plane) and the other node to group the individuals who had been working in Singapore.

The significance of network density, which is here 0.05373530, is that it shows important system characteristics such as the speed at which information diffuses among passengers and the levels of social capital.

The average degree measure is used to describe the structural cohesion of the network. For directed networks, it is the addition of all average degrees divided by the total number of nodes in the network. For undirected networks, the number of edges is calculated twice.

12.5   CONCLUSIONS

Since the data and facts, we are dealing with, are for human beings, most of our investigative processes in this work used deduction. Thus, the validity of our argument cannot be tested. It was indispensable to deal with likelihoods and probabilities, and our results can be challenged and accused of second-guessing or of drawing unfounded conclusions.

Our work involved inductive reasoning and therefore our results can be speculative. There are no general accepted methods for the investigation in our "case" and we have developed probable or acceptable conclusions instead of using the approvable ones. Apparently, our techniques and thinking, when dealing with human factors, must be substantially improved if to use certainty and credibility.

The use of social network analysis—a tool for studying structures of individuals—helped us throughout our analysis of the Flight MH370 passenger network. Our focus was on extracting important mesoscopic and macroscopic features such as network density, diameter, average degree, as well as how these features reflect organizational patterns of the network.

The unknown fate of flight MH370 has become one of the evidences that even when humans are armed with modern technology, they are still unable to solve all puzzles. We are still waiting for more clues to emerge towards solving the Flight MH370 mystery.

KEYWORDS

- **airplane accidents**
- **disasters**
- **flight MH370**
- **social media**
- **social network analysis**

REFERENCES

1. Wisner, B., Blaikie, P., Cannon, T., Davis, I. At Risk: Natural hazards, people's vulnerability and disasters, 2nd ed., London: Routledge, 2003.
2. De Silva, S., Burton, C. Building Resilient Communities: Risk Management and Response to Natural Disasters through Social Funds and Community-Driven Development Operations; Washington, DC: World Bank, 2008.
3. Concern Worldwide: Approaches to Disaster Risk Reduction; London: Concern Worldwide, 2005.
4. Reese, C. D. Accident/Incident Prevention Techniques, 2nd ed., Boca Raton, FL: CRC Press/Taylor & Francis, 2011.
5. IFRC World Disasters Report: Focus on Information in Disasters; Eurospan, London, UK: International Federation of Red Cross and Red Crescent Societies, 2005.
6. Ferry, T. S. Modern Accident Investigation and Analysis, 2nd ed., New York: John Wiley & Sons, 1988.
7. McDonald, R. Introduction to Natural and Man-made Disasters and their Effects on Buildings; London: Routledge, 2003.
8. Crisis and Emergency Management Theory and Practice, 2nd Ed. In Public Administration and Public Policy; Farazmand, A. Ed., Boca Raton, FL: CRC Press/Taylor & Francis, 2014.

9. Global Environment Outlook 3: Past, Present and Future Perspectives. United Nations Environment Program, Global Environment Outlook Series; London: Routledge, 2002.

10. Dynes, R. R. The Importance of Social Capital in Disaster Response. Preliminary Paper No 327. Newark, DE: Disaster Research Center, University of Delaware, 2002.

11. Perry, R. W., Lindell, M. K. Understanding Citizen Response to Disasters with Implications for Terrorism. Journal of Contingencies and Crisis Management,11(2), 49–60, 2003.

12. Victim Assistance: Exploring Individual Practice, Organizational Policy and Societal Response. In Springer Series on Family Violence; Underwood, T. L., Edmunds, C., Eds., New York: Springer Pub, 2003.

13. Mikami, S., Ikeda, K. Human Response to Disasters. International Journal of Mass Emergencies and Disasters, 3, 107–32, 1985.

14. Quarantelli, E. L. Human and Group Behavior in the Emergency Period of Disasters: Now and in the Future. Newark, DE: Disaster Research Center, University of Delaware, 1993.

15. Lai, J. W. M. Utilizing Facebook Application for Disaster Relief: Social Network Analysis of American Red Cross Cause Joiners. M. A. Thesis, University of South Florida, 2010.

16. Von Dawns, B., Fischbacher, U., Kirschbaum, C., Fehr, E., Heinrichs, M. The Social Dimension of Stress Reactivity: Acute Stress Increases Prosocial Behavior in Humans. Psychological Science, 23(6), 651–660, 2012.

17. Frey, B. S., Savage, D. A., Torgler, B. Behavior under Extreme Conditions: The Titanic Disaster. Journal of Economic Perspectives, 2011, 25(1), 209–22.

18. Quarantelli, E. L., Dynes R. R. Response to Social Crisis and Disaster. Annual Review of Sociology, 3(1), 23–49, 1977.

19. Encyclopedia of Group Processes and Intergroup Relations; Levine, J. M; Hogg, M., Eds., London: SAGE, 2009.

20. Hurlbert, J. S., Haines, V. A., Beggs, J. Core Networks and Tie Activation: What Kinds of Routine Networks Allocate Resources in Nonroutine Situations? American Sociological Review, 65, 4; ABI/INFORM Global, 598, 2000.

21. Nelson, C. B., Steckler, B. D., Stamberger, J. A. The Evolution of Hastily Formed Networks for Disaster Response. In Global Humanitarian Technology Conference (GHTC), IEEE, 2011.

22. Reducing Disaster: Early Warning Systems for Climate Change; Zommers, Z., Singh, A., Eds., Dordrecht: Springer Netherlands, 2014.

23. Committee on Public Response to Alerts and Warnings Using Social Media: Current Knowledge and Research Gaps. Public Response to Alerts and Warnings Using Social Media: Report of a Workshop on Current Knowledge and Research Gaps. Washington, D. C.: National Academies Press, 2013.

24. Schellong, A. Increasing Social Capital for Disaster Response through Social Networking Services (SNS) in Japanese Local Governments. National Centre for Digital Government No. 07–005, 2007.

25. American Society for Public Administration. The Use of Social Network Applications in the Japanese Earthquake and Tsunami Response Efforts. http://patimes.org/

the-use-of-social-network-applications-in-the-japanese-earthquake-and-tsunami-response-efforts/ (accessed October 30, 2014).

26. Hossmann, T., Legendre, F., Gunningberg, P., Rohner, C. Twitter in Disaster Mode: Opportunistic Communication and Distribution of Sensor Data in Emergencies. ExtremeCom '11 Proceedings of the 3rd Extreme Conference on Communication, SWID, 7: 1–7:8, 2011.

27. Dhillon, B. S. Transportation Systems Reliability and Safety. Boca Raton, FL: CRC Press/Taylor & Francis, 2011.

28. Oster, C. V., Strong, S., Zorn, C. K. Why airplanes crash: Aviation Safety in a Changing World. New York: Oxford University Press, 1992.

29. Cawthorne, N. Flight MH370: The Mystery. London: John Blake, 2014.

30. Byrne, G. Flight 427: Anatomy of an Air Disaster. New York: Copernicus Books, 2002.

31. Marchman III, J. F. Encyclopedia of Flight. Pasadena, CA: Salem Press, 2002.

32. Handbook of Aviation Human Factors, 2nd ed., Wise, J. A., Hopkin, V. D., Garland, D. J., Eds., Boca Raton, FL: CRC Press/Taylor & Francis, 2010.

33. Childs, D. R; Dietrich, S. Contingency Planning and Disaster Recovery: A Small Business Guide; New York: John Wiley & Sons, 2002.

34. National Transportation Safety Board (NTSB) Aircraft Accident Report-Allegheny Airlines, Inc., Nord 262, Mohawk/Frakes 298, 929824, Benedum Airport, Clarksburg, West Virginia, 1979.

35. Al-Taie, M., Kadry, S. Applying Social Network Analysis to Analyze a Web-Based Community. International Journal of Advanced Computer Science and Applications (IJACSA), Vol. 3, No.2, 2012.

36. Al-Taie, M., Kadry, S. Social Network Analysis: An Introduction with an Extensive Implementation to a Large-Scale Online Network Using Pajek; Sharjah, UAE: Bentham Science Publishers, 2014.

37. Upton, J. Boeing 777. In Airliner Tech Series, Volume 2; North Branch, MN: Specialty Press, 1998.

38. Al-Taie, M. Z., Shamsuddin, S. M., Ahmad, N. B. Flight MH370 Community Structure. *Int. J. Advance. Soft Comput. Appl.*, Vol. 6, No.2, 2014.

INDEX